Hydrophyte Strategies for Heavy Metal Stress Mitigation

Mansi Sharma
https://orcid.org/0000-0001-6536-1075
Sharda University, India

Priyanka Mahajan
https://orcid.org/0000-0002-6307-7975
Chandigarh University, India

Har Mohan Singh
Sharda University, India

IGI Global
Scientific Publishing
Publishing Tomorrow's Research Today

Vice President of Editorial	Melissa Wagner
Director of Acquisitions	Mikaela Felty
Director of Book Development	Jocelynn Hessler
Production Manager	Mike Brehm
Cover Design	Jose Rosado

Published in the United States of America by
IGI Global Scientific Publishing
701 East Chocolate Avenue
Hershey, PA, 17033, USA
Tel: 717-533-8845 | Fax: 717-533-7115
Website: https://www.igi-global.com E-mail: cust@igi-global.com

Library of Congress Cataloging-in-Publication Data

Names: Sharma, Mansi, 1992- editor | Mahajan, Priyanka, 1986- editor | Singh, Har Mohan, 1992- editor
Title: Hydrophyte strategies for heavy metal stress mitigation / edited by Mansi Sharma, Priyanka Mahajan, Har Mohan Singh.
Description: Hershey, PA : IGI Global Scientific Publishing, [2026] | Includes bibliographical references and index. | Summary: "The book aims to explain the fundamentals of hydrophytes underpinning their role in heavy metal remediation, metal uptake mechanisms, and presence of bioactives"-- Provided by publisher.
Identifiers: LCCN 2025032390 (print) | LCCN 2025032391 (ebook) | ISBN 9798337340371 hardcover | ISBN 9798337340388 paperback | ISBN 9798337340395 ebook
Subjects: LCSH: Phytoremediation | Aquatic plants--Industrial applications | Heavy metals--Environmental aspects
Classification: LCC TD192.75 .H93 2026 (print) | LCC TD192.75 (ebook)
LC record available at https://lccn.loc.gov/2025032390
LC ebook record available at https://lccn.loc.gov/2025032391

British Cataloguing in Publication Data
A Cataloguing in Publication record for this book is available from the British Library.

Table of Contents

Detailed Table of Contents

Preface

In recent years, the global community has witnessed an unprecedented rise in environmental challenges—none more pervasive or perilous than heavy metal contamination. This threat extends far beyond soil and water quality, permeating ecosystems and impacting human health, food security, and climate resilience. As editors of *Hydrophyte Strategies for Heavy Metal Stress Mitigation*, we recognize the urgent need for innovative, sustainable, and interdisciplinary approaches to address this growing concern. The chapters that follow reflect this urgency and the collective scientific effort to develop environmentally responsible technologies that restore balance to affected ecosystems.

Aquatic plants, or hydrophytes, have emerged as powerful yet underutilized allies in the bioremediation of heavy metals. Their remarkable morphological and physiological traits—including enhanced metal uptake capacity, rapid biomass accumulation, and adaptability to diverse aquatic environments—position them as natural, scalable tools for environmental restoration. This book explores these characteristics in depth, presenting hydrophytes not only as biological curiosities but as essential components of a new technological frontier.

The structure of the book mirrors the multifaceted nature of hydrophyte research. The initial chapters introduce the foundational principles governing hydrophyte biology, metal uptake pathways, and the rich repertoire of bioactive compounds that enable these plants to thrive under stress. These chapters serve as a comprehensive guide for understanding how hydrophytes mitigate heavy metal toxicity and why they hold such promise for future applications.

Subsequent chapters expand the discussion to include the synergistic roles of microbial partners and ecological interactions that enhance hydrophyte-assisted bioremediation. Further contributions explore the deployment of hydrophytes in soil reclamation, wetland restoration, and strategies for climate adaptation. Collectively, these discussions underscore the book's alignment with key United Nations Sustainable Development Goals, particularly those addressing good health and well-being (SDG 3), responsible consumption and production (SDG 12), and climate action (SDG 13).

Our aim is to provide a reference work that is both scientifically rigorous and accessible to a diverse audience. Whether you are an industry professional seeking sustainable technologies, a policymaker developing environmental regulations, a research scholar or university student exploring plant–environment interactions, or an individual committed to ecological stewardship, we hope this book serves as a valuable and inspiring resource.

As editors, we extend our sincere appreciation to the contributing authors whose expertise and dedication have shaped this volume. It is our hope that *Hydrophyte Strategies for Heavy Metal Stress Mitigation* will not only advance academic discourse but also catalyze practical innovation toward a cleaner, healthier, and more resilient planet.

ORGANIZATION OF THE BOOK

Chapter 1: Effect of Metals on Hydrophyte Growth and Development—Toxicity and Tolerance Mechanisms

In this opening chapter, the contributors lay the scientific foundation for understanding how hydrophytes respond to heavy metal exposure in aquatic environments. The chapter examines in detail the physiological, biochemical, and molecular disruptions caused by metals such as cadmium, lead, mercury, arsenic, chromium, and nickel. Emphasis is placed on the mechanisms through which hydrophytes manage toxicity—ranging from detoxification pathways to structural and metabolic tolerance strategies. By outlining how these plants navigate the stresses imposed by contaminated waters, the chapter offers essential context for the bioremediation potential discussed throughout the book.

Chapter 2: Hydrophytes and Heavy Metal Stress—Mechanisms, Adaptations, and Phytoremediation Potential

Building upon the foundational principles introduced in Chapter 1, this chapter investigates species-specific responses to heavy metal stress, presenting clear examples of both hyperaccumulators and sensitive hydrophytes. The authors highlight the sophisticated defense strategies hydrophytes deploy, such as phytochelatin- and metallothionein-mediated chelation, ROS-scavenging enzymes, and root-associated adaptations like iron plaque formation. Through case studies of species such as *Elodea canadensis*, *Eichhornia crassipes*, *Typha latifolia*, *Sedum alfredii*, and *Lemna minor*, the chapter demonstrates the remarkable efficiency with which hydrophytes absorb, immobilize, and compartmentalize metals. It also underscores knowledge gaps in long-term ecological impacts and genetic regulation, setting a research agenda for future exploration.

Chapter 3: The Role of Microbial Communities in Hydrophyte-Assisted Bioremediation

Recognizing that hydrophytes do not operate in isolation, this chapter explores the integral role of rhizospheric microbial communities in enhancing plant-based remediation. The chapter illustrates how root exudates shape microbial composition and activity, driving processes such as nitrification, denitrification, sulfate reduction, and organic matter turnover. In turn, these microbes bolster plant resilience and remediation efficacy. By integrating advances in metagenomics, transcriptomics, and systems-level ecological analysis, the chapter offers an up-to-date understanding of the synergistic plant–microbe networks that underpin hydrophyte-based treatment systems and identifies their promise for creating low-cost, sustainable pollution-mitigation technologies.

Chapter 4: Hydrophytes as a Source of Bioactives for Heavy Metal Remediation

This chapter shifts focus to the biochemical dimension of hydrophyte resilience, detailing the phytochemical compounds that enable aquatic plants to tolerate, accumulate, and detoxify heavy metals. Highlighting species such as *Pistia stratiotes* and *Lemna minor*, the authors discuss bioactive molecules involved in metal chelation, antioxidation, and stress mediation. The chapter reinforces the idea that the metabolic pathways of hydrophytes offer a rich source of tools for enhancing eco-friendly remediation strategies. By unpacking these biochemical mechanisms, the chapter contributes to the growing interest in nature-based solutions for ecosystem restoration.

Chapter 5: Hydrophytes as a Sustainable Tool for Bioremediation of Heavy Metals in Wetland Ecosystems

Wetlands, as ecological hotspots, face increasing threats from heavy metal contamination, and this chapter examines how hydrophytes serve as effective and sustainable mitigation agents in these habitats. The authors review the uptake, stabilization, and transformation processes facilitated by hydrophytes and their associated microbial partners, emphasizing their contributions to biodiversity preservation and water purification. The chapter also acknowledges inherent challenges—seasonality, invasive tendencies, saturation thresholds, and the need for responsible biomass disposal—while affirming the value of hydrophyte-based remediation as an affordable, ecologically aligned solution for restoring polluted wetlands.

Chapter 6: Hydrophytes for Sustainable Agriculture—Achieving SDG3 by Soil Reclamation

Connecting bioremediation to public health and agricultural productivity, this chapter explores how hydrophytes contribute to soil reclamation and water purification in ways that support SDG 3 (Good Health and Well-being). The authors describe reclamation practices that convert contaminated landscapes into functional agricultural or developmental spaces, highlighting the contributions of floating hydrophytes such as *Pistia stratiotes*, *Lemna gibba*, and *Eichhornia crassipes*. By showcasing their metal-absorbing capabilities and versatility, the chapter underscores the potential of hydrophytes to restore ecological function while promoting food security and human well-being.

Chapter 7: Hydrophytes and Climate Change—Bioremediation of Heavy Metals and SDG13

This chapter situates hydrophyte-based bioremediation within the broader context of climate change adaptation and mitigation, emphasizing its relevance to SDG 13 (Climate Action). As climate-induced stresses intensify contamination and degrade natural resources, hydrophytes provide low-cost, nature-based pathways for building ecological resilience. The chapter outlines how hydrophytes can support climate-adaptive food systems, rehabilitate polluted aquatic environments, and enhance environmental health. It also calls for interdisciplinary collaboration, policy support, and community engagement to integrate hydrophytes into climate strategies globally.

Chapter 8: Hydrophytes and the Circular Economy for Heavy Metal Remediation

Linking environmental restoration with resource efficiency, this chapter explores the alignment between hydrophyte-based remediation and circular economy principles. The authors describe the various cleanup mechanisms—phytoextraction, phytostabilization, and rhizofiltration—through which hydrophytes immobilize or remove heavy metals from aquatic systems. Beyond remediation, the chapter highlights opportunities to repurpose hydrophyte biomass, transforming waste into value while minimizing environmental impacts. This perspective positions hydrophytes not just as remediation tools but as catalysts for regenerative, closed-loop environmental systems.

Chapter 9: Technological Monitoring and Management in Hydrophyte-Based Heavy Metal Stress Management

Advancing into the technological domain, this chapter presents remote sensing as a transformative tool for monitoring heavy metal contamination and evaluating hydrophyte performance. By integrating satellite data, aerial sensors, GIS, and machine learning, researchers can generate precise assessments of pollution patterns and plant health. The chapter demonstrates how these technologies support timely decision-making, enhance remediation planning, and reduce the need for invasive sampling. It underscores the growing importance of data-driven environmental management in supporting hydrophyte-based solutions.

Chapter 10: The Synergistic Role of Artificial Intelligence and Machine Learning in Enhancing Hydrophyte-Based Heavy Metal Stress Management

Concluding the volume, this chapter explores how AI and machine learning can revolutionize hydrophyte-based remediation. From predicting plant performance and optimizing growth conditions to enabling automated stress detection and risk modeling, digital technologies provide powerful tools for scaling nature-based solutions. The authors discuss both the promise and the challenges—such as data limitations, cost barriers, and ethical considerations—associated with AI-driven environmental management. Ultimately, the chapter highlights a future in which biological and computational intelligence converge to strengthen ecological restoration and protect aquatic ecosystems.

Together, these chapters offer a multidimensional exploration of hydrophytes as innovative, sustainable instruments for addressing heavy metal pollution—integrating biology, ecology, technology, and policy to illuminate pathways toward a resilient environmental future.

CONCLUSION

As we conclude this volume, we reflect on the collective insights offered across its chapters—insights that underscore the profound potential of hydrophytes as agents of environmental restoration, technological innovation, and sustainable development. Throughout the book, our contributors have illuminated the remarkable resilience, adaptability, and biological sophistication of these aquatic plants, demonstrating how they confront heavy metal stress through mechanisms that span molecular pathways, ecological interactions, and symbiotic partnerships. Equally compelling is the growing recognition of hydrophytes as not merely components of natural ecosystems, but as vital tools in addressing some of the most pressing challenges of our time.

From their capacity to detoxify polluted waters and rehabilitate degraded soils to their roles in fostering climate resilience, supporting circular economies, and enabling data-driven environmental management, hydrophytes represent a convergence point for science, sustainability, and societal well-being. The integration of advanced technologies—such as remote sensing, AI, and machine learning—further amplifies their promise, opening new opportunities for scalable, precise, and adaptive remediation strategies. At the same time, chapters focusing on microbial partnerships, bioactive compounds, and wetland management remind us that nature-based solutions remain most powerful when they harness the inherent complexity of ecological systems.

As editors, we hope this book inspires researchers, policymakers, industry professionals, and students alike to look more deeply at hydrophytes not only as subjects of scientific study, but as collaborators in achieving global sustainability targets. Their contributions are already evident, yet their full potential remains largely untapped. It is our sincere belief that continued interdisciplinary effort, informed policy frameworks, and public engagement will be essential in advancing hydrophyte-based bioremediation from promising research to widespread application.

We extend our gratitude to the authors whose expertise has shaped this volume, and to the readers who carry this work forward. May the knowledge presented here spark new ideas, foster meaningful collaborations, and ultimately contribute to a healthier, more resilient planet for generations to come.

Chapter 1
Effect of Metals on Hydrophyte Growth and Development Toxicity and Tolerance Mechanisms

Priyanka Devi
https://orcid.org/0009-0004-4931-7858
Lovely Professional University, India

Tanvir Kaur
Graphic Era University, India

Divjot Kour Singh
Chandigarh University, India

ABSTRACT

Abstract Heavy metal pollution in aquatic ecosystems presents a severe environmental hazard, adversely impacting the growth, physiology, and survival of hydrophytes plants specialized for water-saturated habitats. This chapter examines the intricate interactions between heavy metals and hydrophyte systems, emphasizing the physiological, biochemical, and molecular mechanisms that govern metal toxicity and plant tolerance. This chapter provides a comprehensive overview of the effects of heavy metal toxicity on the growth, development, and physiological functions of hydrophytes, with a particular focus on their mechanisms of tolerance and detoxifi-

DOI: 10.4018/979-8-3373-4037-1.ch001

cation. Heavy metals such as cadmium (Cd), lead (Pb), mercury (Hg), arsenic (As), chromium (Cr), and nickel (Ni) are known to disrupt vital metabolic and cellular processes in hydrophytic species.

INTRODUCTION

Contamination by heavy metals has become one of the most serious issues on the environmental front, posing much danger to both aquatic and terrestrial life. High rates of industrialization, mining, and urban sewage, and high-intensity agriculture activities have caused a steep rise in heavy metals, including cadmium (Cd), lead (Pb), mercury (Hg), arsenic (As), and chromium (Cr) concentrations in water (Ali et al., 2013; Nagajyoti et al., 2010). In contrast to organic pollutants, heavy metals are non-biodegradable, as they remain in the water bodies over a long period and are subjected to the processes of bioaccumulation and biomagnification via a food chain (Jaishankar et al., 2014). The high concentrations of these metals pose a significant toxicological danger to marine flora, fauna, and human beings; hence, their cleanup is an urgent environmental concern. The essentiality of plants in supporting the stability of aquatic ecosystems means that hydrophytes, or plants that are able to grow in part or entirely in water, assume a central position. Nevertheless, they are susceptible to their physiological and biochemical disturbances when subjected to high levels of heavy metals, which have an adverse effect on growth, reproduction, and survival (Mishra et al., 2006). Some of the symptoms of toxicosis in hydrophytes include growth retardation, chlorosis, necrosis, compromised photosynthesis, and inhibition of enzyme activities (Singh & Tewari, 2003).

The vulnerability of the hydrophytes to heavy metals is dependent on the type of metal, concentration, and duration of exposure factors with regard to species. Some species exhibit very strong tolerance mechanisms, even allowing them to withstand and flourish in metal-stressed conditions. These defense mechanisms incorporate entrapping metal by vacuoles, production of metal-bound peptides such as phytochelatins and metallothioneins, activation of antioxidants counteract and changes in morphology, including a change in root shape (Sharma & Dubey, 2005; Prasad, 2004). In light of these implications, there is a timely need to tap into the physiological adaptations of hydrophytes to improve the heavy metal remediation process without necessarily affecting the environmental characteristics (realistic, affordable, and viable) (Ali et al., 2013).

1.1 Both Heavy Metal Uptake and Bioavailability

Water bodies contain a wide variety of forms of heavy metals, such as free hydrated metal ions, organic and inorganic ligand complexes, colloids, and adsorbed particles onto the suspended sediment (Pereira et al., 2021a). Bioavailability, mobility, and the toxicity of heavy metals to aquatic plants are, in large part, dictated by the speciation (i.e., chemical form) of heavy metals. Free ionic species are, in general, the bioavailable and toxic among these forms, easily cross plant membranes, and disturb cell functions (Wang & Chen, 2019). The ambiance, in terms of physical and chemical parameters, including pH, redox potential (Eh), the hardness of water, salty water, and dissolved organic matter (DOM), is a major determinant of metal speciation. Incidentally, a decrease in pH might augment metal solubility and hence improve bioavailability, whereas high carbonate or sulfate concentrations can cause the precipitation of the metal and minimize its plant uptake (Turetsky et al., 2020; Huser et al., 2021). Hydrophytes negatively acquire heavy metals mostly via the roots through which there is close contact with sediments, whereby heavy metals are prone to accumulation by adsorption on organic matter and clays (Bonanno & Lo Giudice, 2010). Foliar uptake can also be in the emergent and floating-leaved species when they are submerged or come in contact with contaminated drops of water or aerosol (Mishra et al., 2008).

At the molecular level, there are often uptake mediator metal transporters and ion channels, which typically mediate the movement of essential mineral nutrients such as Fe 2+, Zn 2+, Mn 2+, and Cu 2+. Unfortunately, because of their similarity in chemistry, toxic metals Cd 2 +, Pb 2 +, and Ni 2 + may ride along these transport networks. An example is the Zn and Fe transporters (ZIP, ZRT/IRT-like Protein) that, in some instances, are responsible for transporting cadmium by mistake and NRAMP (Natural Resistance-Associated Macrophage Protein) transporters that allow uptake of analogues of Mn and Fe, such as Pb and Co (Clemens et al., 2013). Aquaporins, membrane proteins that are mainly involved in water transport, also exist in aquatic plants and therefore indirectly could promote the transport of metalloids such as arsenite (As 3+) (Bienert et al., 2008). Upon entry into root tissues, heavy metals may be bound in vacuoles or attached to the cell wall, or complexed with metal-chelating agents like phytochelatins and thereby lessen their toxicity. Nevertheless, uncontrolled exposure to transport can cause excessive accumulation, which disrupts the work of cells and causes oxidative stress (Dal Corso et al., 2019).

1.2 Effects on Physiology and Morphology

When the levels of heavy metals become toxic, numerous physiological and morphological alterations will occur in hydrophytes. These include:

- ✓ Growth inhibition: growth inhibition may result in reduced biomass, chlorosis of roots, and shoot stunting. The use of Pb and Cd concentrations higher than 100 mg/L, for example, revealed a drastic decrease in root and shoot length in Eichhornia crassipes (water hyacinth) (Vestena et al., 2016).
- ✓ Photosynthesis disruption: Owing to heavy metals like Cd and Pb, heavy metals disrupt the production of chlorophyll and degrade photosystem functioning, which hampers photosynthetic rates and energy metabolism (Pushpamala et al., 2014).
- ✓ Nutrient imbalance: Nutrient imbalances due to environmental stress occur when heavy metals compete with Ca^{2+}, Mg^{2+}, Fe^{2+}, and Zn^{2+} essential ions, causing the lack of essential nutrients and deranged metabolic functions (Ali et al., 2020a).

1.3 Antioxidant Responses to Oxidative Stress

Production of reactive oxygen species (ROS) such as superoxide radicals (O_2^-), hydrogen peroxide (H_2O_2), and hydroxyl radicals ($\cdot OH$) is one of the major forms of heavy metal toxicity. When there is excess ROS, there is oxidative stress resulting in protein, lipid, and nucleic acid damage (Pereira et al., 2021b). In response, hydrophytes enable many enzymatic and non-enzymatic antioxidant strategies: Enzymatic antioxidants like superoxide dismutase (SOD), catalase (CAT), ascorbate peroxide (APX), and glutathione reductase (GR) are essential to the scavenging of the ROS. Examples of non-enzymatic antioxidants are ascorbic acid (vitamin C), glutathione (GSH), carotenoids, and flavonoids that trap ROS and safeguard intracellular compounds. Investigations of Salvinia biloba and E. crassipes reported that, in response to metal stress, activities of antioxidant enzymes are greatly increased, possibly as a response mechanism of tolerance (Vestena et al., 2017).

1.4 Species-specific Tolerance

Hydrophytes do not react to exposure to heavy metals equally. Eichhornia crassipes, Salvinia natans, Lemna minor (duckweed), and Hydrilla verticillata are some of the species that have exhibited a significant level of accumulation and tolerance of metal. For instance, the E. crassipes has the potential to reach up to 5.4 percent Pb-accumulation in the roots with little to no translocation to the leaves (Vestena et al., 2016). Salvinia natans is highly tolerant to Zn and has a moderate sensitivity to Cd and Cu (Pushpamala et al., 2014).

1.5 Potential as Phytoremediation

Many hydrophytes are used in phytoremediation, hydrophyte-mediated processes to cleanse polluted water bodies; since many hydrophytes can accumulate and tolerate heavy metals, it is readily applicable, low-cost cost and thus environmentally friendly. Metal removal from wastewater has also been implemented using constructed wetlands composed of such species as E. crassipes and Lemna minor (Ali et al., 2020b). The success of phytoremediation relies on the ability to produce biomass, accumulate metals, and adapt to various environmental circumstances.

1.6 Importance of the study of metal toxicity and tolerance mechanisms in hydrophytes

The accumulation of heavy metals is increasingly proving to be a menace to the aquatic ecosystems of every corner of the globe due to industrialization, mining, urbanization, and intensive farming. The toxic metals such as cadmium (Cd), lead (Pb), mercury (Hg), arsenic (As), and chromium (Cr) are not biodegradable, and consequently form at the base of the sediments and water bodies. The heavy metals stand the test of time compared to organic pollutants that may wear off after some time and are therefore detrimental to the environment and health in the long run. Such ecosystems are based on the hydrophytes, which are aquatic vegetation and are highly sensitive to metal stress. Direct effects of exposure to metal on their growth and development disrupt cellular metabolism, photosynthesis, and nutrient uptake and generate reactive oxygen species (ROS), resulting in oxidative damage. Thus, the ecological health and ecosystem sustainability are significant when researching the toxicity of the metals in hydrophytes. This study is applicable in other fields beyond monitoring ecosystems. Hydrophytes can be employed as natural bioindicators of pollution processes, which help define changes in water quality and environmental stress. Better still, several hydrophyte species possess excellent tolerance and adaptation mechanisms that make them survive in contaminated waters. The strategies include sequestration of the metals in vacuoles, chelation by the phytochelatins and metallothioneins, activation of antioxidant enzymes, and structural changes in the roots and leaves, among others, which provide an upper edge to survival. These are not merely simple processes that focus on the resilience of hydrophytes, but rather an opportunity to utilize them in phytoremediation, particularly rhizofiltration, whereby the lentils absorb and fix the metals in the polluted water.

2. HEAVY METALS IN AQUATIC SYSTEMS

2.1. Natural and Artificial Sources

The pollution of heavy metals into aquatic systems occurs through both natural processes and anthropogenic activities, leading to an increase in total metal loading, although the magnitude, temporal variability, and environmental disturbance differ (Ali et al., 2019).

Natural Sources

The naturally occurring heavy metals include: cadmium (Cd), lead (Pb), arsenic (As), nickel (Ni), and mercury (Hg), whose major sources are geochemical weathering, volcanic eruptions, and the erosion of the soils (Degryse et al., 2009). Oxidation and dissolving release metals to engineers as they get to the immediate surroundings of a soil and water due to the weathering of parent rocks rich in metals to form soils and water, such as sulfide ores (Alloway, 2013). As an example, the weathering of these sulfide minerals may provide Fe, Cu, and Zn to streams, particularly in locations with high rainfall and acidic soils (Hudson-Edwards, 2016). Injections of volcanoes into the atmosphere consist of giant amounts of the metals Hg, Pb, and As; this gets deposited later within the terrestrial and aquatic environment either through the process of precipitation or direct fallout (Zheng et al., 2007). Likewise, the process of soil erosion releases metal-bound particles of terrestrial catchments into rivers and lakes, particularly in river and lake basins with steep slopes or limited covers of vegetation (Bini & Bech, 2014).

Anthropogenic Sources

The loading of heavy metals into the water bodies by the man over the last century has significantly increased, reaching greater levels that are mostly beyond the natural assimilatory capacity of the water bodies (Singh et al., 2011). Anthropogenic key pathways are as follows:

- ✓ **Mining/Smelting processes** - the extraction and processing of metal ores produce high volumes of waste rock and tailings of Pb, Cd, Zn, Cu, and Ni (Lottermoser, 2010). The issue is especially pronounced in developing areas with weak environmental law enforcement (RSC, 2023).
- ✓ **Industrial Discharges** -Electroplating, battery making, dyeing industries, tanneries, and electronic manufacturing produce wastes in the form of wastewater, which contains Cr, Cd, Pb, Hg, and Cu. These effluents are left not to

be treated, and in case they were, they end up contaminating the receiving water bodies, changing the sediment chemistry, and posing chronic risks to the aquatic life (Pereira et al., 2021c; Fu & Wang, 2011).

- ✓ **Agricultural Runoff**- Farming activities can also lead to metal pollution by overusing phosphate fertilizers (in which Cd may be used as an impurity), pesticide mixtures that include Cu or As, and animal manure that contains high levels of Zn and Cu as a result of feed supplements (Ali et al., 2020a). Such metals dissolve into the surface water when irrigated and when there is rainfall.
- ✓ **Atmospheric Deposition-** The combustion of fibrous fuels, waste incineration, and automobile exhausts scatter volatile metals in the atmosphere; Hg, Pb, and As are examples. The resultant metals are later deposited into the water system through either wet deposition (rain, snow) or dry deposition (dust fall out) (Haghshenas, 2020; Pacyna & Pacyna, 2001). As an example, coal-fired electricity producers continue to be one of the biggest anthropogenic sources of atmospheric mercury.

2.2 Common Toxic Heavy Metals

The following heavy metals are of extra concern to the aquatic environment because of their toxicity, persistence, and bioaccumulative potential:

- **Cadmium (Cd):** Cadmium is a non-essential metal that is used in batteries, coatings, and pigments. It gets into the water bodies primarily via industrial wastes and phosphate fertilizers. Cd interferes with the photosynthetic and water absorption of aquatic plants and may cause kidney and bone tissue damage in humans (Pushpamala et al., 2014).
- **Lead (Pb):** Lead poisoning is caused by the recycling of batteries, lead mining, plumbing, and the previous use of petrol lead. Pb is neurotoxic and disrupts the work of enzymes of aquatic plants and animals (RSC, 2023).
- **Mercury (Hg):** Mercury is added through natural outgassing, coal formation, and small-scale gold mining. Its more pernicious variant, methylmercury, is biomagnified in food chains in water, affecting fish and humans (Pereira et al., 2021c).
- **Arsenic (As):** Arsenic is geogenic as well as anthropogenic. It occurs in mining effluents, the combustion of fossil fuels, and the application of pesticides. It disrupts the metabolism of phosphate and results in oxidative stress to the aquatic biota (Vestena et al., 2017).
- **Chromium (Cr):** There are two major forms of chromium, Cr3+, or less toxic, and Cr6+, or highly toxic. They usually occur in industrial effluents,

especially those of the leather tanners and electroplaters. The Cr6+ can damage DNA, as well as cause respiratory distress and the suppression of roots growing in aquatic plants (Pereira et al., 2021a).

- **Nickel (Ni):** Stainless steel and electroplating are made of nickel. Excess Ni alters and disrupts the normal metabolic processes and triggers the oxidative damage of aquatic life, even though it is necessary to trace levels in some enzymes (Ali et al., 2020a).
- **Zinc (Zn):** Zinc is an important micronutrient that proves poisonous at high levels. In galvanization and fertilizers, it is highly sought after. Zn reacts with the enzymes of aquatic vegetables and invertebrates and disrupts membrane functioning (Pushpamala et al., 2014).

2.3 Aquatic Mobility and Bioavailability in the Aquatic Environment and Sediments

Bioavailability is a term used to describe that fraction of the total metal pool that aquatic living things can take up, and this is a key factor that determines their fate, mobility, and ecological risk in the aquatic ecological systems (Luoma & Rainbow, 2008). In addition, sediments are sinks and sources of metals as they bind the metal during specific conditions and release the same in the above water at some other environmental conditions (Burton & Johnston, 2010).

2.4 Metal Speciation, Phase Distribution

This method sorts out metals into five geochemical fractions based on operationally defined distinctions, F1 to F5, which can vary in both mobility and bioavailability:

F1: The Exchangeable / Water-Soluble Fraction

This is a fraction of metals loosely attached to particles of sediment or existing freely in pore waters as hydrated ions. These metals have the capability of the aqueous phase to be easily released under variations in ionic strength or competition of the ionic exchange (Forstner & Wittmann, 2012).

F2: Fraction Carbonate Bound

The metals of this fraction are found either in association with carbonate minerals, usually through adsorption or co-precipitation. They are also pH-stable: their stability in an acidic pH favors carbonate dissolution to release metals bound with them into pore water (Li et al., 2013).

F3: Fraction Bound by Oxidation Fe-Mn Oxides

This fraction is a group of metals that are adsorbed or co-precipitated with iron/manganese oxyhydroxides that are strong trace metal scavengers in oxic media (Tebo et al., 2004).

F4: Organic-Bound Fraction

Metals in this fraction are complexed by natural organic matter in the form of humic and fulvic acids or attached to biogenic detritus. Such complexes can be comparatively inert in oxic environments, but can be freed by microbial breakdown of the organics-particularly in anaerobic environments, binding metals (Burton & Johnston, 2010).

F5: Fraction Residual

That portion contains metals as a constituent of crystal lattices of primary and secondary minerals in the form of silicates. These are typically of geogenic origin and exhibit solid, well-established chemical weathering stages that require bioavailability. The eluted residual fraction is therefore deemed to be environmentally inert in short ecological time (Kabata-Pendias, 2011).

2.5 Bioavailability and Mobility: Environmental Effects

The metal fraction shift is dynamic and highly conceptualized with environmental parameters:

- ✓ As an example, acid mine drainage usually causes high levels of soluble Fe, Mn, and Cu in the downflow of mining locations (Nordstrom, 2011).
- ✓ Redox Potential (Eh). Lower Eh (reducing conditions) increases the dissolution of oxides of Fe/Mn (F3); the breakdown of organic matter (F4), which releases these metals (Tebo et al., 2004).
- ✓ Salinity/Ionic Strength- Male salinity may drive metals out of exchange sites on particles by competitive ion exchange (Du Laing et al., 2009).
- ✓ Temperature-Increasing temperatures increase the metabolic rate of microorganisms and may increase degradation of organic matter and release of metals (Neubauer et al., 2008).

Sediment–Water Interactions: Heavy metals are mainly stored within the sediments in water environments (Forstner & Salomons, 2010). In a steady state, they are long-term sinks, but disruptions to the environment, e.g., dredging, storms, or bioturbation can re-suspend polluted sediments, remobilising metals into the overlying water (Eggleton & Thomas, 2004).

Resuspension Events: Physical processes such as wind action, ship navigation, and floods can resuspend fine sediment particles enriched in fractions of labile metals (Oursel et al., 2014). At such occasions, acute exposure risk appears as dissolved metal, particularly of planktons and fish larvae, due to short-term spikes in dissolved metal concentrations.

Bioturbation: Benthic worms, bivalves, and crustaceans dislodge and mix up sediments by feeding and burrowing. Such an activity transforms the sediment porosity, oxygen penetration, and redox gradient and modifies metal binding and release (Kristensen et al., 2012).

2.6 Metal Mobility and Climate Change

It has been predicted that the effects of climate change, including warming, changed precipitation, and storm intensities, will affect the dynamics of heavy metals in aquatic systems (Noyma et al., 2020).

- **Ecotoxicological Implications**

Acute toxic metals such as Cd 2, Pb 2, and Hg 2 may interfere with the enzyme activity, ion compartment, and provoke nibbling of aquatic organisms as they have high bioavailability. When the level of bioavailable metals is high, plants (and even hydrophytes) might experience retarded photosynthesis, growth, and even nutrient uptake (Sharma & Dubey, 2005; Degryse et al., 2009).

3. TAXONOMIC AND MORPHOLOGICAL IDENTIFICATION OF HYDROPHYTES

Hydrophytes refer to the vascular plants that have undergone morphological, anatomical, and physiological modifications to survive in water. They live in a variety of freshwater and brackish habitats, including rivers, lakes, wetlands, reservoirs, estuaries, and rice paddies in which the water regime is fully submerged to seasonal flooding. Hydrophytes are traditionally classified according to their growth form and contact with the water surface, which defines the roles of a species in an ecosystem, its physiological adaptations, and its fate in regards to contaminant uptake (Lin et al., 2022).

Figure 1. Classification and characteristics of hydrophytes

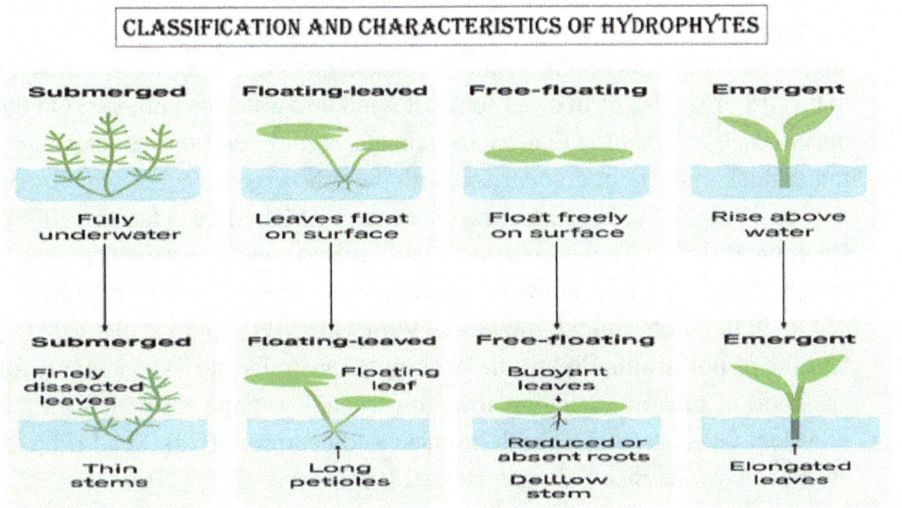

Table 1. Comparative anatomical and functional traits

Hydrophyte Type	Cuticle & Epidermis	Aerenchyma	Stomata Position	Vascular & Mechanical Tissues
Emergent	Thin but functional	Extensive	Both surfaces	Well-developed, robust vascularization
Submerged (rooted/ free)	Thin or absent	Present	Absent	Reduced; stems spongy and flexible
Floating-leaved	Poorly developed; upper only	Present	Only on the upper surface	Moderate; long petioles with air lacunae
Free-floating	Absent or thin	Extensive	Upper surface	Reduced stems and roots

3.1 Emergent Hydrophytes

The phototrophic emergent plants grow at the bottom of the sediment, and then their stems, leaves, and reproductive organs are above the water level. Such plants are Typha latifolia, Phoenix australis, Sagittaria sagittifolia, and Scirpus lacustris. Such plants are predominant in the shallow littoral areas where the resting depth and stability of sediments are enough to have large systems of rhizomes (Lin et al., 2022).

- ✓ **Anatomical and Morphological Characteristic:** The emergent species developed aerenchyma or specialized air space, helping in the movement of oxygen between the aerial organs to the roots, thus helping to survive in the anaerobic sediments (Colmer, 2003; Voesenek et al., 2016). Their stems tend to be stiff or semi-stiff in order to resist wind and water movements. On their upper leaf surfaces, the density of stomata is high, and the cuticles can be rather thick in terms of comparison with the submerged species, which seems to conform to the given partial exposure to air (Maberly & Madsen, 2002).
- ✓ **Ecological Roles:** They also act as biofilters and capture nutrients and heavy metals contained in the inflowing waters, where most sequestration of metals occurs in the roots and the rhizomes (Vymazal, 2011; Alam et al., 2021).
- ✓ **Uptake of pollutants Potential:** Phragmites australis and Typha angustifolia are good at phytoremediation in a constructed wetland, in which they are prevalent because of their high biomass and cadmium (Cd), Lead (Pb), and arsenic (As) tolerance (Alam et al., 2021; Mishra et al., 2023).

3.2 Submerged Hydrophytes

Submerged hydrophyte plants grow completely under the surface. They may be anchored (Hydrilla verticillata, Vallisneria spiralis, Potamogeton crispus) or have no roots/be free-floating (Ceratophyllum demersum, Utricularia vulgaris) (Chambers et al., 2008; Lin et al., 2022).

- ✓ **Morphological and Anatomical characteristics:** The function of most of the root-attached submerged species is to get their nutrients through the sediments, whereas the lack of roots on rootless species diverts them all into the water column (Chambers et al., 2008).
- ✓ **Ecological Roles:** Buried hydrophytes bind sediments, thus stabilizing them, improve water clarity by lessening resuspension, and give cover to young fish and macroinvertebrates (Sand-Jensen, 2003; Bornette & Puijalon, 2011). They also generate oxygen in the daytime, and this sustains aquatic life.
- ✓ One of the parameters, Pollutant Uptake Potential, estimates how absorption of a substance occurs over time. The parameter is adapted to measure the rate of absorption at pH 7.
- ✓ Submerged plants are thus successful since they provide a high surface area to volume and thus can uptake the dissolved metals such as copper (Cu), zinc (Zn), and chromium (Cr) directly in the water (Srivastava et al., 2008).

3.3 Hydrophytes with floating leaves

Floating-leaved species have their roots in the sediments, yet they generate floating leaves on the water surface. Typical representatives of them are Nymphaea nouchali, Nelumbo nucifera, and Nuphar lutea (Lin et al., 2022).

- ✓ **Anatomical and Morphological Attributes:** To reach the surface, these plants grow long and supple petioles to which the leaves grow out horizontally. Stomata are restricted to the upper surface of the leaf to carry out the exchange of gases with outside air (Maberly & Madsen, 2002).
- ✓ **Ecological Roles:** The floating-leaved will create shade, which regulates the temperature of the water and controls algal blooms (Zhao et al., 2021).
- ✓ **Uptake Potential:** Examples of species having the capacities to root (sediment pathway) and leaf (foliar pathway via water) uptake are Nymphaea alba and Nelumbo nucifera, and can accumulate Pb, Cd, and Hg (Zhou et al., 2007a; Ha et al., 2019). They are effective in removing sediments, being large in biomass, and are found in sediments.

3.4 Hydrophytes, free-floating

Free-floating hydrophytes do not grow uprooted into sediments; themselves being floating on the surface, they move along with currents and wind. Typical ones: A typical example is the Eichhornia crassipes (water hyacinth), Pistia stratiotes (water lettuce), Lemna minor (duckweed), and Salvinia molesta (giant salvinia) (Zhou et al., 2007b; Lin et al., 2022).

- ✓ **Morphological and anatomical Characters:** To remain afloat, they have specialized tissues that are buoyant (aerenchyma and spongy mesophyll). The plants have roots freely suspended in a water column, which take up the nutrients and the metals, directing them out of the water (Malik, 2007).
- ✓ **Ecological Roles:** The abundance of free-floating vegetation blocks the window of penetration and changes the dynamics of phytoplankton, which affects the dynamics of the dissolved oxygen (Villamagna & Murphy, 2010).
- ✓ **Potential to gain pollutants:** Eichhornia crassipes and Lemna minor are some of the extensively explored phytoremediators able to remove Pb, Cd, Hg, and nutrient pollutants very effectively (Rezania et al., 2016; Singh et al., 2023).

Table 2. Comparative adaptations across categories

Category	Rooting	Leaf Position	Nutrient Uptake Pathway	Metal Uptake Potential	Notable Examples
Emergent	Rooted	Above surface	Sediment-dominant	High for sediment metals	*Typha, Phragmites*
Submerged	Rooted/ Not	Below surface	Water and/or sediment	High for dissolved metals	*Vallisneria, Ceratophyllum*
Floating-leaved	Rooted	Floating leaves	Sediment and foliar	High for mixed sources	*Nymphaea, Nelumbo*
Free-floating	Not rooted	Floating plant	Water-dominant	High for dissolved metals	*Eichhornia, Lemna*

3.5 Ecophysiological Services in Water Quality & Ecosystem Integrity

Hydrophytes are used in a variety of ecological processes essential to both the health and the cleanup of aquatic environments:

- ✓ **Water Purification Removal of Nutrients:** Water plants, such as E. crassipes and Lemna minor, have also been useful in wastewater treatment due to their ability to efficiently uptake nitrogen, phosphorus, and heavy metals in an aquatic environment and constructed wetlands (Rai, 2008; Vymazal, 2011; Qayoom & Jaies, 2023).
- ✓ **Turbidity (Turbidity Reduction) & Sediment Stabilization:** Species that develop their roots widely (i.e., Typha and Phragmites) help stabilize sediments and minimize erosion processes, increasing water clarity, and promoting submerged plant assemblage (Thomaz & Bini, 2003).
- ✓ **Oxygenation Diaphragm & Redox Conditioning:** The fraction of the dissolved oxygen produced by submerged and floating-leaved macrophytes comes about as a result of photosynthesis, whereas emergent release of root can promote oxidation of the rhizosphere, which encourages the degradation of microbial materials as well as nutrient recycling (Lin et al., 2022; Che et al., 2024).
- ✓ **Uptake & Detoxification of Heavy Metal:** Cd, Pb, Cu, Zn, Ni, and Cr are readily removed by absorption and sequestration by free-floating and rooted hydrophytes. Some of the metals have removal efficiencies by E. crassipes, P. stratiotes, L. minor, and S. polyrhiza as high as 99% (Soda et al., 2012; Qayoom & Jaies, 2023; Sompura et al., 2024). Root biomass tends to accumulate metals, preventing their movement to shoots, which is a critical safety feature in constructed wetlands (Mishra et al., 2008; Lin et al., 2022). Water

hyacinth showed to accumulate up to 80-90% of cadmium and lead in two weeks of exposure in a study carried out close to the Yamuna River, which demonstrates its ability to accumulate cadmium and lead in a rhizofiltration process in the field. Likewise, Typha latifolia (cattail) and Phragmites australis (common reed) have also been used in constructed wetlands in Europe to treat municipal and mining waste water. An example of constructed wetlands in Poland revealed that constructed wetlands containing Phragmites had 70 percent zinc and copper extraction and still had a stable ecosystem.

Table 3. Hydrotypes, types, and Ecophysiological role

Hydrophyte Type	Examples	Primary Ecophysiological Functions
Emergent	Typha, Phragmites	Sediment stabilization; root zone oxygenation; pollutant assimilation
Submerged (rooted)	Hydrilla, Vallisneria	Water oxygenation; metal uptake; nutrient cycling
Submerged (unrooted)	Ceratophyllum, Utricularia	Metal absorption; biodiversity support
Floating-leaved	Nymphaea, Nelumbo	Pollutant interception; habitat complexity
Free-floating	Eichhornia, Lemna, Pistia	Rapid biomass turnover; high metal uptake and sequestration

4. NEGATIVE EFFECT OF THE HEAVY METALS ON AQUATIC PLANTS

4.1 Influences on Seed Germination and Early Growth

Cadmium (Cd), lead (Pb), mercury (Hg), chromium (Cr), nickel (Ni), and arsenic (As) are well-known among the heavy metals that severely affect seed germination and the early development (growth) of the seedling. The uptake of water (imbibition), activation of enzymes, and reactivation of metabolism needed to germinate require the disturbances caused by exposure, even to low to moderate metal concentrations (Osman & Fadhlallah, 2010; Pourrut et al., 2011). Pb and Cd suppress or diminish the percentage of germination, delay the time of germination, and decrease the germination index (Shourie, 2022; Collin et al., 2022). Metal pollution of starch hydrolysis and amylase/protease activity severely reduces the energy release in germination. Although germination does take place, the growth of radicles and shoot growth is inhibited in smaller seedlings that lack vigor (Collin et al., 2022;

Shourie, 2022). Oxidative stress activating germination results in fragmentation of DNA, peroxidation of lipids, and oxidation of proteins, which reduces viability of the embryo (Pourrut et al., 2011). These impacts underline the idea that seed and frond initiation stages present bottlenecks of heavy metal stress.

4.2 Effect on the accumulation of biomass and morphological traits

The stress of heavy metals causes a decrease in root and shoot biomass and modification of plant structure, and impairs the usual morphological growth of the aquatic species. In Eichhornia crassipes, exposure to Pb showed ~50 biomass reduction, and the root tissue had the highest concentration of metal in comparison to leaves (Vestena et al., 2016). Cd impairs ~41%, and Pb impairs ~36% dry biomass, explaining that dose-dependent toxicity is demonstrated ($Ni2+$ effects are less critical), and root elongation, leaf extension, and petiole sizes all shrink in water spinach (Ipomoea aquatica) (Sompura et al., 2024).

Under heavy metal stress, the morphology of roots and shoots undergoes significant changes. The elongation growth of roots is hampered, the number of root hairs decreases, and the formation of roots is deformed due to exposure to heavy metal pollution. Shoot morphology is also impacted, with shorter stems, thinner leaves, and a reduced number of nodes occurring at moderate to high exposures, and chlorotic patches occur at moderate exposures.

4.3 Cellular Injury and Oxidative Stress

Exposure to heavy metals causes the acceleration of reactive oxygen species (ROS) such as superoxide radicals (O 2 -), oxidative hydrogen peroxide (H 2 O 2), hydroxyl radicals (OH), and singlet oxygen. Unless these oxidative agents are scavenged, they lead to extensive cell damage. Overproduction of ROS causes lipid peroxidation, which is indicated by high levels of MDA (malondialdehyde), and this is generally documented at intermediate exposures to metals and then decreases due to cell death (Vestena et al., 2016). Due to peroxidation, the membranes are disrupted and the ion gradients are lost, leading to the event of ion leakage and the damaged ability of nutrient transport. The metabolic disintegration is a result of protein oxidation, loss of enzyme functionality; when the exposure is extended or the doses are high, DNA damage and aberrations in chromosomes are achieved (Pourrut et al., 2011). The activation of antioxidant enzymes is first triggered: SOD, which further gets degraded by CAT, APX, or some other peroxidases. Nonetheless, these defense enzymes decrease at very high levels of metals, indicating an overloaded detox capacity (Vestena et al., 2016).

Figure 2. Negative effect of the heavy metals on aquatic plants

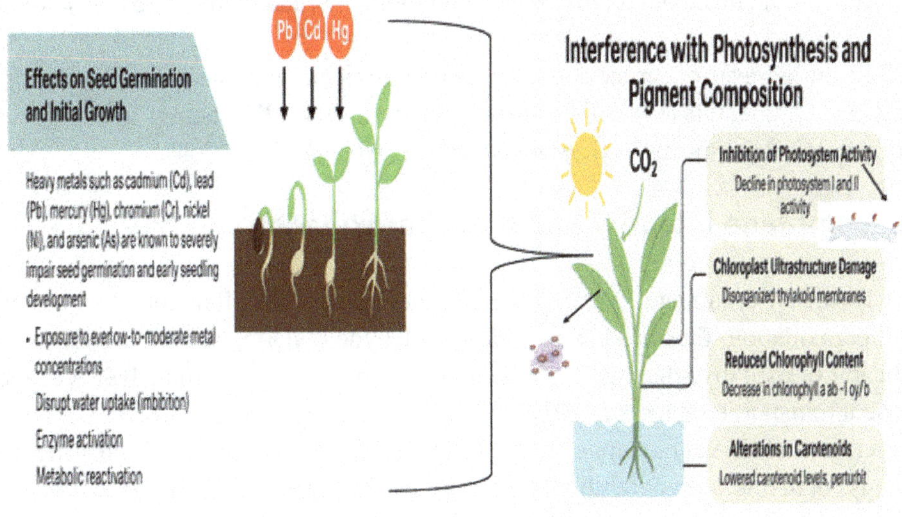

5. BIOCHEMICAL AND MOLECULAR RESPONSES TO STRESS OF HEAVY METALS

5.1 Constitutive and Regulated Generation of Reactive Oxygen Species (ROS)

The heavy metals increase the formation of reactive oxygen species (ROS)/superoxide (O2-) /hydrogen peroxide (H2O2)/hydroxyl radicals (OH)/singlet oxygen (O2) in chloroplasts, mitochondria, and peroxisomes (Mittler et al., 2004; Kosakivska et al., 2020). Even though cadmium (Cd) is not redox-inactive, it indirectly finds its way into the electron transport systems of the energy organelles, increasing the amount of ROS produced (Keunen et al., 2011; Kosakivska et al., 2020).

In hydrophytes, researches indicate:

✓ Cd causes a rise in ROS in Vallisneria spiralis, and this results in decreased leaf and fresh root weight and an increased antioxidative response (Wang et al., 2008).

✓ Long-term cultivation of Eichhornia crassipes sparked SOD activities and MT1 expression after being exposed to low-moderate Cd and assisted ROS homeostasis (Zhang et al., 2024).

✓ In Lemna minor, Cd elevates phytochelatins (PCs) 6 8-fold, which is associated with elevated oxidative stress alleviation (Szalai et al., 2015).

The initial level of protection is against antioxidant enzymes: superoxide dismutase (SOD), catalase (CAT), ascorbate peroxidase (APX), peroxidases (POX), and glutathione reductase (GR), among others (Apel & Hirt, 2004).

5.2 Membrane integrity and Lipid peroxidation

Chronic ROS damages the unsaturated lipids within the cell membrane through lipid peroxidation, thus producing malondialdehyde (MDA), which constitutes an indicator of oxidative damage. This inhibits membrane fluidity, ionic balance, and nutrient transport (Baba et al., 2016). All of E. crassipes exposed to Pb or Cd show an increase in MDA at moderate doses, which decreases at high doses, indicating the collapse of membranes or inhibition of enzymes (Zhang et al., 2024). In Cd, electrolyte permeability and chlorophyll decreases occur in Vallisneria spiralis, which is an indication of damaged membrane permeability (Wang et al., 2008). The oxidation of both protein and DNA damage has been reported and caused downstream effects on mitosis and cell viability (Kosakivska et al., 2020). According to the evidence, the effects of the genetic editing involve both DNA and protein damage, as well as downstream changes in mitosis and cell viability (Kosakivska et al., 2020).

6. TOLERANCE MECHANISMS IN HYDROPHYTES

Hydrophytes resort to the coordinated action of molecular, biochemical, and cellular mechanisms to be resistant to heavy metal exposure.

6.1 Chelation of Metals and Phytochelatins/ Metallothioneins Synthesis

6.1.1 Phytochelatins (PCs)

Phytochelatins (PCs) are oligomers derived from glutathione (e.g., PC 2-PC 7) and are produced with the help of phytochelatin synthase in response to metal ions (Rea, 1999; Wikipedia Phytochelatin). This study reveals that in V. spiralis, PCs and glutathione rapidly accumulate in roots and leaves as the concentration of Cd rises linearly (0.08 to 0.64 μM), and that the leaf-accumulated PCs are greater than those in the roots (Wang et al., 2008). When plants of L. minor undergo Cd stress, PCs increase 6-8 fold, especially the higher polymer forms (Szalai et al., 2015).

6.1.2 Metallothioneins (Mts)

Metallothioneins are rich in cysteine, and the thiol group binds metals. In fish and halophytic plants, the expression of the MT genes increases in the presence of Cd, Cu, Zn, and Pb (Gonzalez Mendoza et al., 2007). In E. crassipes, the prolonged exposure to low concentrations of Cd induces the expression of MT1, which increases tolerance due to the effect of detoxification and antioxidant potential (Zhang et al., 2024).

6.2 Vacuolar Sequestration and Compartmentalization

Complexes of medals are isolated against cells in vacuoles in order to suppress cytosolic interactions. Tonoplast-bound transporters support transport; such transporters include ABC-type (ABCC1-3) transporters, P1B-like ATPases, and cation diffusion facilitators (Kim et al., 2006). ABCC transporters can transport vacuolar PC-metal complexes in Arabidopsis thaliana (Kim et al., 2006). Proteomic data in E. crassipes indicate increased levels of transporters and vacuolar pump proteins, indicating active sequestration in E. crassipes against Cd (Xu et al., 2014). In V. spiralis, one of the effects of low-dose Cd is accumulation in roots rather than shoots and localization and compartmentalization of the response in antioxidant enzymes (Wang et al., 2008).

6.3 Phytoremediation Results and Integration Mechanisms

Hydrophytes combine several detoxification pathways to accept- and store- metals, in the practice of phytoremediation:

- ✓ Uptake of metal through ZIP, NRAMP, and P1B-ATPase transporters.
- ✓ Removal of electron transport complexes and inhibition of enzymes resulted in the production of ROS.
- ✓ Enzymatic defenses: Activation of antioxidants, beginning with SOD and proceeding to CAT, APX, and GR; non-enzymatic defenses.
- ✓ Chelation through a quick PC manufacture and long-term metallothionein expression.
- ✓ The adaptation protects the proteomic and transcriptomic levels and increases detox genes, antioxidant system, transporter expression, and metabolic resilience (Xu et al., 2014; Zhang et al., 2024).

7. AQUATIC PLANT PHYTOREMEDIATION

7.1 Mechanisms: Phytoextraction, Rhizofiltration, and Phytostabilization

Aquatic phytoremediation uses living plants to remove, stabilize, or transform the metals so that they do not exist in high levels in the water. Phytoextraction entails the absorption of soluble metals to the root and shoot tissues, all of which is extracted via the coveted harvesting of biomass (Su et al., 2012; Wang et al., 2008).

7.2 Conspicuous metal-accumulating hydrophytes species

The transportation and elimination of metal by some aquatic species are extraordinary:

✓ Eichhornia crassipes (water hyacinth): Pb, Cd, Cr, Ni, Zn, and As are removed (even >90% of them in days), in most cases (MDPI review, 2021; Cluj County case study, 2023). Bioconcentration levels in roots are high, and the translocation factor is low, which is ideal in phytoextraction and safety (Maurya et al., 2025).

✓ Lemna minor (duckweed): Microorganisms remove Pb (>85%), Ni (~68%), and As (~ 70%) in 7-15 days. The values of bioconcentration factors of Ni are more than 1000, which makes it a hyperaccumulator (Rahman & Hasegawa, 2011).

7.3 Natural Remediation and Engineered Wetlands

The use of the engineered wetland technique aims at implementing potent heavy metal mitigation through a combination of hydrophyte-driven remediation, microbial biodegradation of heavy metals, and sedimentation. In constructing wetlands (CWs), plants such as Typha latifolia, Eichhornia crassipes, and Pistia stratiotes are manipulated in a shallow basin, to which recourse is given for the contaminated water. Heliyon study, 2024 T. latifolia removed 96.7 percent Cr in a tannery wastewater CW with backup by E. crassipes and P. stratiotes (~80%) (Heliyon study, 2024). The multiple-stage wetlands (hybrid flow) combine elements of floating, emergent, and submerged plant species in order to utilise the complementary mechanisms of rhizofiltration, phytoextraction, and phytostabilization (Cluj County case, 2023).

8. CONCLUSION

The widespread pollution of water bodies with heavy metal contamination continues to be one of the most significant environmental issues throughout the world, mostly because of the fastest rates of industrialization, mining, agricultural losses, and urban wastewater pollutants. These are non-biodegradable elements, such as cadmium (Cd), lead (Pb), mercury (Hg), arsenic (As), and chromium (Cr), which remain in the environment, endangering both aquatic life and human life. These distributions comprise some of the most affected biological samples in freshwater ecosystems, such as hydrophytes, which are aquatic plants with specialized functions, including nutrient cycling, sediment stabilization, and water cleansing. Besides, heavy metals stimulate excessive production of reactive oxygen species (ROS), leading to oxidative stress, which causes the oxidation of proteins, nucleic acids, and lipids, eventually affecting the viability of the plants. Nonetheless, hydrophytes have developed complex biochemical and molecular mechanisms of tolerance, which help them survive the metal stress. The leading among them are the synthesis of antioxidant enzymes like the superoxide dismutase (SOD), catalase (CAT), and ascorbate peroxidase (APX), as they counteract the adverse effects of the ROS. The redox homeostasis in the face of stress is further promoted by non-enzymatic antioxidants, e.g., glutathione, ascorbic acid, and phenolic substances. The synthesis of detoxifying metal-binding peptides such as phytochelatins and metallothioneins at the molecular level is an important determinant of detoxification. These chelators combine with metal ions and transfer them into vacuoles, hindering their cytotoxic activities to a minimum. One of the most efficient detoxification mechanisms, Vacuolar sequestration, keeps the free metal ions away from the vital cellular targets. They make good phytoextraction and rhizofiltration agents because many of these species have large biomass, high growth rate, and very large root surface area per plant. The use of engineered wetlands with hydrophytes will provide low-cost maintenance methods of treating metal-contaminated effluents, especially in areas where there is a poor infrastructure for comprehensive wastewater treatment. Finally, hydrophytes constitute an exclusive category of plants that may grow in water bodies polluted by metals. This can be exploited both through their utility in ecological investigations and in environmental remediation processes, as they are able to tolerate, accumulate, and detoxify heavy metals. Leveraging their inherent potential in combination with biotechnological remedial measures, it can be stated that the abusive impact of the heavy metal pollution can be reduced and aid in restoring balance in the environment. To secure water security, biodiversity, and public health during the next several decades, it is essential to consider hydrophyte-based phytoremediation solutions since the level of global pollution is increasing every day.

9. FUTURE PROSPECTS

The future of phyto-remediation with hydrophytes is bright, particularly as the world continues to pressure in search of technologies that are more sustainable to combat the heavy metal pollution at relatively low cost and in an environmentally friendly manner. Research trends indicate that this green technology has had few directions whose implementation would render the technology much more efficient, scalable, and acceptable.

- First, to start with, forward biotechnological approaches, such as genetic engineering, CRISPR/Cas9-based genome editing, and marker-assisted selection, can be integrated to design hydrophytes with increased tolerance and heavy metal accumulation capacity. Altered plants that are engineered to synthesize larger amounts of metal transporters, antioxidant enzymes, and metal-binding peptides (phytochelatins and metallothioneins) have the potential to enhance the capacity of plants to phyto-remediate significantly.
- Second, the additional application of omics technologies (genomics, transcriptomics, proteomics, metabolomics, and ionomics) will be utilized to have a deeper insight into stress-responsive regulatory networks in hydrophytes.
- Third, the implementation of hydrophytes in the design of engineered wetlands and phyto-remediation-based wastewater treatment systems will tend to gain wider acceptance as an alternative to conventional wastewater treatment plants. To ensure a high degree of heavy metal removal, pilot and large-scale field experiments will be necessary to optimize environmental factors, including pH, nutrient levels, and hydraulic retention time.
- Fourth, one of the largest opportunities in the future is the valorization of contaminated biomass. This corresponds to what a circular economy implies, by which such a notion as phytoremediation does not just provide a solution to the environment, but is also a profitable alternative.

REFERENCES

Alam, M. Z., Ahmad, S., Malik, A., & Khan, Z. (2021). Phytoremediation potential of *Typha* and *Phragmites* for heavy metal-contaminated wetlands. *Environmental Science and Pollution Research International, 28*(12), 14527–14539. DOI: 10.1007/s11356-020-11463-1

Ali, H., Khan, E., & Ilahi, I. (2019). Environmental chemistry and ecotoxicology of hazardous heavy metals: Environmental persistence, toxicity, and bioaccumulation. *Journal of Chemistry, 2019*, 1–14. DOI: 10.1155/2019/6730305

Ali, H., Khan, E., & Sajad, M. A. (2013). Phytoremediation of heavy metals—Concepts and applications. *Chemosphere, 91*(7), 869–881. DOI: 10.1016/j.chemosphere.2013.01.075 PMID: 23466085

Ali, M. M., Ali, M. L., Islam, M. S., & Rahman, M. Z. (2020a). Preliminary assessment of heavy metals in water and sediment of Karnaphuli River, Bangladesh. *Environmental Nanotechnology, Monitoring & Management, 14*, 100333. DOI: 10.1016/j.enmm.2020.100333

Ali, S., Rizwan, M., Qayyum, M. F., Ok, Y. S., Ibrahim, M., Riaz, M., & Arif, M. S. (2020b). Application of floating aquatic plants in phytoremediation of heavy metals polluted water: A review. *Sustainability (Basel), 12*(5), 1927. DOI: 10.3390/su12051927

Alloway, B. J. (2013). *Heavy metals in soils: Trace metals and metalloids in soils and their bioavailability* (3rd ed.). Springer., DOI: 10.1007/978-94-007-4470-7

Apel, K., & Hirt, H. (2004). Reactive oxygen species: Metabolism, oxidative stress, and signal transduction. *Annual Review of Plant Biology, 55*(1), 373–399. DOI: 10.1146/annurev.arplant.55.031903.141701 PMID: 15377225

Baba, K.. (2016). Vacuolar sequestration mechanisms of heavy metals in plants. *Plant Science Review, 5*(2), 101–118.

Bienert, G. P., Thorsen, M., Schüssler, M. D., Nilsson, H. R., Wagner, A., Tamás, M. J., & Jahn, T. P. (2008). A subgroup of plant aquaporins facilitates the bidirectional diffusion of $As(OH)_3$ and $Sb(OH)_3$ across membranes. *BMC Biology, 6*(1), 26. DOI: 10.1186/1741-7007-6-26 PMID: 18544156

Bini, C., & Bech, J. (2014). PHEs, environment, and human health. In Sanchez, F., & Alvaro, R. (Eds.), *Environmental geochemistry: Site characterization, data analysis and case histories* (pp. 59–88). Elsevier., DOI: 10.1016/B978-0-08-099364-7.00004-3

Bonanno, G., & Lo Giudice, R. (2010). Heavy metal bioaccumulation by the organs of *Phragmites australis* (common reed) and their potential use as contamination indicators. *Ecological Indicators, 10*(3), 639–645. DOI: 10.1016/j.ecolind.2009.11.002

Bornette, G., & Puijalon, S. (2011). Response of aquatic plants to abiotic factors. *Aquatic Sciences, 73*(1), 1–14. DOI: 10.1007/s00027-010-0162-7

Burton, E. D., & Johnston, S. G. (2010). Impact of sediment geochemistry on trace metal bioavailability. *Environmental Chemistry, 7*(6), 451–464. DOI: 10.1071/EN10075

Chambers, P. A., Lacoul, P., Murphy, K. J., & Thomaz, S. M. (2008). Global diversity of aquatic macrophytes in freshwater. *Hydrobiologia, 595*(1), 9–26. DOI: 10.1007/s10750-007-9154-6

Che, J., Kumar, A., Gupta, S., & Khatri, N. (2024). Comparative performance of emergent macrophytes in treating heavy metal-contaminated water. *Environmental Science and Pollution Research International, 31*(2), 2841–2853. DOI: 10.1007/s11356-023-29235-1

Clemens, S., Aarts, M. G. M., Thomine, S., & Verbruggen, N. (2013). Plant science: The key to preventing slow cadmium poisoning. *Trends in Plant Science, 18*(2), 92–99. DOI: 10.1016/j.tplants.2012.08.003 PMID: 22981394

Collin, A.. (2022). Heavy metal interference with seed germination, water uptake, and early seedling growth. *Frontiers in Ecology and Evolution, 11*, 130894. DOI: 10.3389/fevo.2023.130894

Colmer, T. D. (2003). Long-distance transport of gases in plants: A perspective on internal aeration and radial oxygen loss from roots. *Plant, Cell & Environment, 26*(1), 17–36. DOI: 10.1046/j.1365-3040.2003.00846.x

DalCorso, G., Farinati, S., Maistri, S., & Furini, A. (2019). How plants cope with cadmium: Staking all on metabolism and gene expression. *Journal of Integrative Plant Biology, 61*(6), 653–676. DOI: 10.1111/jipb.12738 PMID: 19017114

Degryse, F., Smolders, E., & Parker, D. R. (2009). Partitioning of metals (Cd, Co, Cu, Ni, Pb, Zn) in soils: Concepts, methodologies, prediction and applications. *European Journal of Soil Science, 60*(4), 590–612. DOI: 10.1111/j.1365-2389.2009.01142.x

Degryse, F., Smolders, E., & Parker, D. R. (2009). Partitioning of metals (Cd, Co, Cu, Ni, Pb, Zn) in soils: Concepts, methodologies, prediction and applications – A review. *European Journal of Soil Science, 60*(4), 590–612. DOI: 10.1111/j.1365-2389.2009.01142.x

Du Laing, G., Rinklebe, J., Vandecasteele, B., Meers, E., & Tack, F. M. G. (2009). Trace metal behaviour in estuarine and riverine floodplain soils and sediments: A review. *The Science of the Total Environment, 407*(13), 3972–3985. DOI: 10.1016/j.scitotenv.2008.07.025 PMID: 18786698

Eggleton, J., & Thomas, K. V. (2004). A review of factors affecting the release and bioavailability of contaminants during sediment disturbance events. *Environment International, 30*(7), 973–980. DOI: 10.1016/j.envint.2004.03.001 PMID: 15196845

Förstner, U., & Salomons, W. (2010). Sediment research, management, and policy. *Journal of Soils and Sediments, 10*(8), 1440–1452. DOI: 10.1007/s11368-010-0310-7

Förstner, U., & Wittmann, G. T. W. (2012). *Metal pollution in the aquatic environment* (3rd ed.). Springer., DOI: 10.1007/978-3-642-69385-5

Fu, F., & Wang, Q. (2011). Removal of heavy metal ions from wastewaters: A review. *Journal of Environmental Management, 92*(3), 407–418. DOI: 10.1016/j.jenvman.2010.11.011 PMID: 21138785

Gonzalez Mendoza, D., Moreno Sanchez, R., & Santamaria, J. M. (2007). Metallothionein induction in aquatic plants under Cd/Cu stress. *Environmental Botany, 43*(4), 415–422. DOI: 10.1016/j.envexpbot.2006.12.005

Ha, N. T. T.. (2019). Heavy metal accumulation in *Nelumbo nucifera* from contaminated water bodies. *Environmental Monitoring and Assessment, 191*(2), 1–10. DOI: 10.1007/s10661-019-7226-4

Haghshenas, S. M. (2020). An exploratory view of water quality and sediment characteristics in Musa Estuary, the Persian Gulf. *arXiv*. https://doi.org//arXiv.2004.12548DOI: 10.48550

Haghshenas, S. S. (2020). Atmospheric deposition of heavy metals in aquatic ecosystems: Sources, processes, and consequences. *Environmental Monitoring and Assessment, 192*(8), 517. DOI: 10.1007/s10661-020-08525-w PMID: 32666386

Heliyon. (2024). Phytoremediation potential of *Pistia stratiotes, Eichhornia crassipes*, and *Typha latifolia* for chromium removal from tannery wastewater. *Heliyon, 10*(2), e13578. https://doi.org/DOI: 10.1016/j.heliyon.2024.e13578

Hudson-Edwards, K. A. (2016). Tackling mine wastes. *Science, 352*(6283), 288–290. DOI: 10.1126/science.aaf3354 PMID: 27081053

Huser, B. J., Futter, M. N., Bärlund, I., & Ecke, F. (2021). Managing sediment and nutrient interactions with metals in aquatic ecosystems. *The Science of the Total Environment, 757*, 143860. DOI: 10.1016/j.scitotenv.2020.143860

Jaishankar, M., Tseten, T., Anbalagan, N., Mathew, B. B., & Beeregowda, K. N. (2014). Toxicity, mechanism, and health effects of some heavy metals. *Interdisciplinary Toxicology*, *7*(2), 60–72. DOI: 10.2478/intox-2014-0009 PMID: 26109881

Kabata-Pendias, A. (2011). *Trace elements in soils and plants* (4th ed.). CRC Press.

Keunen, E.. (2011). Role of ROS and antioxidant systems in heavy metal stress. *Plant Physiology and Biochemistry*, *49*(6), 649–658. DOI: 10.1016/j.plaphy.2011.03.017

Kim, D. Y., Bovet, L., Maeshima, M., Martinoia, E., & Lee, Y. (2006). Plant ABC transporters sequester PC-metal complexes into vacuoles. *The Plant Journal*, *46*(3), 685–697. DOI: 10.1111/j.1365-313X.2006.02731.x

Kosakivska, I.. (2020). Molecular mechanisms of heavy metal tolerance in aquatic plants. *International Journal of Aquatic Plant Biology*, *12*(4), 183–195.

Kristensen, E., Penha-Lopes, G., Delefosse, M., Valdemarsen, T., Quintana, C. O., & Banta, G. T. (2012). What is bioturbation? The a need for a precise definition of fauna in aquatic sciences. *Marine Ecology Progress Series*, *446*, 285–302. DOI: 10.3354/meps09506

Li, X., Shen, Z., Wai, O. W. H., & Li, Y. S. (2013). Chemical forms of Pb, Zn, and Cu in the sediment profiles of the Pearl River Estuary. *Marine Pollution Bulletin*, *47*(1–6), 107–116. DOI: 10.1016/S0025-326X(02)00418-4 PMID: 11381876

Lin, Y.. (2022). Adaptations and functions of aquatic macrophytes in wetland ecosystems. *Wetlands Ecology and Management*, *30*(4), 1–16. DOI: 10.1007/s11273-022-09887-3

Lin, Y., Zhang, T., Liu, S., & Wang, M. (2022). Phytoremediation of aquatic ecosystems: Mechanisms and future perspectives. *Environmental Science and Pollution Research International*, *29*(12), 17301–17319. DOI: 10.1007/s11356-021-16924-y

Lottermoser, B. G. (2010). *Mine wastes: Characterization, treatment and environmental impacts* (3rd ed.). Springer., DOI: 10.1007/978-3-642-12419-8

Luoma, S. N., & Rainbow, P. S. (2008). *Metal contamination in aquatic environments: Science and lateral management*. Cambridge University Press., DOI: 10.1017/CBO9780511805360

Maberly, S. C., & Madsen, T. V. (2002). Freshwater angiosperm carbon concentrating mechanisms: Processes and patterns. *Functional Plant Biology*, *29*(3), 393–405. DOI: 10.1071/PP01181 PMID: 32689484

Malik, A. (2007). Environmental challenge vis-à-vis opportunity: The case of water hyacinth. *Environment International*, *33*(1), 122–138. DOI: 10.1016/j.envint.2006.08.004 PMID: 17010439

Maurya, N. K., Singh, R. K., & Yadav, P. (2025). Phytoremediation of metal contamination using *Pistia stratiotes* and *Eichhornia crassipes*: A comparative analysis. *International Journal of Environment and Climate Change*, *15*(4), 205–221. DOI: 10.9734/ijecc/2025/v15i44804

Mishra, S.. (2023). Phytoremediation of heavy metals by wetland plants: A review. *Environmental Technology & Innovation*, *30*, 103152. DOI: 10.1016/j.eti.2023.103152

Mishra, S., Srivastava, S., Tripathi, R. D., Kumar, R., Seth, C. S., & Gupta, D. K. (2006). Lead detoxification by coontail (Ceratophyllum demersum L.) involves the induction of phytochelatins and the antioxidant system in response to its accumulation. *Chemosphere*, *65*(6), 1027–1039. DOI: 10.1016/j.chemosphere.2006.03.033 PMID: 16682069

Mishra, V. K., Tripathi, B. D., & Srivastava, S. (2008). Efficiency of constructed wetlands for heavy metal removal. *Ecological Engineering*, *34*(2), 123–130. DOI: 10.1016/j.ecoleng.2008.07.004

Mishra, V. K., Upadhyay, A. R., Pandey, S. K., & Tripathi, B. D. (2008). Heavy metal pollution in a tropical river, *Ganga*: A case study in a small stretch at Varanasi. *Environmental Monitoring and Assessment*, *142*(1-3), 419–425. DOI: 10.1007/s10661-007-9938-1

Mittler, R., Vanderauwera, S., Gollery, M., & Van Breusegem, F. (2004). Reactive oxygen species signaling in plants. *Trends in Plant Science*, *9*(10), 490–498. DOI: 10.1016/j.tplants.2004.08.009 PMID: 15465684

Nagajyoti, P. C., Lee, K. D., & Sreekanth, T. V. M. (2010). Heavy metals, occurrence and toxicity for plants: A review. *Environmental Chemistry Letters*, *8*(3), 199–216. DOI: 10.1007/s10311-010-0297-8

Neubauer, S. C., Franklin, R. B., & Berrier, D. J. (2008). Saltwater intrusion into tidal freshwater marshes alters the biogeochemical processing of organic carbon. *Biogeosciences*, *5*(2), 215–221. DOI: 10.5194/bg-5-215-2008

Nordstrom, D. K. (2011). Hydrogeochemical processes governing the origin, transport, and fate of major and trace elements from mine wastes and mineralized rock. *Applied Geochemistry*, *26*(11), 1777–1791. DOI: 10.1016/j.apgeochem.2011.06.002

Noyma, N. P., Cotovicz, L. C., & Brandini, N. (2020). Climate change effects on aquatic ecosystems: Impacts of extreme events on metal mobility. *Environmental Research Letters*, *15*(11), 113006. DOI: 10.1088/1748-9326/abb4cb

Osman, M. E., & Fadhlallah, N. (2010). Impact of heavy metals on seed germination and stress physiology. *Plant Stress Physiology*, *34*, 45–56.

Oursel, B., Garnier, C., Durrieu, G., Mounier, S., Omanović, D., & Lucas, Y. (2014). Dynamics and fates of trace metals in an urbanized coastal zone: The case study of Marseille Bay (France). *Marine Pollution Bulletin*, *84*(1–2), 29–38. DOI: 10.1016/j.marpolbul.2014.05.053

Pacyna, J. M., & Pacyna, E. G. (2001). An assessment of global and regional emissions of trace metals to the atmosphere from anthropogenic sources worldwide. *Environmental Reviews*, *9*(4), 269–298. DOI: 10.1139/a01-012

Pereira, D., Barros, M. L., & Arruda, M. A. Z. (2021a). Evaluation of cadmium bioaccumulation-related physiological effects in *Salvinia biloba*. *Plants*, *10*(12), 2679. DOI: 10.3390/plants10122679 PMID: 34961150

Pereira, D., Cabrera, G. C., Pires-Lopes, K. M., Siqueira-Silva, A. I., de Souza, M. C., & Damatta, F. M. (2021b). Evaluation of cadmium bioaccumulation-related physiological effects in *Salvinia biloba*. *Plants*, *10*(12), 2679. DOI: 10.3390/plants10122679 PMID: 34961150

Pereira, P., Fernando, A. L., & Caeiro, S. (2021c). Assessment of trace metal contamination and bioavailability in sediments: Implications for ecosystem health. *Environmental Pollution*, *277*, 116842. DOI: 10.1016/j.envpol.2021.116842

Pourrut, B.. (2011). Lead and cadmium toxicity: Oxidative stress in plant cells. *Plant Physiology and Biochemistry*, *49*(6), 633–638. DOI: 10.1016/j.plaphy.2011.03.004

Prasad, M. N. V. (2004). *Heavy metal stress in plants: From biomolecules to ecosystems* (2nd ed.). Springer., DOI: 10.1007/978-3-662-07743-6

Pushpamala, E., Dhanalakshmi, S., & Govindarajan, V. S. (2014). Effect of mercury and cadmium on biochemical parameters of *Hydrilla verticillata*. *Life Science–Environmental Biology*, *3*(2), L58–L66.

Qayoom, I., & Jaies, I. (2023). Review on the phytoremediation potential of aquatic macrophytes. *Environment Conservation Journal*, *24*(1), 273–280. DOI: 10.36953/ECJ.12102318

Rahman, M. A., & Hasegawa, H. (2011). Aquatic arsenic phytoremediation using floating macrophytes: A review. *Chemosphere, 83*(5), 633–646. DOI: 10.1016/j.chemosphere.2011.02.045 PMID: 21435676

Rai, P. K. (2008). Heavy metal phytoremediation by aquatic macrophytes. *International Journal of Phytoremediation, 10*(2), 133–160. DOI: 10.1080/15226510801913918 PMID: 18709926

Rai, P. K. (2008). Heavy metal phytoremediation by aquatic macrophytes. *International Journal of Phytoremediation, 10*(2), 133–160. DOI: 10.1080/15226510801913918 PMID: 18709926

Rezania, S.. (2016). Perspectives of phytoremediation using water hyacinth for the removal of heavy metals. *Process Safety and Environmental Protection, 102*, 128–138. DOI: 10.1016/j.psep.2016.03.017

Royal Society of Chemistry. (2023). *Contamination of heavy metals in aquatic media: Transport, toxicity and technologies for remediation*. Royal Society of Chemistry., DOI: 10.1039/9781839165111

Sand-Jensen, K. (2003). Drag forces on common plant species in temperate streams: Consequences for growth. *Hydrobiologia, 481*(1-3), 25–30. DOI: 10.1023/A:1021299014644

Sharma, P., & Dubey, R. S. (2005). Lead toxicity in plants. *Brazilian Journal of Plant Physiology, 17*(1), 35–52. DOI: 10.1590/S1677-04202005000100004

Sharma, P., & Dubey, R. S. (2005). Lead toxicity in plants. *Brazilian Journal of Plant Physiology, 17*(1), 35–52. DOI: 10.1590/S1677-04202005000100004

Shourie, A. (2022). Effect of cadmium and lead stress on seed germination in *Jatropha curcas. Biosciences Biotechnology Research Asia, 19*(3), 671–678. Advance online publication. DOI: 10.13005/bbra/3019

Singh, A., Sharma, R. K., Agrawal, M., & Marshall, F. M. (2011). Health risk assessment of heavy metals via dietary intake of foodstuffs from the wastewater irrigated site of a dry tropical area of India. *Food and Chemical Toxicology, 48*(2), 611–619. DOI: 10.1016/j.fct.2009.11.041 PMID: 19941927

Singh, R.. (2023). Heavy metal removal efficiency of duckweed species in wastewater treatment. *Environmental Nanotechnology, Monitoring & Management, 20*, 100830. DOI: 10.1016/j.enmm.2023.100830

Singh, R. P., & Tewari, R. K. (2003). Cadmium toxicity induced changes in plant water relations and oxidative metabolism of Brassica juncea L. plants. *Journal of Environmental Biology*, *24*(1), 107–112. PMID: 12974420

Soda, S., Mishima, D., Inoue, D., & Ike, M. (2012). Removal of toxic metals using aquatic plants. *Bioresource Technology*, *121*, 432–439. DOI: 10.1016/j.biortech.2012.06.036

Sompura, Y., Patel, K., & Joshi, N. (2024). Combined microplastic and heavy metal effects on biomass and physiology of water spinach. *Journal of Current Opinion in Crop Science*, *5*(1), 48–61. DOI: 10.62773/jcocs.v5i1.233

Sompura, Y., Patel, K., & Joshi, N. (2024). Recent advances in phytoremediation of polluted waters using hydrophytes. *Journal of Current Opinion in Crop Science*, *5*(1), 48–61. DOI: 10.62773/jcocs.v5i1.233

Srivastava, S.. (2008). Phytoremediation of chromium from industrial wastewater by aquatic macrophytes. *Environmental Technology*, *29*(3), 299–307. DOI: 10.1080/09593330802095083

Su, C., Jiang, L., Zhang, W., & Zhou, Q. (2012). Uptake and accumulation of heavy metals in plants: A review focusing on phytoremediation applications. *Environmental Science and Pollution Research International*, *19*(7), 2251–2261. DOI: 10.1007/s11356-011-0679-0

Szalai, G.. (2015). Phytochelatin induction in *Lemna minor* exposed to cadmium. *Ecotoxicology (London, England)*, *24*(2), 304–316. DOI: 10.1007/s10646-014-1375-6

Tebo, B. M., Bargar, J. R., Clement, B. G., Dick, G. J., Murray, K. J., Parker, D., Verity, R., & Webb, S. M. (2004). Biogenic manganese oxides: Properties and mechanisms of formation. *Annual Review of Earth and Planetary Sciences*, *32*(1), 287–328. DOI: 10.1146/annurev.earth.32.101802.120213

Thomaz, S. M., & Bini, L. M. (2003). Ecological responses of aquatic vegetation to nutrient enrichment. *Acta Limnologica Brasiliensia*, *15*(1), 23–29.

Turetsky, M. R., Benscoter, B., Page, S., Rein, G., Van Der Werf, G. R., & Watts, A. (2020). Global vulnerability of peatlands to fire and carbon loss. *Nature Geoscience*, *13*(1), 1–6. DOI: 10.1038/s41561-019-0468-3

Vestena, S., Cambraia, J., & Ribeiro, C. (2016). Lead-induced heavy metal toxicity alters growth and antioxidative enzyme levels in *Eichhornia crassipes*. *Botanical Studies*, *55*(1), 54. DOI: 10.1186/s40529-014-0054-6 PMID: 28597420

Vestena, S. A., Ribeiro, C., & Cambraia, J. (2017). Cadmium-induced oxidative stress and antioxidative enzyme response in water hyacinth and *Salvinia. Brazilian Journal of Plant Physiology*, *23*(2), 131–139. DOI: 10.1590/S1677-04202011000200005

Villamagna, A. M., & Murphy, B. R. (2010). Ecological and socio-economic impacts of invasive water hyacinth (*Eichhornia crassipes*): A review. *Freshwater Biology*, *55*(2), 282–298. DOI: 10.1111/j.1365-2427.2009.02294.x

Voesenek, L. A. C. J.. (2016). Plant responses to flooding stress. *Current Opinion in Plant Biology*, *33*, 64–71. DOI: 10.1016/j.pbi.2016.06.005 PMID: 27322538

Vymazal, J. (2011). Constructed wetlands for wastewater treatment. *Environmental Science & Technology*, *45*(1), 61–69. DOI: 10.1021/es101403q PMID: 20795704

Wang, W.-X., & Chen, X. (2019). Bioavailability and bioaccumulation of heavy metals in aquatic organisms from China. *Environmental Science and Pollution Research International*, *26*(19), 19023–19041. DOI: 10.1007/s11356-019-04988-7 PMID: 31090002

Wang, X., Liu, C., & Huang, Y. (2008). Cadmium responses in *Vallisneria spiralis*: A freshwater submerged macrophyte. *Chemosphere*, *71*(2), 297–303. DOI: 10.1016/j.chemosphere.2007.10.008

Wang, X., McInroy, J., Deakin, G. D., & Atkins, C. A. (2008). Cadmium responses in *Vallisneria spiralis. Chemosphere*, *71*(2), 297–303. DOI: 10.1016/j.chemosphere.2007.10.008

Xu, C., Zhou, T., Wang, D., & Sun, X. (2014). Proteomic insights into Cd tolerance in *Eichhornia crassipes. Environmental Science & Technology*, *48*(12), 7012–7020. DOI: 10.1021/es500914n PMID: 24819143

Zhang, D.. (2024). Cadmium tolerance mechanisms in water hyacinth involving MT1 and ROS balance. *Environmental Pollution*, *315*, 120316. DOI: 10.1016/j.envpol.2022.120316

Zhao, H.. (2021). Floating-leaved aquatic macrophytes: Functional traits and ecosystem services. *Aquatic Botany*, *172*, 103394. DOI: 10.1016/j.aquabot.2021.103394

Zheng, J., Tan, M., Shibata, Y., Tanaka, A., Li, Y., Zhang, G., & Zhang, Y. (2007). Characteristics of lead isotope ratios and elemental concentrations in PM2.5 in Beijing, China. *Atmospheric Environment*, *41*(3), 541–551. DOI: 10.1016/j.atmosenv.2006.08.040

Zhou, Q.. (2007a). Role of aquatic macrophytes in phytoremediation of heavy metals: A review. *Environmental Reviews*, *15*(1), 1–12. DOI: 10.1139/a06-003

Zhou, Q., Zhang, J., Fu, J., Shi, J., & Jiang, G. (2007b). Biomonitoring: An appealing tool for assessment of metal pollution in the aquatic ecosystem. *Analytica Chimica Acta*, *606*(2), 135–150. DOI: 10.1016/j.aca.2007.11.018 PMID: 18082645

Chapter 2
Hydrophytes and Heavy Metal Stress:
Mechanisms, Adaptations, and Phytoremediation Potential

Ajay Kumar
https://orcid.org/0000-0002-5312-9823
University Centre for Research and Development, Chandigarh University, Mohali, India

Sandeep Kaur
https://orcid.org/0000-0002-2426-9173
University Centre for Research and Development, Chandigarh University, Mohali, India

Brahmjot Singh
Department of Pharmaceutical Sciences, Guru Nanak Dev University, Amritsar, India

ABSTRACT

Heavy metals like cadmium (Cd) and arsenic (As) significantly affect hydrophyte growth, disrupting processes like photosynthesis and causing oxidative stress. Some species, such as Elodea canadensis (submerged) and Eichhornia crassipes (floating), act as hyperaccumulators, capturing metals at efficiencies exceeding 72.5% in contaminated environments. In contrast, species less tolerant to these metals exhibit toxicity through chlorosis, inhibited root elongation, and nutrient uptake failure. Hydrophytes employ tolerance mechanisms like chelation via phytochelatins and metallothioneins, antioxidant enzymes like superoxide dismutase, and reducing reactive oxygen species (ROS). Structural adaptations, such as iron plaque formation on Typha latifolia roots, immobilize metals effectively. Species

DOI: 10.4018/979-8-3373-4037-1.ch002

like Sedum alfredii promote metal translocation, while Lemna minor uses vacuolar sequestration and rhizosphere symbiosis to minimize metal bioavailability. Further studies are needed to understand long-term ecological impacts and genetic controls of tolerance mechanisms.

1. INTRODUCTION

As a result of anthropogenic activities, including mining processes, agricultural runoff, and industrial emissions, the increasing accumulation of heavy metals (HMs) in the water environment has become a pressing environmental concern. These metals, including cadmium (Cd), lead (Pb), mercury (Hg), chromium (Cr), nickel (Ni), and arsenic (As), are non-biodegradable and persist in the environment, being a threat to aquatic flora and fauna (P. Sharma & Dubey, 2005). Freshwater aquatic macrophytes, also known as hydrophytes, play a crucial role in maintaining the chemistry and physical habitat of aquatic ecosystems. They do this by retaining water quality, adding oxygen to the water column, and providing a habitat for diverse biotic communities. Heavy metal toxicity, which disrupts cellular homeostasis and metabolism, often has adverse effects on the growth and physiological development of organisms (P. K. Rai, 2009; Sytar et al., 2013). Heavy metals and hydrophytes. They have a complex relationship with heavy metals, including their absorption, translocation, accumulation, and detoxification (Figure 1). Certain hydrophytes have an unusual ability to withstand and even accumulate large quantities of heavy metals without suffering much damage, suggesting that they may be potential candidates for phytoremediation, and others are highly sensitive, as evidenced by reduced photosynthesis, enzyme activity, and biomass (DalCorso et al., 2019; Zhang et al., 2020). Conventional remediation methods are costly and hazardous; therefore, phytoremediation utilises plants to remediate contaminants (Sahoo et al., 2025). The use of hydrophytes (aquatic plants) in cleaning up heavy metals dates back to the 1970s. During the 1990s, phytoremediation gained popularity, particularly with metal-absorbing species such as *Eichhornia crassipes* and *Pistia stratiotes* (Bayuo et al., 2024). In the 2000s, researchers discovered that bioactive compounds, such as flavonoids, play a role in detoxification. Nanomaterials, genetic engineering, and interaction between plants and microbes have recently emerged as a cost-effective environmental cleanup option through phyto-remediation (Díaz-Torres et al., 2021). For instance, some species, including *Hydrilla verticillata*, *Lemna minor*, and Eichhornia crassipes, have demonstrated bioaccumulation potential and tolerance mechanisms (antioxidant systems, metals sequestration in vacuoles, and induction of metal-chelating phytochelatins) as well (Song et al., 2018). The ability of hydrophytes to respond to heavy metal-induced oxidative stress is a crucial component of their metal tolerance mechanisms. These metals often lead to overproduction of reactive oxidative species (ROS), which, if unmanaged, can damage lipids, proteins, and nucleic acids (Shahid et al., 2014). In order to scavenge ROS, tolerant plants induce antioxidant systems, including SOD, CAT, and APX. Nevertheless, metal detoxification processes can involve the production of phytochelatins, metallothioneins, and organic acids that assist in the storage or compartmentation of metals in less reactive areas of the cells (Sytar et al., 2013; Visioli et al., 1998). Upon high exposure to heavy metals, the concentrations of these metals in plants may exceed the level that tolerant species can detoxify, resulting in decreased growth, chlorosis, necrosis, and even plant death, despite these changes (Xie et al., 2022). Furthermore, the contamination level and bioaccumulation efficiency in hydrophytes are closely related to metal speciation and bioavailability, which are influenced by temperature, pH, and the presence of organic matter in the water (Clemens, 2006). Hydrophyte cleanup is economical and less toxic than chemical and microbiological processes (Farid et al., 2023). Chemical treatments are costly, energy-consuming, and wasteful; thus, they cannot be used on a large scale (Ahmed et al., 2021). Microbial solutions are an environmentally friendly option that is not particularly fast in removing metals and requires proper management of different microbial populations (Verstraete et al., 2022). An alternative to this is the use of hydrophytes, which naturally recruit and detoxify heavy metals, making it a lower-cost, low-maintenance, and eco-friendly process (Aryal, 2024). Biomass disposal and seasonal growth changes can cause them to slow down and decrease their effectiveness in highly contaminated regions (Saha et al., 2023). These interactions between plants and metals are essential in understanding how to develop effective methods that not only use aquatic grass to achieve a method in phytoremediation but also in evaluating the risks associated with our environment. This chapter aims at: describing the physiological and molecular processes of tolerance and toxicity, evaluating their phytoremediation and survival in contaminated aqueous eco systems and finally, summarising the nature of the growth and toxic effects of heavy metals on the development of hydrophytes.

Figure 1. Heavy metal effects on the growth of hydrophytes

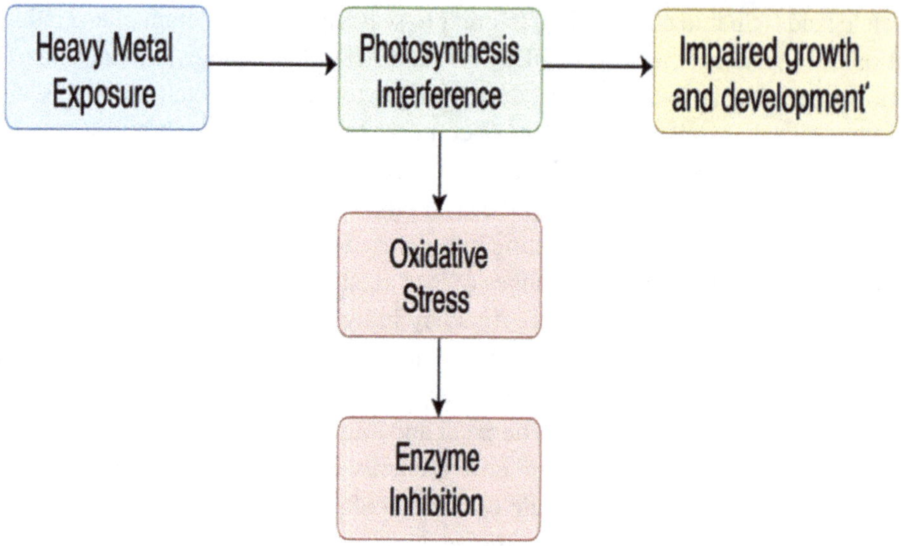

2. EFFECTS OF HEAVY METALS ON HYDROPHYTES

2.1. Toxic Effects of Cadmium and Arsenic on Hydrophyte Growth

Cadmium (Cd) and Arsenic (As) are two examples of heavy metals that have a strong negative impact when it comes to hydrophyte physiology. Namely, they disable photosynthesis, augment oxidative tension and restrain essential metabolic enzymes. These effects are combined to prevent the natural development and growth of the aquatic weeds, hence decreasing their phytoremediation capability as well as their ecological practices (Figure 2).

Figure 2. Toxic effects of heavy metals on hydrophyte growth

Interference with Photosynthesis

Cadmium (Cd) and Arsenic (As) have a major impact on the photosynthesis of hydrophytes. The Cd prevents the formation of chlorophyll by replacing important cations such as Mg 2+ in the chlorophyll molecule, thus leading to the degradation of pigments and damage to the structure of chloroplasts (G. Deng et al., 2014). Exposure to Cd profoundly reduced the chlorophyll concentration in *Ceratopteris pteridoides*, which is a direct outcome of the carbon assimilation and photosystem II (PSII) performance (G. Deng et al., 2014). Looking similar to phosphate and affecting the synthesis of ATP, arsenic, in particular arsenate, can suppress the supply of energy required to perform photosynthesis and metabolic reactions (Wang et al., 2021). Such disruptions lead to the noticeable retardation of the growth in aquatic macrophytes, along with the biomass reduction in their growth (Bali & Sidhu, 2021).

Induction of Oxidative Stress

Cd and As are both strong inducers of oxidative stress. More effective generation of reactive oxygen species (ROS), such as superoxide anion (O 2 -), hydrogen peroxide (H 2 O 2), and hydroxyl radicals, is increased, which are destructive to lipids, proteins, and nucleic acids (Dhir et al., 2004; Santos et al., 2015). Because of higher lipid peroxidation, which is more often measured as higher levels of malondialdehyde (MDA), cell membrane destabilization is also an indication of such oxidative damage. As per Al-Zurfi et al.(2021), Cd exposure in *Hydrilla verticillata* elevated the level of MDA and repressed the activity of catalase (CAT) and peroxidase (POD), which suggested the attenuated antioxidant defence mechanism (Al-Zurfi et al., 2021). Likewise, arsenic stress led to the destruction of chloroplasts, loss of photochemical activity, and increased oxidative damage in rice seedlings (Mishra et al., 2017).

Inhibition of Enzymes

Cd and As interfere with the work of key enzymes in the assimilation and metabolism of nutrients and protection against free radicals. Cd also prevents the effects of the nitrate reductase, glutathione reductase, and sulfur metabolism and detoxification enzymes, which are needed in protein synthesis and detox processes (Aravind et al., 2009; Clijsters & Van Assche, 1985). Both as trivalent species (As^{3+}), specifically, arsenic forms a bond at the sulfhydryl (-SH) functional groups on the proteins, which leads to the inhibition of part of the enzymes (Wang et al., 2021). The mechanisms include redox signalling, metabolic balance, and ion homeostasis, among others, which cause lower enzymatic efficacies in the water system, and this ultimately causes poor plant growth and development challenges. There is a synergistic toxic effect of these mechanisms where Cd and As not only suppress the growth and productivity of the hydrophytes, but they also pose a threat to the ecological stability of these organisms. In order to identify and develop aquatic organisms to be utilized in the bioremediation project and sustain the health of the freshwater ecosystem, there exists a need to know such interactions.

2.2. Metal Accumulation in Hydrophytes

The concentration of heavy metals in hydrophytes varies significantly between species, with noticeable distinctions between floating and submerged aquatic macrophytes. Because a plant's capacity to absorb metals determines its ecological and remediative potential, it is crucial to comprehend this diversity in order to optimise phytoremediation techniques (Figure 3).

Figure 3. In hydrophytes: Metal uptake and accumulation

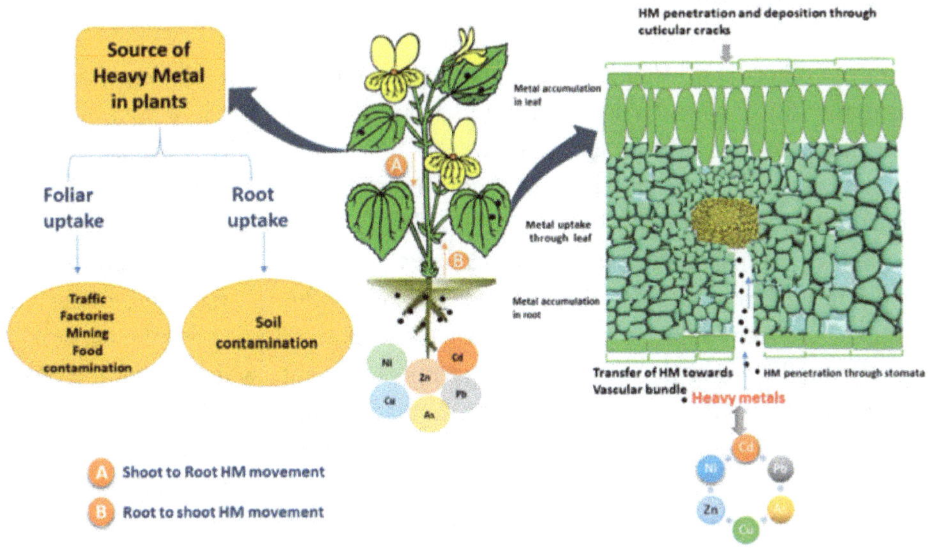

(Hasan et al., 2019).

Low Accumulation in *Elodea canadensis*, a Submerged Species

Generally, the submerged hydrophytes like Elodea are believed to accumulate lesser amounts of heavy metals as compared to floating species. *E. canadensis*, due to its physiology and limited surface contact with the contaminated water column, has been observed in most studies to assimilate metals, including cadmium (Cd), zinc (Zn), and copper (Cu) in relatively low concentrations (Nyquist & Greger, 2007). As a result of a lack of direct interaction between the root and the sediment, the absorption of many of them takes place primarily via the leaves and not the roots. In the experiments, it is found that it is not effective in the collection of the metal due to the inability to accumulate a certain number of metals and the cuticle layers, some of which are negative features of *E. canadensis,* despite its ability to accumulate cadmium (Borisova et al., 2017). Furthermore, the plant is also inclined to control the flow of the metals in order to maintain the process of photosynthesis and reduce accumulation in the plant itself.

Hyperaccumulation in *Eichhornia crassipes* is a characteristic of floating species.

On the other hand, floating water macrophytes like the *Eichhornia crassipes* (popularly known as water hyacinth) are spectacular in taking in a host of heavy metal elements, including cadmium. It was demonstrated by various research that in a relatively brief time, it can absorb more than 72.5 percent of cadmium in contaminated water (Javed et al., 2019). This hyperaccumulation is facilitated by a well-developed root system that occupies a large area and high amounts of metal-binding biomolecules, which include phytochelatins and metallothioneins. In addition, *E. crassina* has ion exchange and sequestration tools with the capability to carbonate by using the vacuolar compartmentalization, which limits the cytotoxic impacts but keeps a high ratio of accumulation capacity (Favas et al., 2016). Moreover, root exudates released by the plant have been demonstrated to change the rhizosphere PH and increase bioavailability, resulting in metal uptake. This is because *E. crassina* can be proposed as an alternative that can be employed in a long-term phyto-remediating initiative that will be implemented on contaminated freshwater systems, because of its capability to generate and recycle biomass. Lots of cadmium-related chemicals, such as lead (Pb), arsenic (As), chromium (Cr), and nickel (Ni), accumulate in large amounts, and in numerous aspects, it is less costly and cleaner than conventional methods of waste treatment via adherence to chemicals or mechanical processes (Zorza et al., 2022). When used together, these species help to demonstrate the extent of the methods available to hydrophytes to store heavy metals. *Eichhornia crassipes* has been shown to be the best at long-scale technology of phytoremediation of heavy metals and thus may rank well on factors including heavy metal pollution as compared to Elodea canadensis, being viable in biomonitoring and less polluted areas.

Emerging Trends in Phytoremediation and Policy Implications

Phytoremediation, a green technology, is increasingly becoming significant in the face of increased environmental issues in the world, especially in the application of hydrophytes in the eradication of heavy metals in water bodies (Abdelaal et al., 2021). This technique can be used as an ecologically friendly substitute for the standard chemical and mechanical remedies that might be expensive, dangerous, and less scalable, especially within the polluted urban and industrial environments (Di Stasio et al., 2025). The sustained focus on sustainable cleanup of environmental pollution in the world is impacting the development and introduction of hydrophytes to serve as an agent of remedial activity. The impact of climatic changes has necessitated the regulatory powers to reduce the industrial waste in our water bodies, and this is one of the ways through which they have managed this situation, and that is why the governmental bodies are on the move to develop low-cost but effective water purification technologies, particularly in parts of the world where there may be few resources (M. Sharma et al., 2024). The probable emergent trends that may yield great benefits to the efficacy of hydrophyte-based remediation are new technologies such as bioengineering and nanotechnology, which CAN promise to yield improvement and maximization of intake and detoxification using nanomaterials and plant tolerance to metals. The studies in the given areas remain in their early stages yet, the first findings are already encouraging, meaning that the given advances within the sector could be utilized to run the cleanup processes faster and more efficiently, using deeply-polluted locations (Mondal et al., 2025). As well, hydrophytes and microbial solution combination or even chemical remedies as part of a hybrid remediation trial could be adopted as a stronger rebuttal to hidden issues of the heavy metal pollution. Using an example, the synergetic effect alongside microbial degradation facilitated in the phytoremediation may be expediently reinforced during the deterioration of the contaminants, such that hydrophytes will eventually thrive in unfavorable circumstances (Khatoon et al., 2024).

An integrated approach to forming the links between phytoremediation techniques and the overall trends on global sustainability and enabled policy schemes could place phytotransformation as a central element in the systems of environmental regulation in the future. The development of these trends and technologies will define the future of phytoremediation as the world manages the ecological impact of industrial activity, which will make the given concept a more helpful and scalable instrument for purifying the polluted aquatic ecosystem.

2.3. Toxicity Symptoms in Non-Tolerant Hydrophytes

Heavy metal pollution of the water bodies takes a serious toll on the physiological and morphological features of the non-tolerant hydrophytes. These symptoms are the immediate effects of the cellular and metabolic deformity of the toxic metal ions such as lead (Pb), cadmium (Cd), mercury (Hg), and arsenic (As). The intolerant species often show premature stress at both biochemical and physical stages, including chlorosis, short root systems, and low nutrient uptake, both at the biochemical level and at the physical level, culminating in death.

Pigment Breakdown and Chlorosis

One of the most noticeable effects of overexposure to heavy metals is chlorosis, whose prevalence is caused by the disintegration of chlorophyll pigments or the loss of the green hue of leaves. This effect is due to the fact that production of chlorophyll is affected by oxidative stress, or the process of pigment degradation is accelerated (P. Sharma & Dubey, 2005). As an example, chromium and cadmium have demonstrated the ability to inhibit the activities of such enzymes as d-aminolevulinic acid dehydratase, required in the synthesis of chlorophyll, which leads to a significant loss of a green pigment (Arif et al., 2016). Moreover, metal interference leads to a reduction of iron and magnesium absorption, through which pigment loss is aggravated and photosynthetic activity and plant vigour are reduced (Yadav et al., 2025).

Impaired Root Development

Early inhibitory responses to heavy metals are root necrosis, darkening, and shortening of intolerant hydrophytes. Because they are the main location for heavy metal contact and deposition, roots are the most affected by cellular toxicity. According to research, lead and cadmium cause the disintegration of root apical meristems, cell wall thickening, and xylem element collapse, which impairs the transmission of water and nutrients (M. Ali & Nas, 2018; Aravind et al., 2009). Cadmium toxicity in *Potamogeton crispus* resulted in considerable root growth suppression and root hair distortion, which restricted nutrient absorption and anchorage (Q. S. Xu et al., 2010).

Interruption of Nutrient Absorption

Heavy metals either compete directly with vital nutrients for absorption sites or interfere with ion channels and transport proteins. By depleting the internal supply of essential macronutrients like potassium (K), calcium (Ca), and magnesium (Mg), as well as micronutrients like iron (Fe) and zinc (Zn), this condition lowers the amount of these nutrients available within the body. The imbalance causes metabolic diseases, slowed development, and decreased defensive responses (Khan & Faisal, 2024; Shanker et al., 2005). For example, even when cadmium is present in the growth medium, it inhibits nitrate reductase and phosphate transporters, resulting in nitrogen and phosphorus deficiencies (Jibril et al., 2017).

In polluted water bodies, these symptoms together significantly impair the survival and ecological function of hydrophytes that are not tolerant. These symptoms must be identified in order to monitor the environment and take prompt action in ecosystems that are experiencing heavy metal stress.

3. CELLULAR PROCESSES

3.1 Cellular Chelation

One of the main defence mechanisms exhibited by hydrophytes is cellular chelation in order to counter the effects of heavy metals (H. de P. Lima & Asencios, 2021). Hydrophytes will initiate the production of designated peptides and proteins when contacted with high levels of metals, like cadmium (Cd), lead (Pb), and arsenic (As); chiefly phytochelatins (PCs) and metallothioneins (MTs) to bind and store the metal ion cellularly (Parida et al., 2024). Low-molecular-weight, cysteine-rich peptides derived via glutathione, known as phytochelatins, form covalently bound complexes within the cytosol with heavy metals and are known to help detoxify heavy metal pollutants (Seregin & Kozhevnikova, 2023). The complexes are subsequently carried into vacuoles, whereby they render the metals inaccessible to the sensitive cellular pathways and make them less cytotoxic (Balzano et al., 2020). Metallothioneins are similarly cysteine-rich and similarly bind metal ions to the thiol groups, playing a role in detoxification and homeostasis of important micronutrients (Yang et al., 2024). Hydrophytes also respond to oxidative stress caused by heavy metals by stimulating effective antioxidant defence systems, besides the response to the oxidative stress involving chelation (Mansoor et al., 2023). The superoxide dismutase (SOD), catalase (CAT), and ascorbate peroxidase (APX) are also upregulated to neutralise the effect of reactive oxygen species (ROS) produced during metal exposure. An example is provided by research on *Elodea canadensis* and *Lemna minor*, where increased activities of SOD and CAT were displayed in cadmium and lead stress, and this situation protects against oxidation of lipids, proteins, and nucleic acids. These antioxidant systems also consist of enzymatic as well as non-enzymatic components, and the executive components or systems are glutathione and ascorbic acid that increase the resilience of the plant further (Seregin & Kozhevnikova, 2023). Chelation and antioxidant defence mechanisms together create a coordinated step towards a response in hydrophytes, which allows them to survive and grow in polluted environments (Mansoor et al., 2023). The efficiency of these mechanisms varies with specific species, and some hydrophytes have greater competencies in metal-binding and ROS-scavenging than others (Seregin & Kozhevnikova, 2023). Such molecular and physiological adaptation understanding will help in the development of phytoremediation technologies and the identification of suitable plant species in remediating heavy metal-polluted water bodies.

3.2 Structural Adaptations

Structural adaptation is important to hydrophytes exposed to heavy metal stress as a form of external defence, in addition to the cellular systems (H. Deng et al., 2004). An example of one of the most visible adaptations is the formation of iron plaques on root surfaces, which may occur on the root of such plants as *Typha latifolia*, oxidized iron (Fe(III)). The plaques are deposited on the epidermis of the root by precipitation when the oxygen emitted by the plant is combined with the iron-containing water (Y. Xu et al., 2018). The plaques employ their physical as well as chemical character to attach and cement heavy-metal cadmium, lead, and arsenic at the root-soil association, hindering their relocation into the plant vascularized tissue (Javed et al., 2019).

Besides the pot development iron plaque, there, cell wall structure changes are observed in the case of HDME exposure on the hydrophytes (Ronzhina et al., 2009). The cell wall has the capability to bind and immobilise metal ions, which restrict their connections into the cytoplasm and hence reduce their toxicity (Cheng et al., 2025). To further add to the strength of this barrier, some species thicken the cell walls or amass more lignin and suberin inside the storage cells and secondary walls. The other adaptation technology is the secretion of root exudates that can potentially mediate changes in the rhizosphere redox potential and pH and influence metal solubility and supply (He et al., 2024). Hydrophytes can also fix metals through acidification of the rhizosphere or by organic acids of chelators released by them (Bhat et al., 2022). Instead, they can endure the metals through blockage in non-root tissues. They are also considered to be significantly important to Rhizosphere microorganisms, with which they are symbionts, in structural adaptation (Alves et al., 2022). Metal immobilising Mycorrhizal fungi and metal-resistant bacteria can increase rhizosphere immobilisation of metals and can decrease plant uptake, resulting in general tolerance (Han et al., 2020). Metal-immobilising Mycorrhizal fungi and metal-resistant bacteria may augment rhizosphere Immunisation of metals and may decline plant Immunisation, leading to general tolerance (Han et al., 2020). These metabolisms may lose enzymes or chelators released by these microorganisms to enable holding of the metals, thus being more immobile and toxic (Priya et al., 2022). All these structural adaptations enable hydrophytes to survive in a polluted ice, which includes the minimization of the degree of metal ingestion coupled with the protection of internal organs. They are of special significance to the species that live in very polluted waters, where external barriers should be strong enough to facilitate survival and ecosystem performance. Knowing this adaptation can aid in the choices and design of hydrophytes to be used in effective phytoremediation in aquatic environments.

3.3 Hyperaccumulation and Ecotype-Specific Adaptations

There is a special adaptation of some hydrophytes called hyperaccumulation, which helps the plant absorb and tolerate concentrations of heavy metals that are extremely high and allows it to exclude the symptoms of toxicity (Skuza et al., 2022). An example of such species is the *Sedum alfredii* and *Eichhornia crassipes* that showed effective uptake, translocation, and sequestration of metals such as cadmium, zinc, and lead between roots and below-ground shoots (Tian et al., 2017). This occurs on the basis of augmented metal transporter protein expression, which helps with the transport of the metal ions across roots and incorporates the shoot (K. K. Rai et al., 2021). On entry into the plant, metals can be chelated with organic acids, peptides, or proteins and excluded into homely places such as vacuoles or cell walls, and this can inhibit interaction with vital metabolic processes (Anjum et al., 2015).

Adaptations specific to ecotypes also narrow down hyperaccumulation. The ability to take up and tolerate metals may therefore differ across the various ecotypes of the same species because these ecotypes have adapted to different environmental conditions (Muszyńska et al., 2020). Indicatively, some ecotypes of *Sedum alfredii* in metal-soil-rich environments have adapted to have stronger metal transporters and chelators and stronger antioxidant systems, tolerating extremely contaminated habitats. These modifications are largely through selective pressure, and thus genetic differentiation and specialization to heavy metal tolerance often occur (Hou et al., 2018).

The ecological importance of hyperaccumulation is not limited to a single plant's survival. Hyperaccumulators may affect the cycling and distribution of nutrients and metals in water bodies by concentrating the metals in the above-ground tissues (Pasricha et al., 2021). Their exclusive adaptations make them important to phytoremediation, which is what they do: extract and concentrate the metal in polluted soil so that it can be safely eliminated (Bhat et al., 2022). Comparative studies have revealed that some hydrophytes do not need to grow and store mass and increase their metal concentration levels, but others do not profit from the hyperaccumulation via large root systems or transporter proteins (Ali et al., 2020).

Altogether, hyperaccumulation and ecotype-specific adaptations are rather advanced evolutionary modules contributing to the enhancement of the resilience of hydrophytes in metal-contaminated sites, entailing convenient ways to solve the environmental clean-up and recovery of lost ecosystems.

3.4 Vacuolar Sequestration and Symbiotic Interactions

Vacuolar sequestration and symbiosis are advanced tolerance approaches that the hydrophytes can accommodate heavy metal toxicity (DalCorso et al., 2019).

Vacuolar sequestration is the process by which metal ions are actively transported into vacuoles, large, membrane-bound structures that make up the bulk of the volume of a plant cell, and in these vacuoles are bound to organic acids, peptides, or other ligands, reducing the activity of the metals and decreasing the probability that they can cause cellular dysfunction (Peng & Gong, 2014). This compartmentalisation lowers the levels of metals in the cytosol, and also the most critical organelles of the cell are guarded against cell damage initiated by the metal, hence providing the cell with its homeostasis and sustenance (Bar-Peled & Kory, 2022; Dumanović et al., 2021).

Lemna minor is one of the examples where vacuolar sequestration is a dominant detoxification mechanism (Xue et al., 2018). *Lemna minor* is not like many land plants that increase the stress accumulation of proline and can instead utilise vacuolar storage as a means to cope with heavy metal stress, with the overall effect of reducing the cytotoxicity and balancing metabolism (Liu et al., 2018). Adaptation enables the plant to grow well in metal-rich environments, and this aspect explains why this plant has become useful in phytoremediation (Montenegro et al., 2010).

Hydrophyte tolerance is additionally improved by symbiotic relationships with rhizosphere microorganisms. Reduction, precipitation, and methylation conversion (Alzate Zuluaga et al., 2024). Metals may also be converted to less toxic forms in the rhizosphere by metal-resistant bacteria and mycorrhizal fungi (Karnwal et al., 2025). These microbes also release extracellular polymers and siderophores, which contain chelate metals, therefore making them less bioavailable to the plant. To compensate, hydrophytes provide root exudates, a source of sustaining the growth of microbes, and a mutually advantageous relationship (Upadhyay et al., 2022).

The resulting interaction of the vacuolar storage and symbiosis both causes the capability of the hydrophytes to survive in the contaminated waters, as well as the wellness of the aquatic system due to moderations around the liquidity of the heavy metal and the effects created on the environment. These processes are essential in the formulation of phytoremediation processes and ensuring that de-contamination of metals in a polluted ecosystem is successful.

4. PHYTOREMEDIATION POTENTIAL OF HYDROPHYTES

Hydrophytes (aquatic macrophytes) are also proving to be highly promising in regard to phytoremediation, which is a green technology that uses plants to remediate, store, or absorb water-contaminating pollutants, including heavy metals (Justin et al., 2022). They are applicable particularly in contexts of countering pollutants that appear in the aquatic system, because, in this case, the traditional remediation process might be too expensive and disruptive to the ecosystem. Such Hydrophytes that have been investigated most and used include: *Eichhornia crassipes* (water hyacinth) and *Lemna minor* (duckweed), that has been found to accumulate and tolerate such high concentrations of metals like cadmium, lead, and arsenic, among others (Mustafa & Hayder, 2021).

Phytoremediation by hydrophytes occurs in various processes, including phytoextraction, rhizofiltration, and phytostabilisation (Kafle et al., 2022). During phytoextraction, plants are used to take up dissolved metals via roots and then transfer them to above-ground tissue; mining plants can be harvested, and the soil metals safely disposed of, or may be processed to recover the metals (Kafle et al., 2022; Paz-Alberto & Sigua, 2013; J. K. Sharma et al., 2023). Rhizofiltration is the precipitation or adsorption of metals on the surfaces of roots; this process is especially efficient in floating and submerged hydrophytes that have large root systems (Mocek-Płóciniak et al., 2023). Phytostabilisation, in its turn, means that metal in the rhizosphere is immobilised, thus its bioavailability is reduced, and it can no longer spread to the rest of the aquatic environment (Mendez & Maier, 2008).

The overall performance of phytoremediation depends on the species of plants, the speciation of metals, the chemistry of water (PH and organic matter), and environmental conditions (Nedjimi, 2021). Indicatively, *Eichhornia crassipes* has been found capable of the extraction of huge amounts of cadmium, lead, and other less significant determinants of the occasions of industrial effluents and mine waters, and in the majority of cases without any composite indications of the existence of injuriousness (Malar et al., 2016). Lemna minor has been known to grow with a high growth rate and good surface-volume ratio, hence making it a viable plant in situations where it is used to store metals like cadmium and copper (Vulpe et al., 2025).

The current studies guarantee that the process of the hydrophytes to store many times more metals than the surrounding water should be termed the best way of eco-restoration and enhancement of water quality (Ziegler et al., 2023). They are economically feasible and they are sustainable in their usage, in addition to building the levels of biodiversity and system strength of the ecosystems. The maximisation of the efficacy of phytoremediation, such as the selection of species, the manipulation of the environment, and the possible genetic improvement of the hydrophytes, should be maximised as it goes through the research to make the hydrophytes more vulnerable to purifying the polluted water bodies.

5. LIMITATIONS AND CHALLENGES OF HYDROPHYTE-BASED REMEDIATION

Despite being effective and environmentally friendly, Hydrophyte-based remediation also faces several major challenges that discourage its widespread use

(Sahoo et al., 2025). Seasonal changes in plant growth are one representation of the biggest problems, as they have a negative impact on the effectiveness of metal uptake, especially during colder seasons when plants are less active (Cubas, 2020). Additionally, there is a risk of heavy metal concentration on the vegetative cover. If biomass decay is not properly carried out, there is a risk of metal reintroduction into the surroundings, which would pose further environmental threats (Singh et al., 2024). Another issue is the contaminated biomass, but it can be resolved by reducing the harvested biomass to biochar or retrieveing metals via thermal or chemical methods (da Silva et al., 2025). The other underlying issue is the invasive species *Eichhornia crassipes*, which may outcompete the native species, which is why the process of recuperation is interrupted. Constant monitoring of the situation should also be done to make sure that the remediation measures work in the long run. To eliminate such limitations, appropriate design measures, utilization of suitable species, and the development of effective post-harvest treatment technologies to curb such limitations are needed in an effort to contribute to the sustainability and scalability of hydrophyte remediation projects.

5.1 Addressing the Under-Emphasis on Socio-Economic and Logistical Factors

To improve the debate on the hydrophyte-based phyto-remediation, the paper would respond by getting deep into the economic barriers and policy challenges. Although phytoremediation is currently cheaper compared to the conventional ways of remediation, it still has indirect expenses, including the management of invasive species and the disposal of the contaminated biomass, which should be addressed to scale it to the required level (Sahoo et al., 2025). These costs can be alleviated through partnerships with industries and government incentives. Also, such logistical factors as monitoring and maintenance are not discussed in depth; continued control implies increasing operation costs, which would imply the use of automated monitoring technology to decrease the workforce costs. Biomass disposal poses significant challenges, and while risks of decay and metal reintroduction are mentioned, a more thorough discussion on alternative management strategies is warranted. Converting biomass to biochar or using it for energy recovery could decrease disposal costs and environmental impacts. The integration of metal recovery technologies could further enhance sustainability and economic viability. Finally, technological advancements in metal uptake, plant tolerance, or hybrid approaches with microbial solutions should be emphasized to improve the cost-effectiveness and scalability of hydrophyte-based remediation (Chai et al., 2022).

5.2 Expanding on Socio-Economic and Regulatory Factors

During phytoremediation, especially through the use of hydrophytes, the interactions of regulatory forces, economic considerations, and modern techniques can ensure the sustainability of any of the various remediation techniques in combating environmental problems (Sahoo et al., 2025). With the trend of being green across the world, governments and organizations are in support of green technologies to avert pollution and climate change, particularly with the pollution of water bodies with heavy metals. Remediation carried out using hydrophytes is environmentally and economically beneficial, but its expansion is largely dependent on financial matters. Although this is cheaper compared to the traditional methods, they have their own problems, including the disposal of biomass, handling of invasive species, and providing monitoring solutions, particularly in a limited resource area (Rodrigo, 2021).

The policymaking contributes significantly because, in case the laws become even stricter, there are practices that will be introduced, which will supposedly be clean and will decrease pollution. These types of regulatory requirements can be solved through remediation of the hydrophytes, which can be further strengthened by the active regulatory frameworks constituted by international treaties, alongside the national laws, which can regulate the water quality. The large-scale implementation concerns logistics, which, however, have been neglected to a greater extent. A proponent of the text is the effectiveness of the systems of monitoring that would see the success of the interventions in their sustainability (Pollack Porter et al., 2018).

The connection to the industries, including wastewater treatment and mining, resulted in huge applications in remote areas. Generation of innovations such as genetic engineering, nanotechnology, and hybrid systems can also be deployed to increase the efficiency of hydrophytes to extract metals. The chapter highlights why these forms of technological advancement should be incorporated into the existing economic and regulatory regulations, where it is suggested that the use of genetically modified plants may lead to much quicker phytremediation levels, although so far, this must continue to align with environmental rules and economic forces to be utilised in large-scale use (J. P. M. de Lima et al., 2025).

A substantial amount should be done to verify the economic and logistical limitations that might hinder the large-scale adoption of hydrophyte-based phytoremediation in order to make the process viable and scalable in the long term. Even though such approaches are typically more affordable in comparison to the conventional remediation strategies, some difficulties related to the invisible expenses of biomass discarding, as well as the exclusion of invasive species, still remain. Creativity in solutions, such as using biomass to produce biochar to cover the health of the soil or to provide energy, can assist in the reduction of some of these issues by generating value-added products. To counterbalance the capitalization of the operating costs of large projects, it is important to consider the use of financial models, subsidies, and public-private partnerships (Haque & Gazi-Khan, 2025). The technology adoption can also be done on regulatory conditions, particularly with the policies that focus on environmental conditions and sustainable production and practices. Recommended is that the chapter should explore how the alteration to regulations can either help or frustrate the use of the genetically modified hydrophytes and how the biomass polluted can be used to be conformed to the international standards of sustainability. The study of these areas will contribute to improving the economic benefit and scalability of phytoremediation using hydrophytes, which will translate into its effectiveness to reduce heavy metal contamination. (Al-Emran & Griffy-Brown, 2023).

The Phytoremediation references used in the chapter are mostly relevant to both the foundational and the current research. It could, however, use some updates to incorporate new research on new applications like genetic engineering and nanotechnology that would help projects of hydrophytes to be more efficient. Furthermore, the use of examples, especially those of developing nations that have serious pollution issues, would provide an international perspective. Error Logging inferences to the policy papers/environmental laws favouring the use of sustainable technologies would reinforce the need to use hydrophyte remediation. On balance, it is necessary to add more innovations, case studies and policy insights to the chapter to make it more comprehensive and to provide applications of policies.

6. CONCLUSION AND FUTURE DIRECTIONS

Hydrophytes are the most suitable species to maintain in order to reduce water body pollution by heavy metals since it has an independent tolerance mechanism, and also have elaborated functions of phytoremediation. The plants play an important role in the preservation of the ecological balance within the eco-friendly volumes of a particular water body, as they have the ability to sequester, absorb, and detoxify harmful substances through diverse processes. This can be attributed to the great capability of their cells to endure excessive metal in the environment, achieved by facilitating the cellular chelation, antioxidants, and adjustments of the organ tissues, giving the body the ability to endure this and survive in the polluted environment. The metals in the water bodies are reduced through absorption by the metals in the surroundings using the hydrophytes, and therefore provide clean and healthy habitats. Moreover, they can also detoxify the heavy metals besides eliminating them via mechanisms such as phytoremediation, which provides a less toxic habitat to other species, as well as the absence of the metals in the food chain.

Eichornia crassipes (water hyacinth) *and Lemna minor* (duckweed) are two of the most typical hydrophytes that have been studied to discover their various possible uses in phytoremediation. They have been used individually to complement the quality of water in contaminated water ecosystems. They are very effective in cleaning polluted water bodies since they have the capacity to accumulate metals such as cadmium (Cd), lead (Pb), and arsenic (As). Factors have been established to prove how these plants have the power to lower heavy metals, especially in an area where the water situation has been grossly contaminated. Their use has been found efficient in not only ensuring better water quality but also in restoring polluted ecosystems. An example is the *Eichhornia crassipes*, which has been employed in treating water (such as in wetlands and water purification ponds) through uptake and concentration of the pollutants. Although hydrophytes are currently well-renowned due to their capacity to considerably lessen the effects of heavy metal pollution, there are a number of shortcomings coupled with ignorance that are yet to be addressed in order to facilitate better application in phytoremediation. Accumulation in the biomass of such plants is one of the major difficulties encountered. The amount of metals in hydrophytes after some time may be so much to the extent of causing a toxic effect, and this effect can make the plants dangerous to the ecosystem. Moreover, the possibility of metals bioaccumulating in the plant biomass should make a person worry about the ecological consequences of their application in phytoremediation in the long term. Another major concern that ought to be dealt with is the fact that the biomass that stores the accumulated metals has to be disposed of, creating secondary pollution. Metal concentrations in plants may give rise to contamination of soil or passage of metals to herbivores and predators within the ecosystem. Moreover, methods of harvesting these plants and disposing of them have to be investigated so that they do not end up polluting the surroundings further. To balance these challenges, in the future, the effort should be geared towards coming up with solutions for the biomass that should be utilized during phytoremediation. It will be necessary to learn how to manage and dispose of the plants, as well as cases of possibly recycling the plants that end up with high metal concentration. Specifically, the creation of long-term and green solutions to phytoremediation will help it become more feasible and useful in the long run. Also, it is desired to research the impacts of metal deposition on food webs and aquatic life, which help themselves through these plants as food sources. Among them is learning how bioaccumulation and biomagnification of heavy metals in aquatic food webs happen, and how the metals can impact the health of organisms in higher food chains. Along with the addressing of the practical use of the metal disposal and metal harvesting, much attention is also aimed at explaining the genetic and molecular factors that make the hydrophytes tolerant. Recent technology in genomics, transcriptomics, and proteomics has led to the development of newer means to study the genetic and molecular mechanisms of metal tolerance in plants. Determination of the molecular routes through which metals are taken in, transported, and detoxified will play a major role in enhancing the use of phytoremediation. To reduce metal toxicity in hydrophytes, several types of advanced ways may be utilised: cellular chelation, usage of antioxidant defence and metal sequestration. Other genes and proteins that govern the operations of metal assimilation, vicinity and disarmament are in place to regulate them. These processes have the potential to find out the regulatory genes which are relevant, and this will be the future of genomic research. In one instance, genes participating in the production of phytochelatin, metallothioneins and other metal-bounding serums are essential to the macrophage of metals in flora. By this finding, the researchers will be at a stand position of developing genetically modified hydrophytes that would be more effective in regards to metal accumulation and detoxification. Such changes would also prevent anything altering, which would lead to the environmentally resistant side of the plants that would withstand drought or too much heat, making the plants more applicable in a wide range of varying conditions. Selective breeding/hybridisation would also have the advantage of improving the performance of hydrophytes alongside genetic modifications. Arguably, it is possible to have a more advanced plant that can achieve phytoremediation with a specific environmental condition achieved through the confinement of the most desirable lines of various ecotypes. This kind of progress in the evolution of molecules might result in superior, more-productive and faster-growing crops endowed with greater accumulation of metals. In addition, scientists would be able to work on such packages as hydrophytes with other organisms or plants that would complement their phytoremediation capacities and make the operation even efficient. In addition to this, the ecological components, molecular biology, and environmental engineering nature and essence will be demanded to establish and scale up the phytoremediation technologies to their maximum efficacy. The nature of interaction between hydrophytes and the surrounding should be researched such that the phytoremediation systems are sustainable and effective by incorporating water, soil, as well as other organisms. Interdisciplinary studies will be important in ensuring the fact that phytoremediation is not only potentially efficient, but also ecologically sound and cost-effective. In summary, despite the performance of the hydrophytes as evidenced by alternatives that are an incredibly useful tool in restoring heavy metals, much more is required to optimise the use of the tool on a large scale. Their genetic, physiological and ecological aspects need further examination that will allow coming up with more effective strategies of phytoremediation. The problems, such as the metals disposal, the biomass disposal, and the environmental ecological implications of the same, would be sorted, and as such, the use of hydrophytes in the environmental cleanup could proceed even further. As the levels of knowledge on the mechanisms of hydrophyte tolerance increase, there exists the likelihood as well that such plants will play the dominant role of restoring and ensuring the sustainability of the health of aquatic systems to be enjoyed by future generations.

REFERENCES

Abdelaal, M., Mashaly, I. A., Srour, D. S., Dakhil, M. A., El-Liethy, M. A., El-Keblawy, A., El-Barougy, R. F., Halmy, M. W. A., & El-Sherbeny, G. A. (2021). Phytoremediation perspectives of seven aquatic macrophytes for removal of heavy metals from polluted drains in the Nile Delta of Egypt. *Biology (Basel)*, *10*(6), 560. DOI: 10.3390/biology10060560 PMID: 34203088

Ahmed, S. F., Mofijur, M., Nuzhat, S., Chowdhury, A. T., Rafa, N., Uddin, M., Inayat, A., Mahlia, T. M. I., Ong, H. C., Chia, W. Y., & Show, P. L. (2021). Recent developments in physical, biological, chemical, and hybrid treatment techniques for removing emerging contaminants from wastewater. *Journal of Hazardous Materials*, *416*, 125912. DOI: 10.1016/j.jhazmat.2021.125912 PMID: 34492846

Al-Emran, M., & Griffy-Brown, C. (2023). The role of technology adoption in sustainable development: Overview, opportunities, challenges, and future research agendas. *Technology in Society*, *73*, 102240. DOI: 10.1016/j.techsoc.2023.102240

Al-Zurfi, S. K. L., Al-Tabatabai, H. H. M., & Abojassim, A. A. (2021). Cadmium metal effects physiological and anatomical variations for aquatic plant Hydrilla verticillata (L. f.) Royle. *Plant Physiology Reports*, *26*(3), 457–465. DOI: 10.1007/s40502-021-00600-5

Ali, M., & Nas, F. S. (2018). The effect of lead on plants in terms of growing and biochemical parameters: A review. *MOJ Ecology & Environmental Sciences*, *3*(4). Advance online publication. DOI: 10.15406/mojes.2018.03.00098

Ali, S., Abbas, Z., Rizwan, M., Zaheer, I., Yavaş, İ., Ünay, A., Abdel-DAIM, M., Bin-Jumah, M., Hasanuzzaman, M., & Kalderis, D. (2020). Application of Floating Aquatic Plants in Phytoremediation of Heavy Metals Polluted Water: A Review. *Sustainability (Basel)*, *12*(5), 1927. Advance online publication. DOI: 10.3390/su12051927

Alves, A. R. A., Yin, Q., Oliveira, R. S., Silva, E. F., & Novo, L. A. B. (2022). Plant growth-promoting bacteria in phytoremediation of metal-polluted soils: Current knowledge and future directions. *The Science of the Total Environment*, *838*, 156435. DOI: 10.1016/j.scitotenv.2022.156435 PMID: 35660615

Alzate Zuluaga, M. Y., Fattorini, R., Cesco, S., & Pii, Y. (2024). Plant-microbe interactions in the rhizosphere for smarter and more sustainable crop fertilization: The case of PGPR-based biofertilizers. *Frontiers in Microbiology*, *15*, 1440978. Advance online publication. DOI: 10.3389/fmicb.2024.1440978 PMID: 39176279

Anjum, N. A., Hasanuzzaman, M., Hossain, M. A., Thangavel, P., Roychoudhury, A., Gill, S. S., Rodrigo, M. A. M., Adam, V., Fujita, M., Kizek, R., Duarte, A. C., Pereira, E., & Ahmad, I. (2015). Jacks of metal/metalloid chelation trade in plantsâ€"an overview. *Frontiers in Plant Science*, *6*. Advance online publication. DOI: 10.3389/fpls.2015.00192 PMID: 25883598

Aravind, P., Prasad, M. N. V., Malec, P., Waloszek, A., & Strzałka, K. (2009). Zinc protects Ceratophyllum demersum L.(free-floating hydrophyte) against reactive oxygen species induced by cadmium. *Journal of Trace Elements in Medicine and Biology*, *23*(1), 50–60. DOI: 10.1016/j.jtemb.2008.10.002 PMID: 19203717

Arif, N., Yadav, V., Singh, S., Singh, S., Ahmad, P., Mishra, R. K., Sharma, S., Tripathi, D. K., Dubey, N. K., & Chauhan, D. K. (2016). Influence of High and Low Levels of Plant-Beneficial Heavy Metal Ions on Plant Growth and Development. *Frontiers in Environmental Science*, *4*. Advance online publication. DOI: 10.3389/fenvs.2016.00069

Aryal, M. (2024). Phytoremediation strategies for mitigating environmental toxicants. *Heliyon*, *10*(19), e38683. DOI: 10.1016/j.heliyon.2024.e38683 PMID: 39430524

Bali, A. S., & Sidhu, G. P. S. (2021). Arsenic acquisition, toxicity and tolerance in plants - From physiology to remediation: A review. *Chemosphere*, *283*, 131050. DOI: 10.1016/j.chemosphere.2021.131050 PMID: 34147983

Balzano, S., Sardo, A., Blasio, M., Chahine, T. B., Dell'Anno, F., Sansone, C., & Brunet, C. (2020). Microalgal Metallothioneins and Phytochelatins and Their Potential Use in Bioremediation. *Frontiers in Microbiology*, *11*, 517. Advance online publication. DOI: 10.3389/fmicb.2020.00517 PMID: 32431671

Bar-Peled, L., & Kory, N. (2022). Principles and functions of metabolic compartmentalization. *Nature Metabolism*, *4*(10), 1232–1244. DOI: 10.1038/s42255-022-00645-2 PMID: 36266543

Bayuo, J., Rwiza, M. J., Choi, J. W., Njau, K. N., & Mtei, K. M. (2024). Recent and sustainable advances in phytoremediation of heavy metals from wastewater using aquatic plant species: Green approach. *Journal of Environmental Management*, *370*, 122523. DOI: 10.1016/j.jenvman.2024.122523 PMID: 39305882

Bhat, S. A., Bashir, O., Ul Haq, S. A., Amin, T., Rafiq, A., Ali, M., Américo-Pinheiro, J. H. P., & Sher, F. (2022). Phytoremediation of heavy metals in soil and water: An eco-friendly, sustainable and multidisciplinary approach. *Chemosphere*, *303*, 134788. DOI: 10.1016/j.chemosphere.2022.134788 PMID: 35504464

Borisova, G. G., Chukina, N. V., Maleva, M. G., & Levchenko, U. A. (2017). Accumulation of heavy metals in leaves of submerged hydrophytes (Elodea canadensis Michx. and Potamogeton perfoliatus L.) and their responses to the effect of the wastewater of a metallurgical plant. *Inland Water Biology*, *10*(2), 176–181. DOI: 10.1134/S1995082917020031

Chai, Y., Chen, A., Bai, M., Peng, L., Shao, J., Yuan, J., Shang, C., Zhang, J., Huang, H., & Peng, C. (2022). Valorization of heavy metal contaminated biomass: Recycling and expanding to functional materials. *Journal of Cleaner Production*, *366*, 132771. DOI: 10.1016/j.jclepro.2022.132771

Cheng, Z., Wang, C., Tang, F., Zhou, Y., Zhu, C., & Ding, Y. (2025). The cell wall functions in plant heavy metal response. *Ecotoxicology and Environmental Safety*, *299*, 118326. DOI: 10.1016/j.ecoenv.2025.118326 PMID: 40403694

Clemens, S. (2006). Toxic metal accumulation, responses to exposure and mechanisms of tolerance in plants. *Biochimie*, *88*(11), 1707–1719. DOI: 10.1016/j.biochi.2006.07.003 PMID: 16914250

Clijsters, H., & Van Assche, F. (1985). Inhibition of photosynthesis by heavy metals. *Photosynthesis Research*, *7*(1), 31–40. DOI: 10.1007/BF00032920 PMID: 24443011

Cubas, P. (2020). Plant Seasonal Growth: How Perennial Plants Sense That Winter Is Coming. *Current Biology*, *30*(1), R21–R23. DOI: 10.1016/j.cub.2019.11.044 PMID: 31910371

da Silva, D. F., Dias, O. B. S., de Carli Souza, A., Parnian, A., Borgo, L., Melo, L. C. A., & Rabêlo, F. H. S. (2025). Transforming contaminated biomass from phytoremediation into biochar. In *Biochar Ecotechnology for Sustainable Agriculture and Environment* (pp. 131–155). Elsevier., DOI: 10.1016/B978-0-443-29855-4.00006-0

DalCorso, G., Fasani, E., Manara, A., Visioli, G., & Furini, A. (2019). Heavy metal pollutions: State of the art and innovation in phytoremediation. *International Journal of Molecular Sciences*, *20*(14), 3412. DOI: 10.3390/ijms20143412 PMID: 31336773

de Lima, J. P. M., Amaral, M. C. S., & de Lima, S. C. R. B. (2025). Sustainable water management in the mining industry: Paving the way for the future. *Journal of Water Process Engineering*, *71*, 107239. DOI: 10.1016/j.jwpe.2025.107239

Deng, G., Li, M., Li, H., Yin, L., & Li, W. (2014). Exposure to cadmium causes declines in growth and photosynthesis in the endangered aquatic fern (Ceratopteris pteridoides). *Aquatic Botany*, *112*, 23–32. DOI: 10.1016/j.aquabot.2013.07.003

Deng, H., Ye, Z. H., & Wong, M. H. (2004). Accumulation of lead, zinc, copper and cadmium by 12 wetland plant species thriving in metal-contaminated sites in China. *Environmental Pollution, 132*(1), 29–40. DOI: 10.1016/j.envpol.2004.03.030 PMID: 15276271

Dhir, B., Sharmila, P., & Saradhi, P. P. (2004). Hydrophytes lack potential to exhibit cadmium stress induced enhancement in lipid peroxidation and accumulation of proline. *Aquatic Toxicology (Amsterdam, Netherlands), 66*(2), 141–147. DOI: 10.1016/j.aquatox.2003.08.005 PMID: 15036869

Di Stasio, L., Gentile, A., Tangredi, D. N., Piccolo, P., Oliva, G., Vigliotta, G., Cicatelli, A., Guarino, F., Guidi Nissim, W., Labra, M., & Castiglione, S. (2025). Urban Phytoremediation: A Nature-Based Solution for Environmental Reclamation and Sustainability. *Plants, 14*(13), 2057. DOI: 10.3390/plants14132057 PMID: 40648066

Díaz-Torres, R. D. C., Alonso-Castro, A. J., Carrillo-Inungaray, M. L., & Carranza-Alvarez, C. (2021). Bioactive compounds obtained from plants, their pharmacological applications and encapsulation. In *Phytomedicine* (pp. 181–205). Elsevier., DOI: 10.1016/B978-0-12-824109-7.00017-0

Dumanović, J., Nepovimova, E., Natić, M., Kuča, K., & Jaćević, V. (2021). The Significance of Reactive Oxygen Species and Antioxidant Defense System in Plants: A Concise Overview. *Frontiers in Plant Science, 11*, 552969. Advance online publication. DOI: 10.3389/fpls.2020.552969 PMID: 33488637

Farid, N., Ullah, A., Khan, S., Butt, S., Khan, A. Z., Afsheen, Z., El-Serehy, H. A., Yasmin, H., Ayaz, T., & Ali, Q. (2023). Algae and Hydrophytes as Potential Plants for Bioremediation of Heavy Metals from Industrial Wastewater. *Water (Basel), 15*(12), 2142. DOI: 10.3390/w15122142

Favas, P. J. C., Pratas, J., Paul, M. S., Sarkar, S. K., & Prasad, M. N. V. (2016). Phytofiltration of metal (loid)-contaminated water: The potential of native aquatic plants. *Phytoremediation: Management of Environmental Contaminants, 3*, 305–343. DOI: 10.1007/978-3-319-40148-5_11

Han, H., Wu, X., Yao, L., & Chen, Z. (2020). Heavy metal immobilizing bacteria combined with calcium polypeptides reduced the uptake of Cd in wheat and shifted the rhizosphere bacterial communities. *Environmental Pollution, 267*, 115432. DOI: 10.1016/j.envpol.2020.115432 PMID: 32841909

Haque, S. E., & Gazi-Khan, L. (2025). Limitations and challenges of bioremediation approach: alternative solutions. In *Biotechnologies for Wastewater Treatment and Resource Recovery* (pp. 147–157). Elsevier., DOI: 10.1016/B978-0-443-27376-6.00023-2

Hasan, Md. M., Uddin, Md. N., Ara-Sharmeen, I., F. Alharby, H., Alzahrani, Y., Hakeem, K. R., & Zhang, L. (. (2019). Assisting Phytoremediation of Heavy Metals Using Chemical Amendments. *Plants*, *8*(9), 295. DOI: 10.3390/plants8090295 PMID: 31438507

He, M., Yin, H., Liu, Z., Luo, F., Zhao, X., Li, H., & Song, X. (2024). Root exudate–assisted phytoremediation of copper and lead contamination using Rumex acetosa L. and Rumex K-1. *Ecotoxicology and Environmental Safety*, *284*, 117036. DOI: 10.1016/j.ecoenv.2024.117036 PMID: 39270415

Hou, D., Lin, Z., Wang, R., Ge, J., Wei, S., Xie, R., Wang, H., Wang, K., Hu, Y., Yang, X., Lu, L., & Tian, S. (2018). Cadmium Exposure-Sedum alfredii Planting Interactions Shape the Bacterial Community in the Hyperaccumulator Plant Rhizosphere. *Applied and Environmental Microbiology*, *84*(12), e02797-17. Advance online publication. DOI: 10.1128/AEM.02797-17 PMID: 29654182

Javed, M. T., Tanwir, K., Akram, M. S., Shahid, M., Niazi, N. K., & Lindberg, S. (2019). Phytoremediation of cadmium-polluted water/sediment by aquatic macrophytes: role of plant-induced pH changes. *Cadmium Toxicity and Tolerance in Plants*, 495–529.

Jibril, S. A., Hassan, S. A., Ishak, C. F., & Megat Wahab, P. E. (2017). Cadmium toxicity affects phytochemicals and nutrient elements composition of lettuce (Lactuca sativa L.). *Advances in Agriculture*, *2017*(1), 1236830. DOI: 10.1155/2017/1236830

Justin, L. D., Olukanni, D. O., & Babaremu, K. O. (2022). Performance assessment of local aquatic macrophytes for domestic wastewater treatment in Nigerian communities: A review. *Heliyon*, *8*(8), e10093. DOI: 10.1016/j.heliyon.2022.e10093 PMID: 36042728

Kafle, A., Timilsina, A., Gautam, A., Adhikari, K., Bhattarai, A., & Aryal, N. (2022). Phytoremediation: Mechanisms, plant selection and enhancement by natural and synthetic agents. *Environmental Advances*, *8*, 100203. DOI: 10.1016/j.envadv.2022.100203

Karnwal, A., Kumar, G., El Din Mahmoud, A., Dutta, J., Singh, R., Mohammad Said Al-Tawaha, A. R., & Malik, T. (2025). Eco-engineered remediation: Microbial and rhizosphere-based strategies for heavy metal detoxification. *Current Research in Biotechnology*, *9*, 100297. DOI: 10.1016/j.crbiot.2025.100297

Khan, H. N., & Faisal, M. (2024). Bioremediation of heavy metals using the symbiosis of hydrophytes and metal-resistant bacteria. In *Membrane technologies for heavy metal removal from water* (pp. 126–142). CRC Press.

Khatoon, Z., Orozco-Mosqueda, M., & Santoyo, G., del C., & Santoyo, G. (2024). Microbial Contributions to Heavy Metal Phytoremediation in Agricultural Soils: A Review. *Microorganisms*, *12*(10), 1945. Advance online publication. DOI: 10.3390/microorganisms12101945 PMID: 39458255

Lima, H. de P., & Asencios, Y. J. O. (2021). Eichhornia crassipes (Mart.) Solms (natural or carbonized) as biosorbent to remove pollutants in water. *SN Applied Sciences*, *3*(8), 750. DOI: 10.1007/s42452-021-04736-9 PMID: 34337324

Liu, C., Gu, W., Dai, Z., Li, J., Jiang, H., & Zhang, Q. (2018). Boron accumulation by Lemna minor L. under salt stress. *Scientific Reports*, *8*(1), 8954. DOI: 10.1038/s41598-018-27343-y PMID: 29895991

Malar, S., Shivendra Vikram, S., & Favas, J. C. (2016). Lead heavy metal toxicity induced changes on growth and antioxidative enzymes level in water hyacinths [Eichhornia crassipes (Mart.)]. *Botanical Studies*, *55*(1), 54. DOI: 10.1186/s40529-014-0054-6 PMID: 28597420

Mansoor, S., Ali, A., Kour, N., Bornhorst, J., AlHarbi, K., Rinklebe, J., Abd El Moneim, D., Ahmad, P., & Chung, Y. S. (2023). Heavy Metal Induced Oxidative Stress Mitigation and ROS Scavenging in Plants. *Plants*, *12*(16), 3003. DOI: 10.3390/plants12163003 PMID: 37631213

Mendez, M. O., & Maier, R. M. (2008). Phytostabilization of Mine Tailings in Arid and Semiarid Environments—An Emerging Remediation Technology. *Environmental Health Perspectives*, *116*(3), 278–283. DOI: 10.1289/ehp.10608 PMID: 18335091

Mishra, R. K., Kumar, J., Srivastava, P. K., Bashri, G., & Prasad, S. M. (2017). PSII photochemistry, oxidative damage and anti-oxidative enzymes in arsenate-stressed Oryza sativa L. seedlings. *Chemistry and Ecology*, *33*(1), 34–50. DOI: 10.1080/02757540.2016.1265516

Mocek-Płóciniak, A., Mencel, J., Zakrzewski, W., & Roszkowski, S. (2023). Phytoremediation as an Effective Remedy for Removing Trace Elements from Ecosystems. *Plants*, *12*(8), 1653. DOI: 10.3390/plants12081653 PMID: 37111876

Mondal, K., Jain, A., & Chakraborty, A. (2025). *Phytoremediation Meets Nanotechnology: A Cutting-Edge Solution for Environmental Detoxification.*

Montenegro, R. C., Araújo, A. J., Molina, M. T., Marinho Filho, J. D. B., Rocha, D. D., Lopéz-Montero, E., Goulart, M. O. F., Bento, E. S., Alves, A. P. N. N., Pessoa, C., de Moraes, M. O., & Costa-Lotufo, L. V. (2010). Cytotoxic activity of naphthoquinones with special emphasis on juglone and its 5-O-methyl derivative. *Chemico-Biological Interactions*, *184*(3), 439–448. DOI: 10.1016/j.cbi.2010.01.041 PMID: 20138029

Mustafa, H. M., & Hayder, G. (2021). Recent studies on applications of aquatic weed plants in phytoremediation of wastewater: A review article. *Ain Shams Engineering Journal, 12*(1), 355–365. DOI: 10.1016/j.asej.2020.05.009

Muszyńska, E., Labudda, M., & Kral, A. (2020). Ecotype-Specific Pathways of Reactive Oxygen Species Deactivation in Facultative Metallophyte Silene vulgaris (Moench) Garcke Treated with Heavy Metals. *Antioxidants, 9*(2), 102. DOI: 10.3390/antiox9020102 PMID: 31991666

Nedjimi, B. (2021). Phytoremediation: A sustainable environmental technology for heavy metals decontamination. *SN Applied Sciences, 3*(3), 286. DOI: 10.1007/s42452-021-04301-4

Nyquist, J., & Greger, M. (2007). Uptake of Zn, Cu, and Cd in metal loaded Elodea canadensis. *Environmental and Experimental Botany, 60*(2), 219–226. DOI: 10.1016/j.envexpbot.2006.10.009

Parida, T., Riyazuddin, S., Kolli, S. K., Chakraborty, A., Srinivas, N., Kundu, P., Bhattacharya, S., Seth, C. S., & Biswas, J. K. (2024). Phytoremediation of heavy metal(loid)s with integral involvement of the endogenous metal-chelators: Present state-of-the-art and future prospect. *Discover Environment, 2*(1), 139. DOI: 10.1007/s44274-024-00170-x

Pasricha, S., Mathur, V., Garg, A., Lenka, S., Verma, K., & Agarwal, S. (2021). Molecular mechanisms underlying heavy metal uptake, translocation and tolerance in hyperaccumulators-an analysis. *Environmental Challenges, 4*, 100197. DOI: 10.1016/j.envc.2021.100197

Paz-Alberto, A. M., & Sigua, G. C. (2013). Phytoremediation: A Green Technology to Remove Environmental Pollutants. *American Journal of Climate Change, 02*(01), 71–86. DOI: 10.4236/ajcc.2013.21008

Peng, J.-S., & Gong, J.-M. (2014). Vacuolar sequestration capacity and long-distance metal transport in plants. *Frontiers in Plant Science, 5*. Advance online publication. DOI: 10.3389/fpls.2014.00019 PMID: 24550927

Pollack Porter, K. M., Rutkow, L., & McGinty, E. E. (2018). The Importance of Policy Change for Addressing Public Health Problems. *Public Health Reports, 133*(1, suppl), 9S–14S. DOI: 10.1177/0033354918788880 PMID: 30426876

Priya, A. K., Gnanasekaran, L., Dutta, K., Rajendran, S., Balakrishnan, D., & Soto-Moscoso, M. (2022). Biosorption of heavy metals by microorganisms: Evaluation of different underlying mechanisms. *Chemosphere, 307*, 135957. DOI: 10.1016/j.chemosphere.2022.135957 PMID: 35985378

Rai, K. K., Pandey, N., Meena, R. P., & Rai, S. P. (2021). Biotechnological strategies for enhancing heavy metal tolerance in neglected and underutilized legume crops: A comprehensive review. *Ecotoxicology and Environmental Safety, 208*, 111750. DOI: 10.1016/j.ecoenv.2020.111750 PMID: 33396075

Rai, P. K. (2009). Heavy metal phytoremediation from aquatic ecosystems with special reference to macrophytes. *Critical Reviews in Environmental Science and Technology, 39*(9), 697–753. DOI: 10.1080/10643380801910058

Rodrigo, M. A. (2021). Wetland Restoration with Hydrophytes: A Review. *Plants, 10*(6), 1035. DOI: 10.3390/plants10061035 PMID: 34063930

Ronzhina, D. A., Ivanov, L. A., Lambers, G., & P'yankov, V. I. (2009). Changes in chemical composition of hydrophyte leaves during adaptation to aquatic environment. *Russian Journal of Plant Physiology: a Comprehensive Russian Journal on Modern Phytophysiology, 56*(3), 355–362. DOI: 10.1134/S102144370903008X

Saha, S., Huang, L., Khoso, M. A., Wu, H., Han, D., Ma, X., Poudel, T. R., Li, B., Zhu, M., Lan, Q., Sakib, N., Wei, R., Islam, M., Zhang, P., & Shen, H. (2023). Fine root decomposition in forest ecosystems: An ecological perspective. *Frontiers in Plant Science, 14*, 1277510. Advance online publication. DOI: 10.3389/fpls.2023.1277510 PMID: 38023858

Sahoo, A., Chhotaray, S. P., Meher, I., Behera, S. P., Pal, A., Meena, M., Swapnil, P., Yadav, A., & Bhardwaj, R. (2025). Phytoremediation for a sustainable future: Integrating plant based strategies in soil and wastewater remediation. *Bioresource Technology Reports, 31*, 102266. DOI: 10.1016/j.biteb.2025.102266

Santos, D., Duarte, B., & Caçador, I. (2015). Biochemical and photochemical feedbacks of acute Cd toxicity in Juncus acutus seedlings: The role of non-functional Cd-chlorophylls. *Estuarine, Coastal and Shelf Science, 167*, 228–239. DOI: 10.1016/j.ecss.2015.10.005

Seregin, I. V., & Kozhevnikova, A. D. (2023). Phytochelatins: Sulfur-Containing Metal(loid)-Chelating Ligands in Plants. *International Journal of Molecular Sciences, 24*(3), 2430. DOI: 10.3390/ijms24032430 PMID: 36768751

Shahid, M., Pourrut, B., Dumat, C., Nadeem, M., Aslam, M., & Pinelli, E. (2014). Heavy-metal-induced reactive oxygen species: Phytotoxicity and physicochemical changes in plants. *Reviews of Environmental Contamination and Toxicology, 232*, 1–44. PMID: 24984833

Shanker, A. K., Cervantes, C., Loza-Tavera, H., & Avudainayagam, S. (2005). Chromium toxicity in plants. *Environment International, 31*(5), 739–753. DOI: 10.1016/j.envint.2005.02.003 PMID: 15878200

Sharma, J. K., Kumar, N., Singh, N. P., & Santal, A. R. (2023). Phytoremediation technologies and their mechanism for removal of heavy metal from contaminated soil: An approach for a sustainable environment. *Frontiers in Plant Science, 14*, 1076876. DOI: 10.3389/fpls.2023.1076876 PMID: 36778693

Sharma, M., Rawat, S., & Rautela, A. (2024). Phytoremediation in sustainable wastewater management: An eco-friendly review of current techniques and future prospects. *AQUA - Water Infrastructure. Aqua (London, England), 73*(9), 1946–1975. DOI: 10.2166/aqua.2024.427

Sharma, P., & Dubey, R. S. (2005). Lead toxicity in plants. *Brazilian Journal of Plant Physiology, 17*(1), 35–52. DOI: 10.1590/S1677-04202005000100004

Singh, R., Singh, P. K., Madheshiya, P., Khare, A. K., & Tiwari, S. (2024). Heavy metal contamination in the wastewater irrigated soil and bioaccumulation in cultivated vegetables: Assessment of human health risk. *Journal of Food Composition and Analysis, 128*, 106054. DOI: 10.1016/j.jfca.2024.106054

Skuza, L., Szućko-Kociuba, I., Filip, E., & Bożek, I. (2022). Natural Molecular Mechanisms of Plant Hyperaccumulation and Hypertolerance towards Heavy Metals. *International Journal of Molecular Sciences, 23*(16), 9335. DOI: 10.3390/ijms23169335 PMID: 36012598

Song, Y., Zhang, L.-L., Li, J., He, X.-J., Chen, M., & Deng, Y. (2018). High-potential accumulation and tolerance in the submerged hydrophyte Hydrilla verticillata (Lf) Royle for nickel-contaminated water. *Ecotoxicology and Environmental Safety, 161*, 553–562. DOI: 10.1016/j.ecoenv.2018.06.032 PMID: 29929131

Sytar, O., Kumar, A., Latowski, D., Kuczynska, P., Strzałka, K., & Prasad, M. N. V. (2013). Heavy metal-induced oxidative damage, defense reactions, and detoxification mechanisms in plants. *Acta Physiologiae Plantarum, 35*(4), 985–999. DOI: 10.1007/s11738-012-1169-6

Tian, S., Xie, R., Wang, H., Hu, Y., Hou, D., Liao, X., Brown, P. H., Yang, H., Lin, X., Labavitch, J. M., & Lu, L. (2017). Uptake, sequestration and tolerance of cadmium at cellular levels in the hyperaccumulator plant species Sedum alfredii. *Journal of Experimental Botany, 68*(9), 2387–2398. DOI: 10.1093/jxb/erx112 PMID: 28407073

Upadhyay, S. K., Srivastava, A. K., Rajput, V. D., Chauhan, P. K., Bhojiya, A. A., Jain, D., Chaubey, G., Dwivedi, P., Sharma, B., & Minkina, T. (2022). Root Exudates: Mechanistic Insight of Plant Growth Promoting Rhizobacteria for Sustainable Crop Production. *Frontiers in Microbiology*, *13*, 916488. Advance online publication. DOI: 10.3389/fmicb.2022.916488 PMID: 35910633

Verstraete, W., Yanuka-Golub, K., Driesen, N., & De Vrieze, J. (2022). Engineering microbial technologies for environmental sustainability: Choices to make. *Microbial Biotechnology*, *15*(1), 215–227. DOI: 10.1111/1751-7915.13986 PMID: 34875143

Visioli, F., Bellomo, G., & Galli, C. (1998). Free radical-scavenging properties of olive oil polyphenols. *Biochemical and Biophysical Research Communications*, *247*(1), 60–64. Advance online publication. DOI: 10.1006/bbrc.1998.8735 PMID: 9636654

Vulpe, C.-B., Toplicean, I.-M., Agachi, B.-V., & Datcu, A.-D. (2025). A Review on Uses of Lemna minor, a Beneficial Plant for Sustainable Water Treatments, in Relation to Bioeconomy Aspects. *Water (Basel)*, *17*(9), 1400. DOI: 10.3390/w17091400

Wang, H., Cui, S., Ma, L., Wang, Z., & Wang, H. (2021). Variations of arsenic forms and the role of arsenate reductase in three hydrophytes exposed to different arsenic species. *Ecotoxicology and Environmental Safety*, *221*, 112415. DOI: 10.1016/j.ecoenv.2021.112415 PMID: 34171691

Xie, P., Zahoor, F., Iqbal, S. S., Ullah, S., Noman, M., Din, Z. U., & Yang, W. (2022). Elimination of toxic heavy metals from industrial polluted water by using hydrophytes. *Journal of Cleaner Production*, *352*, 131358. DOI: 10.1016/j.jclepro.2022.131358

Xu, Q. S., Hu, J. Z., Xie, K. B., Yang, H. Y., Du, K. H., & Shi, G. X. (2010). Accumulation and acute toxicity of silver in Potamogeton crispus L. *Journal of Hazardous Materials*, *173*(1–3), 186–193. DOI: 10.1016/j.jhazmat.2009.08.067 PMID: 19740606

Xu, Y., Sun, X., Zhang, Q., Li, X., & Yan, Z. (2018). Iron plaque formation and heavy metal uptake in Spartina alterniflora at different tidal levels and waterlogging conditions. *Ecotoxicology and Environmental Safety*, *153*, 91–100. DOI: 10.1016/j.ecoenv.2018.02.008 PMID: 29407744

Xue, Y., Wang, J., Huang, J., Li, F., & Wang, M. (2018). The Response of Duckweed (*Lemna minor* L.) Roots to Cd and Its Chemical Forms. *Journal of Chemistry*, *2018*, 1–9. DOI: 10.1155/2018/7274020

Yadav, A., Hasan, M., Kour, S., Chandra, R., Singh, V., Noor, I. M., Yahya, M. Z. A., Singh, H. M., & Tripathi, A. (2025). Heavy Metal Contamination in Plants: Sources, Monitoring, and Data-Driven Insights. *Macromolecular Symposia*, *414*(1), 2400239. Advance online publication. DOI: 10.1002/masy.202400239

Yang, R., Roshani, D., Gao, B., Li, P., & Shang, N. (2024). Metallothionein: A Comprehensive Review of Its Classification, Structure, Biological Functions, and Applications. *Antioxidants*, *13*(7), 825. DOI: 10.3390/antiox13070825 PMID: 39061894

Zhang, H., Zhang, L.-L., Li, J., Chen, M., & An, R.-D. (2020). Comparative study on the bioaccumulation of lead, cadmium and nickel and their toxic effects on the growth and enzyme defence strategies of a heavy metal accumulator, Hydrilla verticillata (Lf) Royle. *Environmental Science and Pollution Research International*, *27*(9), 9853–9865. DOI: 10.1007/s11356-019-06968-0 PMID: 31927739

Ziegler, P., Appenroth, K. J., & Sree, K. S. (2023). Survival Strategies of Duckweeds, the World's Smallest Angiosperms. *Plants*, *12*(11), 2215. DOI: 10.3390/plants12112215 PMID: 37299193

Zorza, L., Bertins, M., Saleniece, K., Kizane, G., Grinbergs, A., Eismonts, U., Reinholds, I., Viksna, A., & Muter, O. (2022). Caesium-133 Accumulation by Freshwater Macrophytes: Partitioning of Translocated Ions and Enzyme Activity in Plants and Microorganisms. *Sustainability (Basel)*, *14*(3), 1132. DOI: 10.3390/su14031132

Chapter 3
The Role of Microbial Communities in Hydrophyte-Assisted Bioremediation

Shagufta Rashid
https://orcid.org/0000-0002-2063-1323
Annamalai University, India

Aadil Farooq Shah
https://orcid.org/0000-0002-2907-851X
Annamalai University, India

Padma Sharma
https://orcid.org/0000-0001-5098-2725
Department of Ecology Environment and Remote Sensing, Paryavaran Bhawan Transport, India

ABSTRACT

Hydrophyte-assisted bioremediation is a prospective green technology for extenuating water pollution, particularly in raised wetlands and other aquatic treatment systems. The microbial populations linked to hydrophytes, particularly in the rhizosphere, are primarily responsible for this approach. Root exudates from hydrophytes not only shape microbial community composition but also enhance microbial activity, leading to increased rates of processes such as nitrification, denitrification, sulphate reduction and organic matter decomposition. In return, microbes support plant health. This chapter provides an in-depth overview of the functional roles and ecological dynamics of microbial communities in hydrophyte-based remediation systems. It also

DOI: 10.4018/979-8-3373-4037-1.ch003

explores current advances in microbial community profiling, such as metagenomics and transcriptomics, which offer new insights into the plant–microbe interactions underpinning these systems. Understanding these symbiotic relationships is key to designing, low-cost, and sustainable treatment technologies for contaminated aquatic environments.

INTRODUCTION

Aquatic pollution from industrial discharge, agricultural runoff, and domestic sewage poses severe threats to freshwater ecosystems. One of the primary contributors of pollution in the soil and aquatic environment is pollution from various industrial activities. The manufacturing process releases various kinds and levels of heavy metals. Pesticides, fertilizers, and pesticides used in agriculture emit contaminants released into the environment, including lead, arsenic, copper, zinc, nickel, and aluminum (Ayilara & Babalola 2023). Heavy metal-containing waste is inappropriately dumped into aquatic and soil ecosystems which results in chronic diseases among humans as well as livestock. Therefore, these contaminants must be treated using physical, chemical, or biological methods. Physical and chemical procedures have been used for several decades, but they have limitations. Chemical bioremediation requires specific tools and proficiency, while physical bioremediation is affluent (Bakhshoodeh et al., 2020). Plants stimulate physicochemical and biological processes which aid in purify wastewater by eliminating metals from its effluents (DalCorso et al., 2019). Hydrophytes, such as *Eichhornia crassipes*, *Pistia stratiotes*, *Lemna gibba* and *Phragmites australis* are capable of up taking, stabilizing, or transforming pollutants. According to Ayaz et al. (2020), other emergent species classes, such Typha latifolia, can also absorb metals at higher concentrations in their roots and shoots. Because they are ecologically adapted, regional hydrophyte species can be used in artificial wetlands for wastewater treatment, according to a 2009 study by Khan et al. Utilizing biological organisms to transform potentially hazardous contaminants into innocuous substances, bioremediation is therefore one of the most efficient and cost-effective alternatives to conventional approaches (Akhtar et al., 2020). Both plants and microorganisms can be used in biological remediation; hydrophyte-assisted bioremediation, sometimes referred to as engineered wetlands or phytoremediation in aquatic systems, provides a sustainable and environmentally beneficial approach. Rhizospheric and endophytic microbial communities, which are essential for biogeochemical cycling and pollutant degradation, significantly enhance the remediation efficiency of hydrophytes. Through this approach, the microbial communities may enhance the water's quality by eliminating organic carbon, nitrogen, phosphorus, suspended particles, heavy metals, phenols, COD, and BOD.

Exudates from hydrophyte roots release carbon and nutrients that promote microbial activity, pesticides, etc., from wastewater while also preventing algal bloom (Bhatt et al., 2020) there by enhancing processes like nitrogen cycling, sulphur transformation, and organic matter decomposition. Conversely, microbes support plant health and function, creating a mutually beneficial system that drives the effectiveness of bioremediation (Sales da Silva *et al.* 2020). When bioremediation was employed to treat one acre of Pb-polluted soil found that it was 50–65% more cost-effective than traditional approaches like excavation and landfilling (Chibuike and Obiora 2014). Endophytes colonize the internal tissues of hydrophytes without causing harm. They can degrade pollutants directly within the plant and increase its stress tolerance. Epiphytes live on plant surfaces, forming biofilms that helps to breakdown pollutants and protect against disease invasion. The large number of sewages by industries are released to surface water, resulting in aquatic pollution. It causes water to be unsuitable for irrigation, drinking or any other purpose due to accumulation of heavy metals and other toxic substances. Hypertension, cancer, neurological diseases, bone damage, lung insufficiency, and embryotoxic consequences are only a few of the health problems that can result from heavy metal pollution, which has become a serious threat to living organisms in the environment. There are several other approaches that might be utilized to deal with this issue, but they are costly and adversely impact the environment. It is not possible to eradicate heavy metals in underdeveloped nations using such methods. Hydrophyte species are therefore anticipated to play an integral part in reclaiming wastewater (Figure 1).

Figure 1. Impact of Hydrophytes on Humans health and environment

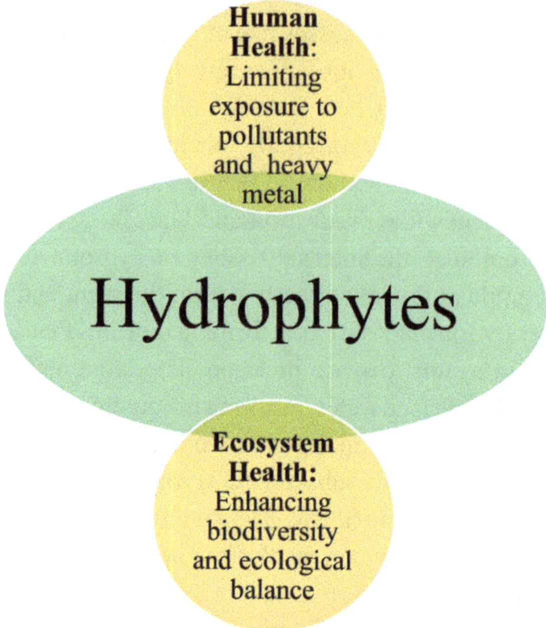

1. HYDROPHYTES: THEIR ECOLOGICAL IMPORTANCE

The term "hydrophytes" describes aquatic plant species that may thrive freely in various aqueous environments, partially or completely submerged. These plants adapt to thrive in an aquatic environment by modifying their leaves, stems, and roots owing to variations in their cellular components (Kumara et al., 2020). The hydrophyte plants can thrive in diverse habitats, including lotic, lentic, and marine water sources. The two scientists Warming and Raunkiaer were the first to coin the term hydrophyte. They have evolved particular traits such as flexible stems to withstand being driven along by turbulent waters, thin leaves to allow nutrients and gasses to diffuse, little or no cuticle, and aerenchyma tissue to give buoyancy and gaseous exchange (Cronk and Fennessy, 2016). In terms of physiology, hydrophytes usually exhibit a high degree of adaptability, which includes low lignin content, effective photosynthetic under low light, and root zone anoxia resistance. Due to these adaptations, they can survive in a variety of freshwater habitats, including lakes, rivers, wetlands, and artificial ponds. With the exception of Antarctica, hydrophytes are found on all continents. They are limited by climate zones, hydrological regimes, and nutrient availability (Chambers et al., 2008). Hydrophytes are essential for management of weeds, shoreline stabilization, and the removal of harmful water contaminants like phosphorus and nitrogen. Moreover, it prevents erosion and traps sediments. it has been demonstrated that these plants are a good source of energy, fiber, paper, fish feed, cattle feed, and biofertilizer. Furthermore, they have the capacity to purify the wastewater through the uptake of dissolved nutrients, including trace metals.

2. OVERVIEW OF POLLUTION IN AQUATIC ENVIRONMENT

2.1. Sources of Aquatic Pollution

Numerous kinds of contaminants that originate from both natural and anthropogenic activities are absorbed by aquatic environments. The primary sources include Industrial effluents discharged from textile, tanning, electroplating, mining, chemical, and paper industries introduces heavy metals (Cr, Pb, Cd, Hg), dyes, solvents, and phenolic compounds. Agricultural runoff fertilizers and pesticides leach into rivers and lakes, contributing nitrogen (N), phosphorus (P), and persistent organic pollutants (POPs). Municipal wastewater, mining and metallurgy, urban stormwater and natural processes (Yang, 2025).

2.2. Types of Pollutants in Aquatic Environments

Excess nutrients (Nitrogen and Phosphorus) from agriculture and sewage lead to eutrophication, resulting in algal blooms, hypoxia, and fish kills. Metals like Cd, Pb, Cr, Ni, Zn, and Hg are non-biodegradable and bioaccumulate through food chains, causing chronic toxicity in aquatic life and humans (Chandel et al., 2023). Organic Pollutants, Pesticides, herbicides, Phenols, hydrocarbons, dyes and Pharmaceuticals are the other category of pollutant that directly or indirectly impact on the human health and environment.

2.3. Impacts of Aquatic Pollution

- **Ecological Impacts**

Aquatic ecosystems are highly subtle to pollutants, which can drastically alter their structure and functioning. The ecological consequences of aquatic pollution are eutrophication of water bodies which occur with excessive nutrients, mainly nitrogen (N) and phosphorus (P), typically from agricultural runoff, sewage, and industrial effluents. Bioaccumulation and Biomagnification, habitat degradation, disruption of food web are the other ecological impacts of aquatic pollution.

- **Health Impacts**

Human health is intimately linked to water quality. Contaminated water can directly or indirectly lead to serious diseases. According to WHO, over 2 billion people worldwide use contaminated water, leading to millions of deaths annually. Wastes from industries, pesticides, and pharmaceuticals in water can cause reproductive disorders and developmental abnormalities, neurotoxicity, chronic diseases, including cancers and kidney or liver damage.

- **Economic impact**

Polluted water bodies create direct and indirect economic burdens such as reduced fish stocks due to toxic pollution or oxygen depletion harm commercial and subsistence fisheries. Coastal oil spills and algal blooms reduce tourist visits, affecting local economies. To ensure safe drinking water municipalities must invest in water purification (Thakur et al., 2022).

- **Environmental Consequences**

Climate change can be exacerbated by the production of greenhouse gases from polluted water bodies, such as methane from eutrophic lakes. Toxic and persistent contamination can cause a species to eventually become extinct regionally or worldwide. In aquatic ecosystems, pollution may disrupt with the availability of oxygen, carbon storage, and nutrient cycling.

3. HYDROPHYTES IN BIOREMEDIATION

Hydrophytes are essential to the bioremediation of contaminated water bodies. They are particularly effective at eliminating contaminants because of their physiological, morphological, and biochemical traits, as well as their unique adaptation to grow in water-saturated conditions.

3.1. Mechanisms of Bioremediation by Hydrophytes

To eliminate contaminants from aquatic ecosystems, the hydrophytes exploit a wide range of physiological, biochemical, and ecological processes. Hydrophyte-assisted systems are safe and effective for treating a number of pollutants because of their mechanisms, which may function separately or in combination.

- **Phytoextraction**

In phytoextraction, dissolved contaminants mainly nutrients and heavy metals are absorbed and incorporated into the plant tissues from the water or sediment. The most hazardous metals that can hyperaccumulate in hydrophytes such as Lemna minor (duckweed) and Eichhornia crassipes (water hyacinth) are cadmium (Cd), lead (Pb), arsenic (As), and zinc (Zn). Metals are absorbed by roots, stored in vacuoles, sometimes transferred to leaves or shoots. According to the study conducted by Tang (2024) the *Eichhornia* can remove up to 90% of Cd within 7 to10 days in controlled conditions.

- **Phytofiltration**

The term "phytofiltration" describes the adsorption and precipitation of dissolved contaminants onto extensive roots of hydrophytes. *Pistia stratiotes* (water lettuce) and *Lemna gibba* are examples of free-floating plants that work particularly well because of their fine roots, which provide a considerable amount of surface area for binding contaminants (Rahman et al., 2016)

- **Phytostimulation**

Phytostimulation is the process by which exudates from plant roots increase microbial activity to degrade organic contaminants. While ethylene inhibits DNA synthesis and cell division at high concentrations, it stimulates root elongation at low concentrations. However, this can be avoided by utilizing the enzyme 1-aminocyclopropane-1-carboxylase deaminase to reduce ethylene levels in plants. This balances the synthesis of ethylene in plants, so reducing abiotic stress (Gaiero et al., 2013). PGPR, a protein linked to plant roots produces this enzyme, which degrade metal pollutants by using plant exudates as a carbon source and energy (Tak et al., 2012).

- **Phytostabilization**

Plant roots are used to remove contaminants from the earth and retain them within the rhizosphere then segregate and stabilize them, rendering them innocuous and stopping their spread throughout the environment. Because to metal valence reduction, root sorption, complexation, and precipitation in the area surrounding plant roots, heavy metals are less accessible or mobile in the environment (Choudhary & Varma, 2016). Phytostabilization plants must have a wide root system and less metal movement from roots to shoots. By altering the pH and amount of organic with the addition of biochar or compost, the capacity of plants for phytostabilization can be enhanced (Abbaszadeh-Dahaji et al., 2017). This will absorb the metals and increase plant productivity (Islam et al., 2013).

- **Phytovolatilization**

This deals to the method by which plants eliminate contaminants from the soil that are easily converted to vapors before being released into the atmosphere. The highly poisonous methyl mercury that tobacco plants may accumulate from mercury-contaminated sites are converted into the least lethal elemental mercury as a volatile substance that can then release into the atmosphere through leaves (Rayu et al., 2012). During phytovolatilization, contaminants are converted into volatile forms by the metabolic abilities of the plants and microbes within the rhizosphere (Tak et al., 2012).

- **Phytodegradation**

The process by which plant enzymes convert organic contaminants into non-hazardous substances is known as phytodegradation. Plants use certain enzymes, such as nitroreductases and dehalogenases, to break down organic pollutants. For optimum breakdown of pollutants, these enzymes must be employed at ideal pH and temperature. By means of the rhizodegradation process, rhizospheric bacteria may also promote the breakdown of organic contaminants present in the soil. Due to the rhizospheric region, the plants have large quantities of nutrients released from the roots that entice large number of microbes to aid degradation of the pollutants as compared to the bulk soil which has a least amount of organic compounds (Favas e al., 2014; Ogunmayowa, 2015).

- **Rhizofiltration**

Rhizofiltration is the process by which plant roots remove harmful materials or contaminants from groundwater. The mechanism of plant rhizospheric accumulation serves as a basis for the rhizofiltration technique. Rhizofilteration is more effective in terrestrial plants than in aquatic plants because the former uses solar energy pumps to obtain specific elements from their environment. Rhizofiltration operates well with plants that can absorb and tolerate large levels of harmful metals, such as hyperaccumulators. Plants are more competent to eliminate heavy metal pollutants and protect themselves from environmental stress when PGPR is introduced to a contaminated location due to the bioavailability of these metals drops. However, there are several limitations to phytoremediation technique, such as retention and accumulation of pollutants in the plant tissue and a decrease in bioremediation rate, which commonly proves inadequate at the contaminated sites where there are many pollutants (López-Chuken, 2012, Ma et al., 2011).

4. ASSOCIATION OF MICROBES WITH HYDROPHYTES FOR BIOREMEDIATION

Bioremediation of contaminated environment can also be accomplished through the interaction of microbes and hydrophytes. The microbes and rhizosphere from specific microbial communities have a symbiotic relationship at the ground level, assisting plants in removing pollutants, particularly heavy metals, from the environment. Several microorganisms, such as mycorrhizae, rhizobacteria and yeast, have been suggested as target inoculants to increase plant growth adaptations, performance, and the entire bioremediation process (Yan et al., 2020, Liu et al., 2016, Raklami et al., 2019). The microbial communities associated with hydrophytes include Bacteria (*Pseudomonas spp., Bacillus spp., Klebsiella spp., Enterobacter spp*), Fungi (*Aspergillus spp., Penicillium spp., Trichoderma spp*) and Actinomycetes. As an external barrier, mycorrhizae trap HMs on fungal mycelium and immobilize them in the environment by spreading, which reduces their transport, bioaccumulation, and bioavailability in plant tissues. This helps with the phytoremediation process. According to some studies, fungal spores, arbuscules, and vesicles may have a role in the storage of heavy metals, providing further resistance against metal toxicity. The synthesis of chelating agents, metal transformation, and passive or active transport from fungal cells to plant cells are the primary mechanisms of mycorrhizae-assisted phytoextraction.

Table 1. Case studies examples of hydrophytes-associated microbial species

Hydrophyte	Associated Microbe	Pollutant	Mechanism	Outcome
Eichhornia crassipes (Water hyacinth)	*Pseudomonas aeruginosa, Bacillus subtilis*	Oil hydrocarbons, Cr, Pb	Biodegradation & biosorption	>75% removal of hydrocarbons, 80% metal removal
Typha latifolia (Cattail)	*Bacillus cereus, Nitrosomonas* spp.	Nitrate, phosphate	Nitrification–denitrification, nutrient uptake	Significant N & P removal in constructed wetlands
Lemna minor (Duckweed)	*Aspergillus niger, Penicillium chrysogenum*	Textile dyes	Enzymatic degradation (laccase, peroxidase)	60–90% dye decolorization
Phragmites australis (Common reed)	*Enterobacter spp., Klebsiella spp.*	Heavy metals, phenols	Metal transformation & phenol degradation	Detoxification in industrial effluent treatment

5. FUNCTIONAL ROLES OF MICROBES IN HYDROPHYTE-ASSISTED BIOREMEDIATION

5.1. Organic Pollutant Degradation

Multiple microbial species working together often accomplish greater degradation of pollutants like PAHs than individual strains alone. For example, a consortium labeled H6 (comprising *Rhodococcus rhodochrous* Q3, *Mycobacterium* species. *Mycobacterium gilvum* and ZL7 degraded 59% of a PAH mixture with 22–32% higher efficiency than single strains. *Sphingobium* and *Pseudomonas*, attained 100% removal of multiple PAHs in wastewater treatment systems. Metagenomic analyses of bioaugmented soils revealed functional shifts toward increased abundance of aromatic-ring-hydroxylating dioxygenases and related enzyme genes, facilitating 84% PAH mineralization in 30 days. The association of microalgae and bacteria demonstrate enhanced degradation of PAHs (e.g., phenanthrene and anthracene). The algae supply oxygen through photosynthesis, while bacteria degrade pollutants more efficiently within this symbiotic environment. *Novosphingobium pentaromativorans* is distinguished for its ability to degrade high molecular weight PAHs through specific enzymes located on a plasmid (pLA1), including ring-hydroxylating dioxygenases that feed into the TCA cycle. Indigenous microbial communities, such as those dominated by *Alcanivorax* and *Cycloclasticus*, showed high degradative potential ~95% degradation of crude oil and ~63% of PAHs in coastal sediment environments (Singh et al., 2016). Microorganisms in hydrophyte systems degrade large number of organic pollutants, which includes hydrocarbons, pesticides, phenols and pharmaceuticals. Enzymes like oxygenases and peroxidases, which degrade complex substances into less hazardous components, are produced by essential bacterial groups like *Pseudomonas*, *Bacillus,* and *Sphingomonas*. These degradation processes often occur in close proximity to plant roots, where oxygen and root exudates support microbial metabolism.

5.2. Degradation of Liquid Waste

Treatment of wastewater can be categorized as primary, secondary and tertiary depending on the level of pollution mitigation.

i) Preliminary Treatment

The "gatekeeper" of sewage treatment systems is preliminary treatment. In order to prevent the purification plate from being clogged, this initial stage of wastewater treatment entails clearing off debris and other harmful substances (Mani et al., 2020). (Mani et al.,2020). Its main objectives are to remove coarse solids, that clog downstream equipment, extract heavy inorganic. Preliminary treatment methods include screening, grit removal, shredding, pre-aeration, and chemical inclusion.

ii) Primary treatment:

By eliminating organic materials, suspended and floatable solids, primary treatment aims to reduce and eradicate them (Singh 2013). BOD can be reduced by 25% to 30%, settling solids by 95%, and total suspended solids by 50% to 60% at each stage of primary treatment. Clearing tanks that are spherical or rectangular might be used for this procedure (Mani et al., 2020).

iii) Secondary treatment:

After primary treatment, the wastewater is transferred to a secondary treatment facility where the main objective is to eliminate organic debris that decomposes naturally. The secondary treatment breaks down the organic content in the wastewater and generates inorganic end products using aerobic microorganisms, primarily bacteria (CO2, NH3, and H2O). For wastewater treatment to be effective, these microbes require a constant supply of oxygen. Secondary treatment can be broadly divided into two categories, such as fixed-film systems and suspended growth systems. In a fixed-film system, A combined film device uses biomass of microbes or slime injected to a medium, which could be made of redwood, stone or any other synthetic material. The bacteria oxidize and degrade the biological components of the wastewater as soon as the slime comes in contact with it. In contrast, suspended growth mechanism, utilizes biological growth mixed directly with wastewater ((Machineni 2019 The fundamental elements of the suspended-growth activated sludge process are a clarifier, secondary settling basin, and aeration tank. The sewage is aerated by constant mechanical agitation (Samstag et al., 2016). The wastewater is periodically released into a tank for aeration with the appropriate concentration of microbes following primary settling. The partially filtered water is moved to another sedimentation tank to eliminate pathogens. To remove pathogenic microorganisms, before being released into collecting water, the effluent is often disinfected with chlorine in the sedimentation tank (Singh 2013).

iv) Tertiary Treatment

The final stage of sewage treatment, tertiary treatment, aims to enhance effluent quality prior to its released into the atmosphere (Naidoo and Olaniran 2014). This treatment eliminates color and other impurities that secondary treatment procedures could not effectively remove. It also reduces BOD and removes inorganic compounds such as phosphates and nitrates (Gunatilake, 2015).

5.3. Degradation of Solid Waste

Biological composting, sometimes referred to as "composting" or "nature's means of recycling," is the most traditional and widely utilized method for disposing of solid waste. It is a biological approach that uses a variety of microbes (bacteria, fungi, etc.) to aerobically break down complex organic waste from many sources, including food scraps, paper, coffee grounds, and plant/garden waste, into a product that resembles humus (Diaz et al., 2020). Because of its tremendous carbon and nitrogen content, the corresponding material so referred to as manure is an appropriate compost for plants (Suyal et al., 2021). Anaerobic and aerobic composting are two other categories for composting. A mesophilic microbe initiates the compositing process by efficiently dissolving the waste's soluble component. When temperatures exceed 40 ∞C, mesophilic bacteria are replaced by thermophilic microbes, and complex carbohydrates, proteins, and lipids break down rapidly. The temperature gradually drops as the compositing process approaches its conclusion, and mesophilic bacteria regulate for the last stage of "curing," or maturing, of the residual organic content. The most sustainable way of managing solid waste is composting, which also enhances the soil's texture and capacity to retain moisture (Banerjee et al., 2019). Composting reduces the demand for artificial fertilizers by balancing the acidic/alkaline composition of the soil and preserving nutrients.

5.4. Degradation of heavy metals

Heavy metals pose a serious ecological and environmental vulnerability on a global scale due to their bioaccumulation and non-biodegradability (Igiri et al., 2018; Yuan et al., 2021). It is imperative that removal of heavy metals (Cr, Hg, Cd, and Pb) techniques become more economical, innovative, effective, and environmentally friendly in order to restore the environment. According to this perspective, recent advances in heavy metals that are dependent on bacteria may spur bioremediation as a potential replacement for conventional methods. Researchers have recently shown a variety of bioremediation techniques; however, due to the prevalence of various kinds of contaminants in the environment, no single bioremediation technique can be utilized as a solution to restore contaminated areas. However, because of their enormous potential, bacteria have been biochemically found to be an emerging agent that can sustain heavy metals. As a result, microbial remediation has emerged as the primary focus of recent research in bioremediation technology. As long as the environment is suitable for their metabolic and developmental processes, indigenous microbes inhabiting contaminated areas are capable of overcoming all the challenges related to bioremediation and biodegradation. The detoxification and toxicity mechanisms utilized by microorganisms are diverse and include enzymes, metal oxidation, the production of exopolysaccharides and metal chelators, such as metallothioneins, methylation, extrusion, biotransformation, precipitation, volatilization, surface complex action, electrostatic interaction, demethylation, and intracellular and extracellular redox processes. Microorganisms may bind to metal cations because of the negative charge that is generated by the anionic structures (sulfonate, sulfydryl, carboxyl, alcohol, amine, hydroxyl, phosphoryl, thioester, thiol, and ester groups) that are present on their cell surfaces (Shuaib et al., 2021).

(a) **Bio-accumulation**: Bio-accumulation is a toxicokinetic process that affects the sensitivity of living organisms to chemicals. It occurs when the absorption of contaminants exceeds the degree of loss (Mishra and Malik 2013; Diaconu et al., 2020). The ability of bioaccumulation organisms to biotransform hazardous substances into harmless molecules and their advanced tolerance level to several contaminants enable them to reduce the toxicity of heavy metals.

(b) **Biotransformation:** With the aid of their enzymatic system, microbes can transform contaminants through biochemical processes which include oxidation, reduction, methylation, and de-methylation. These reactions may occur on the cell surface, in extracellular environments, or in vacuoles (Mohd et al., 2019). This involves the reduction of substantial metal ions, such as the conversion of highly noxious Cr (VI) to less toxic Cr (III) by isolated Gram-positive bacteria found in tannery sewers, it is significant.

(c) Methylation

A key aspect in the removal of heavy metals is microbial methylation. Methylation increases the toxicity of metals by improving their lipophilicity, which eventually results in increased the permeability across the cell membrane. The majority of methylated compounds are volatile by nature. For example, certain bacterial species, including *Clostridium, Escherichia, Bacillus, and Pseudomonas,* can produce methyl-Hg by methylating Hg (II); similarly, methylation of Se and As can produce dimethyl selenide and arsines, respectively (Ma et al., 2019).

Table 1. Significance of microbes in degradation of heavy metals

S. No	Heavy metals	Microbial Agents	Remediation Mechanism	References
1	Cu, Cr	*Acidithiobacillus ferrooxidans, Spirogyra sp.*	Bioleaching of metal sulfides copper extraction via oxidation cycles	Mane and Bhosle (2012)
2	Cr (VI)	*Arthrobacter sp., Bacillus sp.*	Biosorption, bioreduction of Cr (VI) → Cr (III)	Chakraborty et al. (2025); Wróbel et al. (2023)
3	U, Hg, V (vanadium)	*Geobacter metallireducens*	Extracellular reduction of heavy metals	Jafari et al. (2015).
4	Fe, Pb	*Shewanella oneidensis MR-1, Staphylococcus sp, Cellulosimicrobium sp*	Anaerobic reduction of metal ions; electron transfer to metals	Bharagava and Mishra (2018). Kumaran et al., (2011).
5	Zn	*Bacillus firmus*	Plasmid-driven resistance to heavy metals; bioremediation in harsh environments	Salehizadeh and Shojaosadati (2003).

5.5. Degradation of Agro-chemicals

Through the reduction of disease, pathogen, and weed losses, agricultural pesticides are utilized for increasing agricultural output. Among a number of agrochemicals used in contemporary agriculture, pesticides, herbicides, fungicides, and fertilizers are now contributing to environmental pollution. Agrochemical field pollution is a rising global issue that requires immediate attention. (Liu et al.,2019).

6. FACTORS AFFECTING MICROBIAL BIOREMEDIATION

The concentration of metal ions, redox potential, metal chemical forms, microbe competition, pH, temperature, soil structure, oxygen content, moisture content, soil type, and the heavy metal's solubility in water are all factors that affect the ability of microorganisms to bioremediate heavy metals. At acidic pH, heavy metals form free ionic species, which increase the number of protons available to saturate its binding site. The optimal pH range for bioremediation is 6.5 to 8.5 since it affects the pollutant's structure and shows how long the microbe can survive in that environment (Kharangate-Lad & D'Souza, 2021). As a source of energy, microbes interact with each other for carbon, a limited resource. Because the bacteria are more competitive, their natural ability to eliminate heavy metal contaminants affect the rate of biodegradation. The C:N:P ratio must be balanced to promote biodegradation in situations where these essential elements are restricted because microbes that perform biodegradation require both carbon and nitrogen (N) and phosphorus (P) to enhance microbial activity, they can be added (Bala et al.,2022). The rate of a bioremediation process is determined by the form and diversity of microorganisms. For example, in an artificial environment, a certain heavy metal may be bioremediated by a strain of organism but this would be difficult in a natural context where a group of bacteria would be required (Patel et al., 2022). Further factors affecting the rate of microbial biodegradation comprise the sorption location, the structure of the microbial cell wall, and the ionization of the cell wall's chemical constituents (Mahmoud et al., 2021). The amount of moisture in an environment influences the solubility, availability, pH, and osmotic pressure of heavy metals in water. With a high moisture level, microbial biodegradation is extremely slow. The reason for this could be that an anaerobic environment is established, which inhibits aerobic microorganisms from surviving. An optimal microbial biodegradation process requires an appropriate moisture content because microorganisms could not be able to endure at low moisture levels. Only psychrophiles can survive in harsh conditions where bacteria break down heavy metals slowly. Because the microbial pathways are immobilized by the freezing water, metabolic activity is reduced. While majority bioremediation processes require high temperatures, the temperature at which each molecule degrades vary (Ren et al., 2018). Increases in temperature cause heavy metals to become more soluble, which speeds up their availability and the pace at which they are broken down by microbes (Mahmoud et al., 2021). Similarly, the rate of microbial biodegradation is determined by the chemical structure, bioavailability, concentration, toxicity, and stability of pollutants. The two processes of biodegradation are anaerobic and aerobic. By decomposing a specific mineral, an organism may be able to thrive in such an environment, depending on its nature (Jacob et al., 2018). The rate of microbial biodegradation depends on an array of soil properties, such as soil area, particle size and texture (Alvarez et al., 2020). Microorganisms are diverse and more prevalent in the top 0–10 cm of the soil. This is due to the reason that there are more available organic matter and oxygen in comparison to sedimentary soils (Ndeddy Aka and Babalola, 2017). The presence of salt impacts the hydrocarbon clastic activity of halophilic and halotolerant bacteria and exposes soil microbes to stresses. The more saline an environment is, greater is the osmotic pressure of microorganisms. There is an instant effect on the heavy metals' solubility, their mode of transportation, and the bacterial metabolic activities (Imron et al., 2020, Huang et al., 2019).

7. MECHANISM OF MICROBIAL BIOREMEDIATION

The method used for eliminating pollutants from the ecosystem is bioremediation. It eliminates harmful contaminants and restores the ecosystem to its natural state through biological processes that involve microbes and plants. The main principles of bioremediation are decreasing the solub ility of these pollutants by their adsorption from areas of contamination, redox reactions, and pH changes (Jain and Arnepalli 2016). According to Tandon and Singh (2016), Bioremediation processes are based on redox processes, with the objective to change the physical characteristics and microorganisms of contaminated water by adding specific reagents. This facilitates the breakdown and extraction of various contaminants using in situ chemical oxidation/reduction reactions. Bioremediation is carried out by the use of microorganisms, which rely on their metabolic capacity to eliminate ecological contaminants and transform them to harmless molecules via redox reactions (Tak et al., 2013). The enzyme oxidoreductase detoxifies certain pollutants by converting toxic molecules into harmless ones, and several types of microbes secrete different enzymes that can degrade organic toxic elements present in environment. Toxic compounds such as metals and phenolics are primarily eliminated from the environment by oxidoreductase enzymes. Mycelium enabled filamentous fungus to readily absorb soil pollutants and release the detoxifying enzymes laccase and peroxidase (Khatoon et al., 2021). Assimilation, biosorption to membranes, immobilization, complex formation, precipitation, and release agents are all processes that certain bacteria accomplish. The purple non-sulfur bacteria *Rhodobacter sphaeroides* and *Rhodobium marinum* may reduce the effects of heavy metals like cadmium, copper and zinc (Li et al., 2016). Functional groups like hydroxyl, carbonyl, or phosphate groups integrate heavy metals into the cell walls of soil microbiota. This causes metabolic processes to change to prevent metal uptake, and different enzymatic reactions turn toxic metals into harmless compounds. Utilizing an ion exchange method, yeast bioremediates the toxic water. Aerobic bacteria rely on microalgae for food, oxygen and enzyme release, these microbes decompose industrial waste and oil. According to Sutherland and Ralph (2019), conjugated consortia of microalgae-bacteria break down ibuprofen and caffeine more quickly than individual bacterial consortia. By lowering metal binding potential, yeast like S. cerevisiae helps to reduce environmental effluents. This enables radioactive metals to be blocked in biomass, removed, repurposed, and converted into biologically innocuous substances. While exopolysaccharides strengthen the ability cell wall for biosorption, the cell wall of yeast uses electrostatic interactions to bind to heavy metals (Abbas et al., 2014).

8. FUTURE RESEARCH DIRECTIONS

8.1. Plant–Microbe Interaction Mechanisms

Despite evidence of hydrophytes supporting microbial activity, the molecular signaling pathways between roots and microbes remain poorly understood. Research should focus on root exudate composition under different pollution stresses, plant hormonal regulation of microbial colonization and microbial quorum sensing and its role in biofilm formation on plant roots. Understanding these interactions will enable us to select or engineer hydrophyte–microbe partnerships optimized for specific pollutants (Mitsch and Gosselink, 2015).

8.2. Enhancement of Phytoremediation via Genetic Improvement

The next phase of hydrophyte-based phytoremediation research should concentrate on genetically improving biomass production, tolerance, and heavy metal uptake efficiency. Hydrophyte cultivars with increased pollutant binding capacity, faster growth, and improved resilience to environmental challenges could be produced using advances in plant genomics, transcriptomics, and CRISPR-Cas9 gene editing. To increase the effectiveness of the phytoremediation process, iron transporter metal transporters (such as the ZIP and NRAMP families) and antioxidant defense technologies may be controlled (Ali et al., 2013). Furthermore, integrating bioengineering and ecological considerations will be important in preventing genetically modified hydrophytes from affecting indigenous species or altering ecological equilibrium.

9. CONCLUSION AND FUTURE PERSPECTIVE

Microbial communities play an important role in hydrophyte-assisted bioremediation by mediating crucial metabolic reactions that results in pollution elimination. Bioremediation is a simple, cost-effective, and sustainable approach to eliminate contaminants. Developing genetically modified bacteria to break down contaminants is a major concern; this is an alternate sustainable method to accomplish. It has a permanent and adaptive impact on the environment, making it an effective approach to mitigate environmental pollutants. As our knowledge of enzyme structure, degradation pathways, and metabolic profile of microorganisms evolves, in future we will be able develop genetically engineered microbes. A more comprehensive understanding of the genetic composition and metabolic processes of bacteria might support bioremediation. There should be a complete computerized database that may be used to detoxify toxins from a particular area. Through enzymatic processes such elimination, hydrolysis, isomerization, oxidation, and reduction at a particular pH and temperature, microbes are able to identify and break down pollutants. Thus, it can be stated that the biodegradation of toxic and hazardous contaminants with microbial aid harmless to the environment and safe, and able to restore the physicochemical activities of soil and groundwater. It is evidently demonstrating how more research into microbes and bioremediation techniques can support in the management of both aquatic and terrestrial waste. The degrading potential of microorganisms can be increased by appropriate environmental conditions, including temperature, pH, oxygen, and nutrition. Significant effort on enzyme characterization and microbial consortium development is required to achieve complete improvements and create new bioremediation solutions. Microbes and hydrophytes interact to form a dynamic, self-sustaining system capable of effectively and environmentally addressing complex environmental issues. These interactions may now be exploited and optimized for versatile and economical water treatment systems because to advances in molecular biology and system engineering. It is possible to maximize the potential of these aquatic plants through research, governance, and community involvement in order to restore degraded ecosystems, preserve water resources, and establish a strong future that will support future generations. subsequently, harnessing hydrophytes provides a synergistic avenue to address pollution management, biodiversity conservation, and climate resilience, highlighting their potential as an essential means in establishing a sustainable and climate-resilient future.

REFERENCES

Abbas, S. H., Ismail, I. M., Mostafa, T. M., & Sulaymon, A. H. (2014). Biosorption of heavy metals: A review. *J Chem Sci Technol, 3*, 74–102.

Abbaszadeh-Dahaji, P., Omidvari, M., & Ghorbanpour, M. (2017). Increasing phytoremediation efficiency of heavy metal-contaminated soil using PGPR for sustainable agriculture. In *Plant-microbe interaction: An approach to sustainable agriculture* (pp. 187–204). Springer Singapore.

Akhtar, F. Z., Archana, K. M., Krishnaswamy, V. G., & Rajagopal, R. (2020). Remediation of heavy metals (Cr, Zn) using physical, chemical and biological methods: A novel approach. *SN Applied Sciences*, *2020*(2), 1–14. DOI: 10.1007/s42452-019-1918-x

Ali, H., Khan, E., & Sajad, M. A. (2013). Phytoremediation of heavy metals—Concepts and applications. *Chemosphere*, *91*(7), 869–881. DOI: 10.1016/j.chemosphere.2013.01.075 PMID: 23466085

Alvarez, L. M., Ruberto, L. A. M., Gurevich, J., & Mac Cormack, W. P. (2020). Environmental factors affecting reproducibility of bioremediation field assays in Antarctica. *Cold Regions Science and Technology*, *169*, 102915. DOI: 10.1016/j.coldregions.2019.102915

Ayaz, T., Khan, S., Khan, A. Z., Lei, M., & Alam, M. (2020). Remediation of industrial wastewater using four hydrophyte species: A comparison of individual (pot experiments) and mix plants (constructed wetland). *Journal of Environmental Management*, *255*, 109833. DOI: 10.1016/j.jenvman.2019.109833 PMID: 31747629

Ayilara, M. S., & Babalola, O. O. (2023). Bioremediation of environmental wastes: The role of microorganisms. *Frontiers in Agronomy*, *5*, 1183691. DOI: 10.3389/fagro.2023.1183691

Bakhshoodeh, R., Alavi, N., Oldham, C., Santos, R. M., Babaei, A. A., Vymazal, J., & Paydary, P. (2020). Constructed wetlands for landfill leachate treatment: A review. *Ecological Engineering*, *146*, 105725. DOI: 10.1016/j.ecoleng.2020.105725

Bala, S., Garg, D., Thirumalesh, B. V., Sharma, M., Sridhar, K., Inbaraj, B. S., & Tripathi, M. (2022). Recent strategies for bioremediation of emerging pollutants: A review for a green and sustainable environment. *Toxics*, *10*(8), 484. DOI: 10.3390/toxics10080484 PMID: 36006163

Banerjee, P., Hazra, A., Ghosh, P., Ganguly, A., Murmu, N. C., & Chatterjee, P. K. (2019). Solid waste management in India: a brief review. In Ghosh, S. (Ed.), *Waste management and resource efficiency*. Springer. DOI: 10.1007/978-981-10-7290-1_86

Bharagava, R. N., & Mishra, S. (2018). Hexavalent chromium reduction potential of Cellulosimicrobium sp. isolated from common effluent treatment plant of tannery industries. *Ecotoxicology and Environmental Safety*, *147*, 102–109. DOI: 10.1016/j.ecoenv.2017.08.040 PMID: 28841524

Bhatt, P., Rene, E. R., Kumar, A. J., Kumar, A. J., Zhang, W., & Chen, S. (2020). Binding interaction of allethrin with esterase: Bioremediation potential and mechanism. *Bioresource Technology*, *315*, 13845. DOI: 10.1016/j.biortech.2020.123845 PMID: 32707504

Chakraborty, A., Bhakat, K., Islam, E., & Murmu, R. (2025). Arthrobacter sp. mediated chromium remediation and copper accumulation from leached liquor for E-waste management. *The Microbe*, *6*, 100277. DOI: 10.1016/j.microb.2025.100277

Chambers, P. A., Lacoul, P., Murphy, K. J., & Thomaz, S. M. (2008). Global diversity of aquatic macrophytes in freshwater. *Hydrobiologia*, *595*(1), 9–26. DOI: 10.1007/s10750-007-9154-6

Chandel, S., Dar, R. A., Singh, D., Thakur, S., Kaur, R., & Singh, K. (2023). Plant assisted bioremediation of heavy metal polluted soils. In *Bio-Inspired Land Remediation* (pp. 85-114). DOI: 10.1007/978-3-031-04931-6_4

Chibuike, G., & Obiora, S. (2014). Heavy metal polluted soils: Effect on plants and bioremediation methods. *Applied and Environmental Soil Science*, *2014*, 2014. DOI: 10.1155/2014/752708

Choudhary, D. K., & Varma, A. (Eds.). (2016). *Microbial-mediated induced systemic resistance in plants* (pp. 147–162). Springer. DOI: 10.1007/978-981-10-0388-2

Cronk, J. K., & Fennessy, M. S. (2016). *Wetland plants: Biology and ecology* (2nd ed.). CRC Press. DOI: 10.1201/9781420032925

DalCorso, G., Fasani, E., Manara, A., Visioli, G., & Furini, A. (2019). Heavy metal pollutions: State of the art and innovation in phytoremediation. *International Journal of Molecular Sciences*, *20*(14), 3412. DOI: 10.3390/ijms20143412 PMID: 31336773

Diaconu, M., Pavel, L. V., Hlihor, R. M., Rosca, M., Fertu, D. I., Lenz, M., Corvini, P. X., & Gavrilescu, M. (2020). Characterization of heavy metal toxicity in some plants and microorganisms-a preliminary approach for environmental bioremediation. *New Biotechnology*, *56*, 130–139. DOI: 10.1016/j.nbt.2020.01.003 PMID: 31945501

Diaz, L. F., Golueke, C. G., Savage, G. M., & Eggerth, L. L. (2020). *Composting and recycling municipal solid waste*. CRC Press. DOI: 10.4324/9781315150444

Gaiero, J. R., McCall, C. A., Thompson, K. A., Day, N. J., Best, A. S., & Dunfield, K. E. (2013). Inside the root microbiome: Bacterial root endophytes and plant growth promotion. *American Journal of Botany*, *100*(9), 1738–1750. DOI: 10.3732/ajb.1200572 PMID: 23935113

Gunatilake, S. K. (2015). Methods of removing heavy metals from industrial wastewater. *Methods (San Diego, Calif.)*, *1*, 14.

Huang, Y., Pan, H., Wang, Q., Ge, Y., Liu, W., & Christie, P. (2019). Enrichment of the soil microbial community in the bioremediation of a petroleum-contaminated soil amended with rice straw or sawdust. *Chemosphere*, *224*, 265–271. DOI: 10.1016/j.chemosphere.2019.02.148 PMID: 30825852

Igiri, B. E., Okoduwa, S. I., Idoko, G. O., Akabuogu, E. P., Adeyi, A. O., & Ejiogu, I. K. (2018). Toxicity and bioremediation of heavy metals contaminated ecosystem from tannery wastewater: A review. *Journal of Toxicology*, *1*, 16. DOI: 10.1155/2018/2568038 PMID: 30363677

Imron, M. F., Kurniawan, S. B., Ismail, N. I., & Abdullah, S. R. S. (2020). Future challenges in diesel biodegradation by bacteria isolates: A review. *Journal of Cleaner Production*, *251*, 119716. DOI: 10.1016/j.jclepro.2019.119716

Islam, M. S., Ueno, Y., Sikder, M. T., & Kurasaki, M. (2013). Phytofiltration of arsenic and cadmium from the water environment using Micranthemum umbrosum (JF Gmel) SF Blake as a hyperaccumulator. *International Journal of Phytoremediation*, *15*(10), 1010–1021. DOI: 10.1080/15226514.2012.751356 PMID: 23819292

Jacob, J. M., Karthik, C., Saratale, R. G., Kumar, S. S., Prabakar, D., Kadirvelu, K., & Pugazhendhi, A. (2018). Biological approaches to tackle heavy metal pollution: A survey of literature. *Journal of Environmental Management*, *217*, 56–70. DOI: 10.1016/j.jenvman.2018.03.077 PMID: 29597108

Jafari, S. A., Cheraghi, S., Mirbakhsh, M., Mirza, R., & Maryamabadi, A. (2015). Employing response surface methodology for optimization of mercury bioremediation by Vibrio parahaemolyticus PG02 in coastal sediments of Bushehr, Iran. *Clean (Weinheim)*, *43*(1), 118–126. DOI: 10.1002/clen.201300616

Jain, S., & Arnepalli, D. (2016). Biominerlisation as a remediation technique: A critical review; *Proceedings of the Indian Geotechnical Conference (IGC2016)*; Chennai, India. 15:17.

Khan, S., Ahmad, I., Shah, M. T., Rehman, S., & Khaliq, A. (2009). Use of constructed wetland for the removal of heavy metals from industrial wastewater. *Journal of Environmental Management, 90*(11), 3451–3457. DOI: 10.1016/j.jenvman.2009.05.026 PMID: 19535201

Kharangate-Lad A., D'Souza N. C. (2021). Current approaches in bioremediation of toxic contaminants by application of microbial cells; biosurfactants and bioemulsifiers of microbial origin. Rhizobiont Bioremediat. Hazard. Waste, 217–263

Khatoon, H., Rai, J. P. N., & Jillani, A. (2021). Role of fungi in bioremediation of contaminated soil. In: Fungi bio-prospects in sustainable agriculture, environment and nano-technology, pp 121-156 DOI: 10.1016/B978-0-12-821925-6.00007-1

Kumara, M., Singhab, M., & Mukherjeec, P. (2020). Economic importance of Hydrophytes. *RE:view*.

Kumaran, N. S., Sundaramanicam, A., & Bragadeeswaran, S. (2011). Adsorption studies on heavy metals by isolated cyano bacterial strain (Nostoc sp.) from Uppanar estuarine water, southeast coast of India. *Journal of Applied Sciences Research, 7*(11), 1609–1615.

Li, X., Peng, W., Jia, Y., Lu, L., & Fan, W. (2016). Removal of cadmium and zinc from contaminated wastewater using *Rhodobacter sphaeroides. Water Science and Technology, 75*(11), 2489–2498. DOI: 10.2166/wst.2016.608 PMID: 28617267

Liu, L., Bilal, M., Duan, X., & Iqbal, H. M. (2019). Mitigation of environmental pollution by genetically engineered bacteria—Current challenges and future perspectives. *The Science of the Total Environment, 667*, 444–454. DOI: 10.1016/j.scitotenv.2019.02.390 PMID: 30833243

Liu, W., Wang, B., Wang, Q., Hou, J., Wu, L., Wood, J. L., Luo, Y., & Franks, A. E. (2016). Characteristics of metal-tolerant plant growth-promoting yeast (Cryptococcus sp. NSE1) and its influence on Cd hyperaccumulator Sedum plumbizincicola. *Environmental Science and Pollution Research International, 23*(18), 18621–18629. DOI: 10.1007/s11356-016-7041-2 PMID: 27306207

López-Chuken, U. J. (2012). Hydroponics and environmental clean-up. *Hydroponics–a standard methodology for plant biological researches*, 181.

Ma, M., Du, H., & Wang, D. (2019). Mercury methylation by anaerobic microorganisms: A review. *Critical Reviews in Environmental Science and Technology, 49*(20), 1893–1936. DOI: 10.1080/10643389.2019.1594517

Ma, Y., Prasad, M. N. V., Rajkumar, M., & Freitas, H. J. B. A. (2011). Plant growth promoting rhizobacteria and endophytes accelerate phytoremediation of metalliferous soils. *Biotechnology Advances, 29*(2), 248–258. DOI: 10.1016/j.biotechadv.2010.12.001 PMID: 21147211

Machineni, L. (2019). Review on biological wastewater treatment and resources recovery: Attached and suspended growth systems. *Water Science and Technology, 80*(11), 2013–2026. DOI: 10.2166/wst.2020.034 PMID: 32198320

Mahmoud, M., Ramadan, M., Naher, S., Pullen, K., & Olabi, A. G. (2021). The impacts of different heating systems on the environment: A review. *The Science of the Total Environment, 766*, 142625. DOI: 10.1016/j.scitotenv.2020.142625 PMID: 33077224

Mani S, Chowdhary P, Zainith S (2020). Microbes mediated approaches for environmental waste management. Microorg Sustain Environ Health 17

Mishra, A., & Malik, A. (2013). Recent advances in microbial metal bioaccumulation. *Critical Reviews in Environmental Science and Technology, 43*(11), 1162–1222. DOI: 10.1080/10934529.2011.627044

Mitsch, W. J., & Gosselink, J. G. (2015). *Wetlands* (5th ed.). Wiley.

Mohd, S., Kushwaha, A. S., Shukla, J., Mandrah, K., Shankar, J., Arjaria, N., Saxena, P. N., Khare, P., Narayan, R., Dixit, S., Siddiqui, M. H., Tuteja, N., Das, M., Roy, S. K., & Kumar, M. (2019). Fungal mediated biotransformation reduces toxicity of arsenic to soil dwelling microorganism and plant. *Ecotoxicology and Environmental Safety, 176*, 108–118. DOI: 10.1016/j.ecoenv.2019.03.053 PMID: 30925326

Naidoo, S., & Olaniran, A. O. (2014). Treated wastewater effluent as a source of microbial pollution of surface water resources. *International Journal of Environmental Research and Public Health, 11*(1), 249–270. DOI: 10.3390/ijerph110100249 PMID: 24366046

Ndeddy Aka, R. J., & Babalola, O. O. (2017). Identification and characterization of Cr-, Cd-, and Ni-tolerant bacteria isolated from mine tailings. *Bioremediation Journal, 21*(1), 1–19. DOI: 10.1080/10889868.2017.1282933

Ogunmayowa, O. T. (2015). *Coupling bio/phytoremediation with switchgrass to biofuel feedstock production in mixed-contaminant soils* (Master's thesis, Tennessee State University).

Patel A. K., Singhania R. R., Albarico F. P. J. B., Pandey A., Chen C.-W., Dong C.-D. (2022). Organic wastes bioremediation and its changing prospects. Sci. Total Environ. 824, 153889.

Rahman, M. A., Reichman, S. M., De Filippis, L., Tavakoly Sany, S. B., & Hasegawa, H. (2016). Phytoremediation of toxic metals in soils and wetlands: concepts and applications. *Environmental remediation technologies for metal-contaminated soils*, 161-195.

Raklami, A., Oufdou, K., Tahiri, A. I., Mateos-Naranjo, E., Navarro-Torre, S., Rodríguez-Llorente, I. D., Meddich, A., Redondo-Gómez, S., & Pajuelo, E. (2019). Safe cultivation of Medicago sativa in metal-polluted soils from semi-arid regions assisted by heat-and metallo-resistant PGPR. *Microorganisms, 7*(7), 212. DOI: 10.3390/microorganisms7070212 PMID: 31336693

Rayu, S., Karpouzas, D. G., & Singh, B. K. (2012). Emerging technologies in bioremediation: Constraints and opportunities. *Biodegradation, 23*(6), 917–926. DOI: 10.1007/s10532-012-9576-3 PMID: 22836784

Ren, X., Zeng, G., Tang, L., Wang, J., Wan, J., Wang, J., Deng, Y., Liu, Y., & Peng, B. (2018). The potential impact on the biodegradation of organic pollutants from composting technology for soil remediation. *Waste Management (New York, N.Y.), 72*, 138–149. DOI: 10.1016/j.wasman.2017.11.032 PMID: 29183697

Salehizadeh, H., & Shojaosadati, S. A. (2003). Removal of metal ions from aqueous solution by polysaccharide produced from Bacillus firmus. *Water Research, 37*(17), 4231–4235. DOI: 10.1016/S0043-1354(03)00418-4 PMID: 12946905

Sales da Silva, I. G., Gomes de Almeida, F. C., Padilha da Rocha e Silva, N. M., Casazza, A. A., Converti, A., & Asfora Sarubbo, L. (2020). Soil bioremediation: Overview of technologies and trends. *Energies, 13*(18), 4664. DOI: 10.3390/en13184664

Samstag, R. W., Ducoste, J. J., Griborio, A., Nopens, I., Batstone, D. J., Wicks, J. D., & Laurent, J. (2016). CFD for wastewater treatment: An overview. *Water Science and Technology, 74*(3), 549–563. DOI: 10.2166/wst.2016.249 PMID: 27508360

Shuaib, M., Azam, N., Bahadur, S., Romman, M., Yu, Q., & Xuexiu, C. (2021). Variation and succession of microbial communities under the conditions of persistent heavy metal and their survival mechanism. *Microbial Pathogenesis, 150*, 104713. DOI: 10.1016/j.micpath.2020.104713 PMID: 33387608

Singh, A., Prasad, S. M., Singh, S., & Singh, M. (2016). Phytoremediation potential of weed plants' oxidative biomarker and antioxidant responses. *Chemistry and Ecology, 32*(7), 684–706. DOI: 10.1080/02757540.2016.1182994

Singh R (2013). Microbial waste management. Lap lambert Academic Publishing

Sutherland, D. L., & Ralph, P. J. (2019). Microalgal bioremediation of emerging contaminants– opportunities and challenges. *Water Research, 164*, 114921. DOI: 10.1016/j.watres.2019.114921 PMID: 31382151

Suyal, D. C., Soni, R., Singh, D. K., & Goel, R. (2021). Microbiome change of agricultural soil under organic farming practices. *Biologia, 76*(4), 1315–1325. DOI: 10.2478/s11756-021-00680-6

Tak, H. I., Ahmad, F., & Babalola, O. O. (2013). Advances in the application of plant growth-promoting rhizobacteria in phytoremediation of heavy metals. *Reviews of Environmental Contamination and Toxicology, 223*, 33–52. PMID: 23149811

Tandon, P. K., & Singh, S. B. (2016). Redox processes in water remediation. *Environmental Chemistry Letters, 14*(1), 15–25. DOI: 10.1007/s10311-015-0540-4

Tang, K. H. D. (2024). Phytoremediation of petroleum hydrocarbons: An update of its recent progress. *Tropical Environment, Biology, and Technology, 2*(2), 106–123. DOI: 10.53623/tebt.v2i2.532

Thakur, S., Mathur, S., Patel, S., & Paital, B. (2022). Microplastic accumulation and degradation in environment via biotechnological approaches. *Water (Basel), 14*(24), 4053. DOI: 10.3390/w14244053

Wróbel, M., Śliwakowski, W., Kowalczyk, P., Kramkowski, K., & Dobrzyński, J. (2023). Bioremediation of heavy metals by the genus Bacillus. *International Journal of Environmental Research and Public Health, 20*(6), 4964. DOI: 10.3390/ijerph20064964 PMID: 36981874

Yan, A., Wang, Y., Tan, S. N., Mohd Yusof, M. L., Ghosh, S., & Chen, Z. (2020). Phytoremediation: A promising approach for revegetation of heavy metal-polluted land. *Frontiers in Plant Science, 11*, 513099. DOI: 10.3389/fpls.2020.00359 PMID: 32425957

Yang, K. M. (2025). Recent trend in phytoremediation of petroleum hydrocarbon contaminated soil: A bibliometric review. *International Journal of Phytoremediation*, •••, 1–9. DOI: 10.1080/15226514.2025.2539429 PMID: 40735952

Yuan, X., Xue, N., & Han, Z. (2021). A meta-analysis of heavy metals pollution in farmland and urban soils in China over the past 20 years. *Journal of Environmental Sciences (China), 101*, 217–226. DOI: 10.1016/j.jes.2020.08.013 PMID: 33334517

Chapter 4
Hydrophytes as a Source of Bioactives for Heavy Metals Remediation:
Phytochemical Potential and Mechanisms in Aquatic Plant-Based Detoxification

Sandeep Kaur
https://orcid.org/0000-0002-2426-9173
Chandigarh University, India

Ajay Kumar
https://orcid.org/0000-0002-5312-9823
Chandigarh University, India

Vaseem Raja
Chandigarh University, India

ABSTRACT

Heavy metal (HM) contamination is one of the major contributors to environmental pollution and the subsequent low quality of life of the affected populace on a global scale. Specifically, the threat of HM in the groundwater poses a heavy risk to human health and disrupts the stability of aquatic ecosystems. Phytoremediation is a less expensive and green technology, and bioremediation, sustainable systems involving the use of plants to mitigate air, water and soil contaminated with toxic pollutants such as hydrocarbons, heavy metals, industrial wastes, and pesticides, have come up as an eco-friendly solution. Plants living in the water, or the hydrophytes, Pistia stratiotes (water lettuce) and Lemna minor (duckweed), exhibit high levels of tol-

DOI: 10.4018/979-8-3373-4037-1.ch004

erance and accumulation to the toxic metals. In the current chapter, we discuss the key biomolecules that are found in the various hydrophytes and are involved in the metabolic pathway in capturing or strengthening the eco-friendly phytoremediation approaches, and to come up with a low-cost, nature-based ecosystem cleaning up approach.

INTRODUCTION

Environmental pollution, especially by heavy metals, has become an alarming situation in recent days because of the non-biodegradability, toxicity, and persistence of heavy metals in ecosystems. The occurrence of an outbreak of heavy metals in the waters is a serious ecological and health issue to the population of the world. When at low concentrations, heavy metals like cadmium (Cd), lead (Pb), arsenic (As), mercury (Hg), and chromium (Cr) are profoundly toxic and capable of inflicting significant ecological and health risks in the form of neurotoxicity, organ failure, carcinogenic factors, and loss of biodiversity (Ali et al., 2019; Jaishankar et al., 2014). In contrast with organic pollutants, heavy metals are not degraded and remain in the environment as they get deposited on the sediments or accumulate through the food chain (i.e., aquatic organisms) and are magnified in the food chain (Tchounwou et al., 2012). The general sources of heavy metal contamination are urban runoff effluents, mining, metal plating, battery production, textile industry, tanneries, industrial releases, as well as overuse of agrochemicals (Verma et al., 2020a; Fu & Wang, 2011). In the fresh and salt water environment, the contamination of HM is discovered to be centred around mining, waste release by industries and run off discharge by agriculture, which is a serious human health and environmental risk. The main threat that is associated with the contamination of HM is its toxicity, persistence and accumulation in both terrestrial and aquatic ecosystems. Currently, measures being undertaken in the elimination of HM pollution in waste water are ineffective, expensive, tedious and most times the elimination is equally hazardous to the surroundings as the contamination itself. The traditional S4 technology to eliminate the heavy metals like chemical S4, reverse osmosis, Ion exchange and electrical S4 treatment tends to be very ineffective, costly and produce secondary waste, hence not applicable in large-scale treatment (Fu & Wang, 2011; Barakat, 2011).

Phytoremediation is a sustainable, cheap, environmentally-friendly methodology of removing, stabilizing or detoxifying pollutants observed in the environment through the aid of plants, as well as heavy metals (Tangahu et al., 2011). In this plan, there is the use of hydrophytes; aquatic plants that are specifically suited to grow in or on water, which play an imperative role in the cleansing of used-up water resources. These plants are either submerged, emergent or floating and have unique morphological, physiological and biochemical characteristics to survive and tolerate high concentrations of toxic metals and also have the ability to detoxify (Mishra & Tripathi, 2008; Rai, 2009). Although the metal-accumulation potential of hydrophytes is widely reported, an increasing number of studies point to the fact that they have an abundant biochemical arsenal. Hydrophytes produce various secondary metabolites like flavonoids, phenolics, alkaloids, terpenoids, and tannins that perform different protective functions such as metal ions chelation, to prevent its toxicity; reactive oxygen species (ROS) scavenging that generates during metal-stress conditions; activation of antioxidant enzymes like superoxide dismutase (SOD), catalase (CAT), and peroxidase (POD); and stabilization of cellular membranes and detoxification routes (Khan et al., 2009; Singh et al., 2020; Hasan et al., 2009).

These bioactive compounds not only defend the plants against metal-stress related oxidative damages, but also assist in rhizofiltration (root-mediated adsorption and uptake), phytoextraction (matrix accumulation in plant tissues), phytostabilization (entrapment in sediments) (Verma et al., 2020b). A number of studies on the hydrophytic species have shown that *Eichhornia crassipes* (Water hyacinth) has a high-rate production of biomass and strong metal accumulation, and flavonoids and polyphenols produced by the species are involved in detoxification processes (Zhou et al., 2022; Malik, 2007). Duckweed *Lemna minor* shows good results in removing arsenic and cadmium that are facilitated by antioxidant enzyme systems and phenolic compounds (Sekomo et al., 2012; Zhao et al., 2015a). *Pistia stratiotes* (Water lettuce) takes up mercury and chromium and contains a high concentration of tannins and alkaloids (Wolverton & McDonald, 1981; Singh & Chandra, 2014). *Typha latifolia* (Cattail) is a very strong emergent plant with long root systems that have the ability to stabilise metals in the sediments (Marchand et al., 2010; Rai, 2008). The mechanisms of action (biochemical) and involved bioactives are, however, not well characterized. Traditional remediation techniques include being expensive, ineffective or not eco-friendly.

These species are relevant in two ways, i.e., to remove metals by physical absorption and detoxify them chemically by synthesizing bioactive molecules (Zhou et al., 2022). Hydrophytes have diverse microbial communities in the rhizosphere which include metal resistant bacteria, fungi and the actinomycetes. Such microbes have the ability to increase phytoavailability of compounds, degrade organic wastes, and even create their own bioactive molecules (Ali et al., 2019; Glick, 2010). The effectiveness in remedying heavy metals is further increased by the synergy between plant and microbe that rely on the exudation of bioactive compounds by roots (Ma et al., 2011).

Hydrophytes have successfully been incorporated in constructed wetlands, floating treatment wetlands (FTWs), and aquatic biofilters to clean the industrial and municipal wastewater (Vymazal, 2011). Their bioactive profiles have a considerable effect on their efficacy in response to the metal toxicity and long-term remediation performance (Rai et al., 2022). Integrated use of the hydrophytes, which involves the potential of metal accumulation, as well as the phytochemical weapon, has opened a new field in environmental biotechnology. A way to further the concept of bioengineering better phytoremediation systems is through the understanding of the role specific bioactive compounds have as part of the metal detoxification process (Khan et al., 2009). Using the metabolic power of hydrophytes, one can not only clean up polluted water but also obtain useful bioactive materials in terms of therapeutic and commercial value (Verma et al., 2020c; Singh et al., 2020). Such a combination of phytoremediation and valorization of bioresources can be characterized as a strategic direction of transition to a circular bioeconomy and environmentally friendly biotechnology.

The current chapter discusses the novel role of hydrophytes as depositories of bioactive agents with inherent dual ability, i.e., direct uptake of heavy metals and their remediation and biochemical detoxification, as well as the challenges and opportunities of their applicability in practice. By means of biochemical, phytochemical and molecular analysis, this chapter will explore the processes by which heavy metals can be taken up, accumulated and detoxified in some favorite hydrophytic species. It represents major plant species, speculative physiological principles and practices of using plant-based systems in the territory of environmental biotechnology. It uses case studies and scientific knowledge to reveal the potential that is still untapped in terms of using aquatic plants to suppress heavy metal toxicity, as well as providing green innovation and supporting the resilience of the ecosystem.

Historical Timeline and Trend Analysis

Heavy metal contamination has followed an evolution process that has been significant since the early times of industrialization. Industrialization of the 19th century resulted in progressive release of large amounts of effluents rich in metals (Pb, Cd, Hg and Cr) without proper treatment into the aquatic environment (Nriagu, 1996). Mining, smelting, as well as, the large usage of agrochemicals and industrial dyes released heavy metal contaminants into the environment that increased the risk of ecological and health problems in the second and last half of the 20th century (Alloway, 2013). The global awareness of metal toxicities developed in the 1970s-1980s along with environmental legislations around the world starting with the U.S. Clean Water Act (1972) or similar regimes. This timeframe witnessed the initial development of phytoremediation as a green and economical substitute to traditional physical and chemical treatment methods (Salt et al., 1995). Research started with terrestrial hyperaccumulators such as *Brassica juncea* and *Alyssum spp.*; four years later and the agenda fully shifted to hydrophytes, such as *Eichhornia crassipes* (water hyacinth) and *Lemna minor* (duckweed), in the case of the former, effectively removing dissolved metals from the water column (Rai, 2009).

With the advent of molecular biology, metabolomics, and nanotechnology, hydrophyte-based remediation has extended to the 21st century. Hypothetically more recent investigations have been carried out on the bioactive compounds with regards to their importance in metal chelation and detoxification of hydrophytes which provide dual benefits of mitigation of pollution and discovery of natural products (Verma et al., 2020c; Ma et al., 2020). With a changing landscape of environmental conditions, the current trends are leaning towards integrated phytoremediation systems using combinations of hydrophytes with microbial consortia and/or nanomaterials to improve removal efficiency and resilience (Kumar et al., 2022).

Still, the way research is done and technology is applied mirrors the constraints of climate change, regulation, and the economics of the situation. Heavy metal remediation from the era of unrestricted emission to biotechnological development, the history reflects the enhancement of interconnected worldwide activities of the adoption of sustainable technology for recovery of aquatic ecosystems.

Heavy Metal Pollution Effect on Aquatic Ecosystems

The pollution of aquatic life with heavy metals has become an issue of environmental concern because the metals are non-biodegradable, toxic and persistent, thereby giving rise to environmentally critical problems. In contrast to organic pollutants, therapeutic metals like mercury (Hg), cadmium (Cd), lead (Pb), arsenic (As), chromium (Cr), nickel (Ni), and zinc (Zn) remain stable during the course of time and they have the tendency to bioaccumulate and be biomagnified in fish and along the food chain (Ali et al., 2019).

This type of pollution is mainly concentrated in freshwater and marine environments, which comprise fisheries, drinking water supply, agriculture and biodiversity. The continuous and extensive application of heavy metals in industries, agriculture and in the household has extensively by far led to raised levels of heavy metals in the natural water bodies and in most cases, the concentration levels have well surpassed the allowed limits that have been laid down by environmental authorities such as the World Health Organization (WHO) and the United States Environmental Protection Agency (USEPA) (Akratos & Tsihrintzis, 2007).

Major Sources of Heavy Metals in Aquatic Environments

Naturally and anthropogenically, heavy metals are introduced into aquatic ecosystems. Natural sources involved the weathering of the metal-rich rocks and volcanic activity, erosion and transportation of the soil, as well as the natural erosion by leaching. But the anthropogenic factors include the industrial wastes from mining, metal plating, the battery industry, tanning, and paint industries, which have mainly been involved in causing heavy metal pollution (Fu & Wang, 2011). The phosphate fertilizers and pesticides that contain metals such as arsenic and cadmium, which are mostly used in agricultural runoffs, are likely to cause a deposition of these metal compounds in the adjacent waters. Untreated or partially treated municipal waste with contents such as lead, chromium and mercury mostly comes from municipal wastewater, such as the discharge of sewage. Road dust, vehicle emissions and storm drainage contain metals (lead, zinc, copper and so on) in urban runoff, which are deposited in rivers and lakes. Deposition to the atmosphere that involves the combustion of fossil fuel and the burning of waste may lead to the deposition of metals in the air to the aquatic systems (Tangahu et al., 2011).

Environmental Impacts of Heavy Metals

Heavy metals are toxic to a number of biological and ecological levels. a. (Aquatic life: Metals disrupt the performance of enzymes, cellular respiration, and procreation in fish, invertebrates, and aquatic plants (Tchounwou et al., 2012). Biomagnification can occur in human eaters of fish in the case of lower trophic organisms bioaccumulating toxins (e.g., plankton). b. Loss of biodiversity: High concentrations of heavy metals decrease the species richness and the species abundance. Metal-tolerant organisms replace sensitive species, and this causes ecological imbalance (Varol, 2011). c. Contamination of sediments: Metals are deposited on organic and inorganic material on the sediments, which forms an extended source of pollution as it may be remobilized due to physical mobilization (Singh et al., 2005). d. Health risks to humans: Eating polluted fish and water is dangerous and is associated with risks such as neurotoxicity, carcinogenicity, kidney and liver diseases, and child developmental conditions (Jaishankar et al., 2014). The pollution of aquatic ecosystems by heavy metals is a long-term and increasing threat that has far-reaching environmental and health impacts. This problem should be addressed with the help of a multi-faceted approach incorporating strict regulation, sustainable industrial activities, ongoing surveillance and the creation of new remediation technologies.

Comparison between Hydrophytes and Terrestrial Phytoremediators

In groundwater systems, terrestrial plants have received much attention in remediation of heavy metals from soils but hydrophytes (aquatic plants) appear to have several advantages over terrestrial plants in treating contaminated waters. Hydrophytes including *Eichhornia crassipes*, *Lemna minor*, and *Typha latifolia* are the species which are adapted by nature to submerged or semi submerged conditions and are well suited to contact with water in a directly manner that dissolved metals present in water. Includes high surface-area-to-volume ratio, large lateral root system, and fast growth rates to promote the uptake, accumulation, and sequestration of metals including cadmium, lead, and nickel. In comparison, terrestrial plants are frequently constrained by root contact with polluted soil or water and lower metabolic rates, which may mitigate the efficiency of metal uptake from aquatic environments. Additionally, different hydrophytes can be incorporated into constructed wetlands, floating mats, and other applied engineered systems to decontaminate water providing a practical, sustainable and affordable method towards alleviating waterborne heavy metal pollution (Yadav et al., 2012; Sharma et al., 2020). This highlights the main reason for using hydrophytic plants in remediation and phytoremediation strategies in water systems that have been polluted. Awareness of these benefits can assist researchers and practitioners in selecting plant species for specific phytoremediation activities on a given site (**Table 1**).

Table 1. Represents the Comparison between Hydrophytes and Terrestrial Phytoremediators

Parameter	Hydrophytes (Aquatic Plants)	Terrestrial Phytoremediators (Land Plants)
Environment	Submerged, floating, or semi-aquatic	Soil-based or land environments
Direct Interaction with Metals	Direct contact with dissolved metals in water	Limited to metals in soil; indirect contact with waterborne metals
Root System	Highly fibrous, dense, and often floating; enhanced surface area for metal absorption	Deep or fibrous roots, primarily adapted for soil nutrient uptake
Uptake Efficiency	High efficiency for dissolved metals (Cd, Pb, Ni, Hg, Cr)	Moderate; metal availability depends on soil properties
Growth Rate	Rapid growth; can cover water surfaces quickly	Usually slower growth; seasonal limitations
Biomass Production	High biomass generation; facilitates large-scale removal	Moderate biomass; may limit metal accumulation capacity
Adaptability to Contaminated Sites	Can tolerate high metal concentrations in water; suitable for constructed wetlands, ponds, and floating mats	Tolerates metals in soil; effectiveness depends on soil type and remediation strategy
Cost & Maintenance	Low-cost, easy to harvest and deploy in water treatment systems	Moderate cost; may require soil amendments and longer remediation periods
Eco-Friendly Applications	Can be integrated into wetlands, floating treatment systems, and water purification setups	Used in soil stabilization and land-based remediation projects
Examples	*Eichhornia crassipes, Lemna minor, Typha latifolia, Pistia stratiotes*	*Brassica juncea, Helianthus annuus, Arabidopsis thaliana*

In addition, these hydrophytes can be incorporated in engineered systems such as floating treatment wetlands (FTWs) and hydroponic-based filtration systems, also contributing efficient and low-cost technology for industrial effluents, mine waters and urban wastewaters. On the other hand, terrestrial organisms are more appropriate for the remediation of heavy metal contaminated soils, tailings, and solid waste sites where he has an active role in soil stabilization and revegetation.

Thus, an integrated eco-engineered system containing both terrestrial chlorophytes and hydrophytes can feasibly provide a whole approach to both soil and water contamination processes. Such a strategy combines synergistically the various sustainable phytoremediation technologies that will inform its future development.

HYDROPHYTES: CLASSIFICATION AND ECOLOGICAL ROLES

Hydrophyte is classified into various types (emergent, floating, submerged) according to their habit and activities in the ecological business. Aquatic macrophytes, or hydrophytes, are those plants which have become adapted to life in or on water. Such species form an important part of any freshwater ecosystem and their presence provides a form of contribution to the great balance of functions of a freshwater ecosystem, both in terms of physical, chemical and biological processes. Currently they have morphological and physiological adjustments, where they endure in conditions that are saturated with water, usually with low oxygen levels, altered nutrients, and fluctuating water depths.

Besides having a major role in primary production in aquatic food webs, hydrophytes also help maintain normal nutrient activities, stabilise deposition, and provide oxygen and create habitat for aquatic life organisms. Moreover, they are the only organisms that have the special capacity to absorb and detoxify pollutants, especially heavy metals, making them invaluable towards applications in phytoremediation and water purification research and applications (Mishra et al., 2008; Brix, 1997).

Classification of Hydrophytes- Based on Growth Habit

Hydrophytes are broadly readily divided into three primary classes depending on growth form and section of the plant dealing with the water column: Emergent Hydrophytes have roots in the sediment, but shoots, leaves, and flowers grow above the water. These are the species that are adapted to shallow water edges and wetlands whose water level changes frequently. Examples *Typha latifolia* (cattail), *Phragmites australis* (common reed) and *Sagittaria* spp. (arrowhead). They play beneficial ecological functions that stabilise the shorelines, and they proved useful in erosion prevention and rhizosphere oxygenation by aerenchyma tissues. They are also sources of breeding facilities for amphibians and birds, and play the role of biofilters of surplus nutrients and contaminants (Brix, 1997)

Floating hydrophytes are either free-floating or rooted plants with leaves floating in the water. The free-floating forms float along the water surface and are rooted floaters, which anchor sediment in the substrate and leave that which floats on top of the water. Some are free-floating, such as *Eichhornia crassipes* (water hyacinth) and *Pistia stratiotes* (water lettuce), and others are rooted floating-leaved, such as *Nymphaea* spp. (water lilies). They perform ecological functions of shading the water bodies and decreasing the blooms of excessive algae. They also shelter fish and invertebrates as well as take up nutrients and heavy metals in the water column and consequently take part in the oxygen exchange through photosynthesis (Rezania et al., 2016a).

Submerged plants include submerged hydrophytes (whose meristems grow entirely below the water surface, either as rooted plants or as free-floating), which have their leaves delicately and highly dissected to maximize nutrient uptake and turbulent transfer of gases. Such species are *Hydrilla verticillata*, *Ceratophyllum demersum* and *Elodea canadensis*. Their ecological functions include enhancing the clarity of water through stabilisation of the sediments and providing habitats and spawning grounds to aquatic life. This was also achieved through absorption of dissolved nutrients and metals in the water column and also producing oxygen during the day, thus boosting aquatic life support (Chambers et al., 2008).

Ecological Functions and Ecosystem Services of Hydrophytes

Hydrophytes provide numerous ecosystem services that help to sustain the health of the water column, including Water Purification and Detoxification: Hydrophytes take up surplus nitrogen, phosphorus and heavy metals to limit eutrophication and metal poisoning. Their root systems are extensive and give them large areas where they form their biofilms, and this increases the degradation of the pollutants (Brix, 1997; Vymazal, 2011). Habitat Provision and Biodiversity Support: Aquatic plants provide homes to various communities of microbes, invertebrates and vertebrates since they provide the spawning places, habitats, refuges and nursery (Carpenter & Lodge, 1986; Thomaz & Cunha, 2010). Erosion Control/Sediment Stabilization: Emergent and submerged plants assist in holding the sediments together with their roots and rhizomes, reducing the water turbidity, therefore limiting erosion of the shores (Doyle, 2001; Madsen et al., 2001). Carbon Sequestration and production of oxygen: Hydrophytes provide carbon cycling and also produce oxygen through photosynthesis, and this plays a crucial role in sustaining the aerobic conditions within the water bodies (Wetzel, 2001; Mitsch et al., 2013). Moderation of micro-habitats and climate regulation: Shaded floating plants also help in the moderation of microhabitats because they minimise algal overgrowth on different plants in the water, which regulates ecosystem metabolism (Chambers et al., 2008; Jeppesen et al., 2012).

The health and functionality of aquatic ecosystems rely on hydrophytes; you have emergent, floating and submerged hydrophytes. The variety of different ecological functions is attributed to them, including nutrient circulation and sequestration of heavy metals, which makes them useful in restoring ecosystems and sustainable forms of water management. It is possible to integrate them better into constructed wetlands, phytoremediation systems, and conservation techniques implemented to clean up polluted and degraded aquatic setups due to this understanding of their functional classification and habitat-specific adaptations (Kadlec & Wallace, 2009; Vymazal, 2014).

PHYTOREMEDIATION MECHANISMS OF HYDROPHYTES

Phytoremediation is a science based on using green plants to remove the contamination of the soil, air and water with harmful pollutants, heavy metals, organic compounds and radionuclides. The multiple families of phytoremediators vary in the degree to which they grow well in metal-rich aquatic environments, as well as utilizing complicated biochemical and physiological detoxification methods. One well-represented family is that of the hydrophytes (aquatic macrophytes). Four main mechanisms of phytoremediation: phytoextraction, phytostabilization, phytodegradation and rhizofiltration are used to remove, or to convert the heavy metals in the water with help of hydrophytes (Figure 1). All these mechanisms have unique processes and pathways which may either sequester, immobilize or degrade the contaminants.

1. Phytoextraction (Phytoaccumulation)

Phytoextraction is a process whereby hydrophytes take out heavy metals by absorbing them and transferring them to the parts growing above the ground, like stems and leaves, where they are accumulated. The plants can then be harvested to eliminate the contaminants from the system. Water metal ions are either actively taken in by roots using metal ion transporters or passively using the water pathways. Phytochelatins and metallothioneins, as well as organic acids, are the organic molecules that chelate the metals. The resulting Chelated metals are exuded with xylem to aerial plant parts in order to be sequestered. It is in the leaves of *Eichhornia crassipes* (water hyacinth) that Cr, Cd, Pb and As are accumulated. *Lemna minor* (duckweed) with a rapid growth and a high ability to absorb Cd and Cu. The method is also employed, especially in eutrophic waterbodies, where high metal concentrations can be eliminated through the quick and frequent extraction of biomass. According to Rezania et al. (2016b), with controlled coverage, *Eichhornia crassipes* may reach the phytoextraction capacity of up to 85 percent of the contamination in Cd and Pb contained in wastewater (Figure 1).

2. Phytostabilization

Phytostabilisation refers to a process in which heavy metals in the form of suspended materials are immobilised in the root zone so that they are not mobilised into the water column or eaten by aquatic organisms. Root exudates and mucilage substances hold metals. Inspection into the formation of insoluble complexes in the rhizosphere. There are root barrier functions that lessen translocation to the aerial parts. Root activity alters soil pH and redox conditions to bind metals. Cattail (*Typha latifolia*) and bulrush (Scirpus spp.). The strong root systems of such species facilitate the minimisation of leaching and erosion of the sediments, consequently entrap the metals. This method will be most suited for long-term and pollutant containment in wetlands and riverbanks with high mobility activity of metals, where the downstream contamination should be prevented. In discussing the remediation of contaminated sediments occurring in wetlands, Salt et al. (1998) stated that phytostabilisation was a viable technology since methyl Fe^{2++} and methyl Fe^{2} complex ligands attached themselves to the plant roots to influence the rhizosphere chemistry to reduce the availability of metals (Figure 1).

3. Phytodegradation (Phytotransformation)

Phytodegradation is the transformation or degradation of pollutants (even including some complexes of organic and heavy metals) by plant enzymes to substances that are less toxic overall. Hydrophytes excrete enzymes, which include peroxidases, dehalogenases and laccases. These enzymes conflagrate complex molecules, making metal-organic compounds non-toxic. Part of the metals becomes less mobile or bioavailable. The common reed (*Phragmites australis*) and lotus (*Nelumbo nucifera*) are the appropriate plants that can be used in Phytodegradation. Although phytodegradation has been addressed most often when it comes to organic pollutants (e.g., pesticides, hydrocarbons), it is applicable to metal complexes as well as organometallic compounds (e.g., methylmercury or arsenicals). Such a mechanism is important in constructed wetlands, where it depends on plants that have active biochemical activity in the detoxification of complex metal compounds. Pilon-Smits, (2005) observed that some water plants had the capability to convert selenate and arsenate to elements or some less toxic compounds enzymatically, making the environment less toxic (Figure 1).

4. Rhizofiltration

Rhizofiltration entails the adsorption or absorption of heavy metallic substances on the roots of plants, mainly with fast-growing fibrous roots. The metal ions either attach to the root surface due to ion exchange, precipitation or electrostatic forces. There are metals that are absorbed into root cells and retained without translocation. Binding capacity is boosted by the roots exudates like phenolics and mucilage.

Among the plants suited to rhizofiltration are *Salvinia molesta* (giant salvinia), *Hydrilla verticillate* and *Pistia stratiotes* (water lettuce). This strategy is especially useful in industrial wastewater treatment, mine effluents or contaminated irrigation ditches, whereby hydrophytes may be used to "filter" pollutants in situ. Dushenkov et al. (1995) developed rhizofiltration to be used to treat metals in contaminated waters using a scale-down approach and proved that sunflower roots were a viable strategy to remove lead and uranium from water, a concept also extrapolated to hydrophytes. Hydrophytes have been known to use more than one mechanism at the same time. To give an example, *Eichhornia crassipes* is able to adsorb metals with roots (rhizofiltration), accumulate them in leaves (phytoextraction) and excrete antioxidative enzymes to degrade parts of the organo-metal complexes (phytodegradation). These multifaceted tactics ensure the aquatic macrophytes are very dynamic in the field of remediation.

Hydrophytes also exhibit a wide array of highly effective phytoremediation processes, such as phytoextraction, phytostabilisation, phytodegradation, and rhizofiltration, to counter pollution of heavy metals in water bodies. These methods are successful, depending upon plant species, type of metal, water chemistry and the environment. New opportunities that exist in the extensive, environmentally friendly cleansing of water bodies polluted with industrial waste could be opened with the connection of similar processes when plant selection, biostimulation, and the engineering of constructed wetlands are combined (Figure 1).

Figure 1. Schematically representing the mechanisms of different phytoremediation processes

BIOACTIVE COMPOUNDS IN HYDROPHYTES: NATURE AND FUNCTION

Secondary metabolites (e.g., polyphenols, flavonoids, and alkaloids) are extremely important to the detoxification and stress response (Figure 2). Hydrophytes, or so-called aquatic macrophytes, are efficient phytoremediators under the influence of certain characteristics of these and other hydrophytes; that is, the plants can easily absorb and accumulate contaminants such as heavy metals and hold them from the ecosystems. Besides that, these plants also synthesise a wide variety of bioactive secondary metabolites. These are phenolics, flavonoids, alkaloids, terpenoids, tannins and saponins and play vital roles in their physiological robustness and ecorelations. These compounds are biochemical defence agents of a heavy metal, whose individual roles vary according to the nature of exposure. As biochemical defence agents, these compounds minimise toxicity, scavenge reactive oxygen species (ROS) and mediate metal chelation or sequestration under heavy metal stress. Such kinds of bioactive molecules production can be viewed as an evolutionary adaptation, allowing the plants, known as hydrophytes, to endure the changing conditions of an aquatic habitat, frequently under toxic and nutrient-limited conditions (Verma et al., 2020a). These bioactives not only work internally within their own plant system but also communicate with the microbes/contaminants in the rhizosphere, thus increasing the efficacy of phytoremediation.

1. Phenolic Compounds

Phenolics form a widespread group of secondary metabolites in which the key feature is one or more hydroxyl groups bound to aromatic units. They encompass simple phenols, hydroxycinnamic acids and complex polyphenols (lignin and tannins) in hydrophytes. The primary role of their detoxification of metals is that it serves as an antioxidant to scavenge ROS that is produced following oxidative stress caused by the presence of heavy metals. They can chelate to remove the reactivity and toxicity of metal ions (e.g., Fe^{2+}, Cu^{2+}, Pb^{2+}). They have also reinforced cell lignification to the walls, a process aiding in metal entrapment in apolar space. *Eichhornia crassipes* accumulates phenolic acids of the ferulic and caffeic acid under the stress of Cd and Cr stress (Rezania et al., 2016a). There is an elevated level of polyphenols in *Lemna minor* with copper and zinc, and this helps in the reduction of oxidative stress. As explained by Michalak (2006), phenolic compounds are also induced by heavier metal stress in plants, and they play the underlying role of providing biochemical cover against metals in plants (**Table 2**).

2. Flavonoids

Flavonoids are polyphenolic compounds that contain the phenylpropanoid pathway and occur in aquatic plants in large quantities. The most typical subclasses are flavones, flavonols, flavanones and anthocyanins. The most important role of the Flavonoids on metal detoxification is that they are metal chelators in the sense of binding free metal ions, which was achieved through their involvement in ROS scavenging, antioxidant enzyme activity regulation, membrane stabilisation and decreasing lipid peroxidation in the stress intervals. *Pistia stratiotes* accumulates derivatives of quercetin and kaempferol under the stress of Pb and Ni. Hg exposure induces the biosynthesis of flavonoids in the *Hydrilla verticillata* and increases its antioxidative ability. According to Pietta (2000), flavonoids are potent radical scavengers and they have the ability to minimise oxidative damage brought about by metal toxicity (**Table 2**).

3. Alkaloids

Alkaloids refer to the heterocyclic compounds that contain nitrogen, which, on the one hand, have various physiological and ecological functions in plants. Whereas they have received less attention in hydrophytes than phenolics, they have come to be known more so due to their roles in stress. They play a significant role in the detoxification of metals, such as complexation and detoxification of metals by nitrogenous ligands, and also protect the cellular structures by regulating membrane integrity, as well as pH control, and induce the signalling pathways connected to the expression of metal-stress genes. *Nelumbo nucifera* accumulates nuciferine, which is regulated in response to metal stress, hence conducive to cellular homeostasis. In the Cd-contaminated water, *Pistia* strain produces a higher number of total alkaloids. According to Verma et al. (2020b), alkaloids are essential, yet unexplored, metabolites in the management of metal stress, which have the potential to respond to both organic and inorganic pollutants (**Table 2**).

4. Terpenoids

The terpenoids comprise a very large family of isoprene-based compounds. They comprise monoterpenes, diterpenes, triterpenes and carotenoids. The primary role in metal detoxification is as membrane stabilisers and antioxidants, signal transduction and metal stress responses. Certain terpenoids (e.g. phytochelatins and squalene derivatives) have direct metal-binding ability. The mechanism that Typha angustifolia uses to produce triterpenoids that have antioxidative and stress alleviation properties is under Cr stress. In response to metal-induced oxidative stress, *Eichhornia crassipes* upregulates carotenoid biosynthesis as a means of shielding the chloroplast. According to Gershenzon & Dudareva (2007), Terpenoids have been identified as having multiple and probably inconsistent functions in the defence of plants, especially the arrays of tissues with cellular oxidative systems (**Table 2**).

5. Tannins and Saponins

Tannins are polyphenolic compounds that precipitate proteins and are known to bind metals. Saponins are glycosidic products of mixed nature with amphiphilic qualities capable of adjusting the permeability of the membrane. Their key role in the metal detoxification process is to combine the metal ions through the hydroxyl and carboxyl groups (tannins), change the membrane transports to regulate the intake and efflux of metal ions (saponins), and have antimicrobial effects in promoting rhizosphere microbial communities to participate in the bioremediation. Under Ni stress, *Phragmites australis* is accumulating tannins in the roots and rhizomes. *Salvinia molesta* makes saponins that become heightened in contact with arsenic and control how ion channels act. Haslam (1996) indicated a high affinity of tannins to metal ions, particularly iron and aluminium, which is taken advantage of by plants in cases of metal overload (**Table 2**).

The working of bioactive compounds is usually synergistic and not isolated. As an example, the presence of flavonoids could be complemented by the action of phenolics and terpenoids to sustain the redox balancing and the protection of the photosynthetic apparatus. Also, these chemicals are capable of conditioning the rhizosphere microbial community in a way that stimulates metal-tolerant and metal-transforming bacteria, therefore facilitating enhanced remediation ability. Secondly, injury signals released in bioactives during exudation into the water and sediment promote detoxification of the ecosystems, functioning as natural chelators and signalling molecules. Use of such hydrophytes in constructed wetlands and floating treatment wetlands (FTWs) can be optimised through the use of species that have rich bioactive profiles.

Figure 2. Represents the bioactive components of the hydrophytes

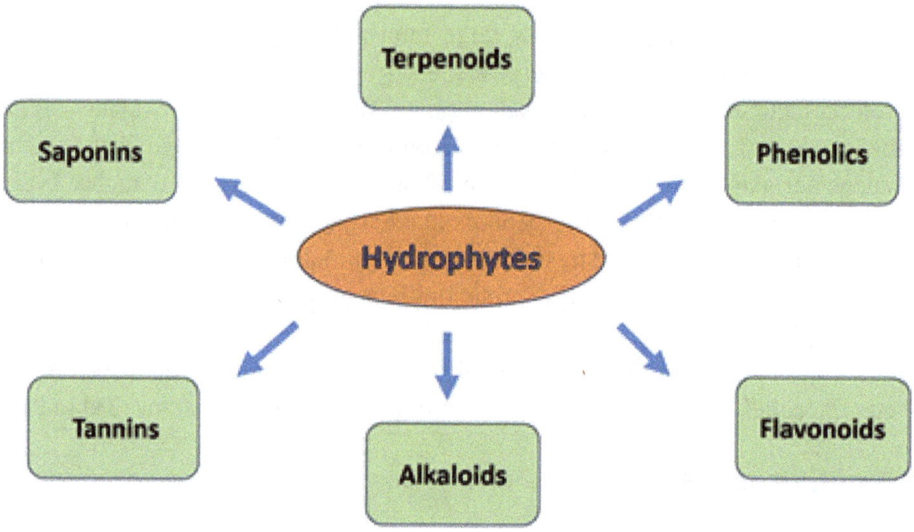

The huge bioactive arsenal of the hydrophytes permits them to withstand, endure and deactivate aquatic heavy metals. Specifically, these metabolites, i.e. phenolics, flavonoids, alkaloids, terpenoids, tannins and saponins, not only provide protection to these plants but they also play a substantial role towards the overall success of phytoremediation strategies (**Figure 2**). Finding out the biosynthesis, activity, and interactions of these bioactives is central towards the development of better and more nature-based technologies for cleaning up polluted water bodies more efficiently.

Table 2. Summarize the known hydrophyte-derived bioactive compounds, their functions and associated plant species

Species (common name)	Representative bioactive compound(s)	Functions	Plant part(s) reported	References
Nelumbo nucifera (sacred lotus)	Alkaloids (nuciferine), flavonoids, saponins, polyphenols, polysaccharides	Antioxidant, anti-inflammatory, anti-cancer, hepatoprotective, cardioprotective	Seed, rhizome, leaf, flower	Paudel & Panth, 2015; Bakrim et al., 2022
Eichhornia crassipes (water hyacinth)	Phenolics, flavonoids, tannins, saponins, terpenoids, fatty acids, volatile oils	Antioxidant, antimicrobial, cytotoxic, insecticidal, allelopathic	Whole plant	Bakrim et al., 2022; Alharthi et al., 2024
Nymphaea nouchali / Nymphaea spp. (water lily)	Flavonoids (quercetin derivatives), phenolic acids, tannins, glycosides	Antioxidant, anti-inflammatory, anticancer, hepatoprotective	Tuber, rhizome, flower	Sarma et al., 2024; Mao et al., 2025
Hydrilla verticillata (hydrilla)	Terpenoids (phytol), fatty acids, flavonoids, phenolics, alkaloids	Antibacterial, antifungal, antioxidant	Whole plant	Sarma et al., 2019; Prabha et al., 2020
Ipomoea aquatica (water spinach)	Flavonoids, phenolic acids, carotenoids (β-carotene, lutein), sterols	Antioxidant, hepatoprotective, nutritional, anti-inflammatory	Leaf, stem	Saikia et al., 2023; Rehman et al., 2024
Trapa natans / T. bispinosa (water chestnut)	Tannins, phenolics, flavonoids, starches, polysaccharides	Antioxidant, hepatoprotective, nutritive, antimicrobial	Seed/fruit, kernel	Rajput and Singh, 2023; Koley et al., 2024
Lemna minor (duckweed)	Proteins, polyphenols, flavonoids, carotenoids, sulfated polysaccharides	Antioxidant, anti-inflammatory, immunomodulatory	Whole frond	Sosa et al., 2024; Alharthi et al., 2024
Pistia stratiotes (water lettuce)	Phenolics, flavonoids, saponins, alkaloids, steroids	Antimicrobial, antioxidant, allelopathic, insecticidal	Leaves, roots	Alharthi et al., 2024
Azolla spp. (mosquito fern)	Phenolics, flavonoids, carotenoids, essential amino acids	Antioxidant, nutritive, antimicrobial	Whole plant	Saikia et al., 2023; Alharthi et al., 2024
Vallisneria natans	Flavonoids, phenolic acids, sterols	Antioxidant, anti-inflammatory	Leaves, roots	Alharthi et al., 2024
Potamogeton spp. (pondweeds)	Phenolics, phenolic acids, flavonoids	Antioxidant, allelopathic	Whole plant	Alharthi et al., 2024; Paudel & Panth, 2015
Ceratophyllum demersum (hornwort)	Polyphenols, fatty acids, terpenoids	Antioxidant, allelopathic, antimicrobial	Whole plant	Alharthi et al., 2024

continued on following page

Table 2. Continued

Species (common name)	Representative bioactive compound(s)	Functions	Plant part(s) reported	References
Egeria densa (Brazilian water weed)	Flavonoids, phenolics, fatty acids	Antioxidant, antimicrobial	Whole plant	Alharthi et al., 2024; Koley et al., 2024
Myriophyllum spicatum (watermilfoil)	Phenolics, tannins, flavonoids	Antioxidant, allelopathic	Whole plant	Alharthi et al., 2024; Sarma, 2019
Salvinia molesta (giant salvinia)	Phenolics, flavonoids, terpenoids, tannins	Antioxidant, allelopathic, insecticidal	Whole plant	Alharthi et al., 2024
Typha spp. (cattails)	Flavonoids, phenolics, terpenoids, essential oils	Antimicrobial, antioxidant, wound-healing	Rhizome, pollen, leaf	Alharthi et al., 2024; Koley et al., 2024

ADVANCED ANALYTICAL APPROACHES FOR CHARACTERIZING BIOACTIVES IN HYDROPHYTES FOR HEAVY METAL REMEDIATION

Hydrophytes play essential roles in the hydrological cycle and are also widely studied for remediation of heavy metals in contaminated soils and water systems; however, less attention has been given to their bioactive components and the related specific analytical approaches employed for their characterization. Several hydrophytes like water hyacinth (*Eichhornia crassipes*), cattail (*Typha latifolia*), duckweed (Lemna minor), and water lettuce (*Pistia stratiotes*) have proven to have a huge potential in decontaminating bodies of water that are contaminated with heavy metals. Identification and characterization of bioactive compounds responsible for metal uptake, sequestration and detoxification is essential to exploit their capabilities. More advanced analytical techniques such as metabolomics, proteomics, and molecular docking studies are valuable tools to uncover the complex biochemical pathways and molecular interactions that are involved in these processes.

1. Metabolomics: Profiling Small Molecule Changes

Metabolomics is the systematic investigation of these small molecules (metabolites) within a part of, or whole organism, allowing for the identification and quantification of these metabolites enabling us to understand the changes that occur on metabolic level once organisms are exposed to an environmental stressor such as heavy metals. Hydrophytes also provide a unique opportunity for metabolomic profiling that can help discover important metabolites that confer tolerance and detoxification functions to metals. For example, changes to metabolite composition, such as increased amino acid and organic acid concentrations associated with metal stresses have been shown for the aquatic macrophyte *Eichhornia crassipes* in response to cadmium exposure (Singh, et al. 2015). With the help of an aquatic macrophyte species, *Elodea canadensis*, as a model organism, the simultaneous impact of temperature and trace metals (specifically Co and Ni) was investigated on bioaccumulation of metals in plants and their viability. The results emphasised the changes in metabolism of plant during metal stress and the prospects of phytoremediation (Polechońska et al., 2024).

2. Proteomics: Identifying Protein Markers and Pathways

Proteomics is the large-scale study of proteins (particularly their functions and structures). Proteomic analysis can provide information on key proteins involved in metal uptake, transport, and detoxification in hydrophytes. For example, metallothioneins and phytochelatins are among the different proteins involved in the processes of heavy metals such as cadmium and lead sequestration in proteomic studies of *Typha latifolia* (Raza, A., et al., 2022). A study on *Pistia stratiotes* was carried out to examine its potential in remediating a heavily polluted drain located in Egypt, rich in macronutrients and heavy metals. El-Liethy et al. (2022) characterized the plant proteome responses to metal stress and implicated two important metal stress-related pathways that may help this plant be good at phytoextraction of metals.

3. Molecular Docking Studies: In Silico Prediction of Molecular Interactions

Molecular docking studies of small molecules with proteins at the atomic level reveal mechanistic insights and binding affinities. This methodology may be used to predict the interaction of metal ion and associated proteins of the hydrophyte-derived compounds. For example, some hydrophilic types of heat shock protein 70 (Hsp70) showed very high binding affinities toward cadmium and lead in the molecular docking study, which indicates Hsp70 may play a role in protecting animals during metal stress responses (El-Sappah et al., 2024). El Awady et al., (2021) investigated the efficiency of silver nanoparticles modified *Azolla filiculoides* in removing heavy metals from contaminated water. The evaluation offered valuable information about plant molecular interaction with metals and its relatively higher efficiency in phytoremediation.

For Comprehensive Understanding of Omics Approaches as Integrated Suite

This research offers a comprehensive perspective on the biochemical and molecular strategies used by hydrophytes for heavy metal remediation by integrating metabolomics, proteomics, and molecular docking studies. These multi-omics approaches allow for the discovery of metal stress tolerance biomarkers and strategies to improve phytoremediation performance (Khan, et al. 2025). *Alternanthera philoxeroides, Canna indica* L. and *Nymphaea tetragona* are widely spread emergent macrophytes that were investigated in this study for their application in the remediation of heavy metal-contaminated sediments. This study sheds lights on the multi-omics responses of these plants to metal stress as well as their phytoremediation potential.

The application of these advanced analytical techniques makes it possible for researchers to gain greater insight into the bioactive compounds within hydrophytes, thereby facilitating the development of improved phytoremediation strategies for combating aquatic heavy metal pollution.

HEAVY METAL TOLERANCE AND ACCUMULATION STRATEGIES IN HYDROPHYTES

As aquatic macrophytes, hydrophytes are regularly subjected to the contamination of toxic heavy metals, including cadmium (Cd), lead (Pb), mercury (Hg), chromium (Cr), and, of course, arsenic (As). Hydrophytes do not possess the same soil buffer to counter metal-contaminated ground and water so potential harsh tolerance processes must also be very strong to enable survival of a hydrophyte and continue to perform ecologically. These strategies are complex cellular and molecular mechanisms that control their capacity to accumulate, detoxify, and tolerate heavy metals in many ways, such as the involvement of transporters of metals, chelation mechanisms, sequestration into intracellular compartments, and stimulation of antioxidant systems (Yadav, 2010; Verma et al., 2020c).

Cellular Adaptations

Cell Wall Binding and Metal Immobilisation: The cell wall is a physical and chemical barrier through which heavy metals may be adsorbed and immobilised in addition to preventing penetration into the cytoplasm. The composition of the cell wall of the hydrophytes under metal stress changes in such a way that enhances the amount of pectins, lignin, and hemicelluloses with carboxyl, hydroxyl, and sulfhydryl groups that have heavy metal binding capabilities (Krzesłowska, 2011; Hall,

2002). In case of Cd and Pb, in *Eichhornia crassipes*, there is an accumulation of the heavy metals at roots, epidermis and at the cell wall, which is evidence of its accumulation in roots as a tentative defense mechanism (Zhao et al., 2015b; Rai, 2009). Vacuolar Sequestration: Metal ions can commonly be compartmentalised within large vacuoles, membrane-bound organelles, which act as storage units once they are inside the cell. This vacuolar detention protects metals against sensitive organelles such as the mitochondria and chloroplasts. It is facilitated by tonoplast transporters e.g. H^+/metal antiporters (CAX, MTP families), which take part in ionic homeostasis (Martinoia et al., 2007; Gustin et al., 2009; Ricachenevsky et al., 2013).

Uptake and Distribution of Metal Ions in Plants

Specialized transporter proteins regulate plant metal homeostasis, tolerance, and detoxification, and play pivotal roles in metal ion uptake and distribution in plants. These transporters are localized in cell membranes for the uptake and efflux of essential metals (Fe, Zn, Cu) and non-essential or toxic metals (Cd, Pb) from cells or organelles. P1B-type ATPases: A family of transmembrane proteins which involved in effluxing divalent and trivalent metals such as Cu^{2+}, Zn^{2+}, and Cd^{2+}; Accomplishes this by using ATP hydrolysis These are of great importance in terms of metal detoxification and intracellular sequestration (Williams & Mills, 2005). ZIP (ZRT/IRT-like Proteins) family: ZIP transporters mediate uptake of Zn, Fe and, to a lesser extent, Mn from extracellular environments through the plasma membrane and thus play important roles in controlling metal levels in the cytosol. They are important for the acquisition of metals in situations of deficiency (Guerinot, 2000). NRAMP (Natural Resistance-Associated Macrophage Proteins): NRAMP transporters are involved in the uptake and intracellular delivery of divalent metals, including Fe^{2+}, Mn^{2+}, and Cd^{2+}. Commonly found in roots, these are indicative for metal mobilization and storage (Nevo & Nelson, 2006). Heavy Metal ATPases (HMAs): HMAs export metals to vacuoles, or out of the cell for detoxification. Example of HMA2 and HMA4 in Arabidopsis as play significant role in the translocation of Zn and Cd from roots to shoots (Hussain et al., 2004). Function in phytoremediation: In hyperaccumulator plants and also in some aquatic macrophytes in phytoremediation, these transporter proteins play important roles in the uptake, translocation and sequestration of heavy metals from contaminated water or soils (Wang et al., 2012; Zhou et al., 2010) (**Table 3**). To conclude, metal transporter proteins are responsible for the site-specific uptake, translocation and targeting of metals and are important for metal nutrition and also play a key role in plant tolerance to metal stress and metal immobilization within the rhizosphere and subsequent removal from contaminated sites for environmental clean-up programs.

The uptake and distribution of metal ions are facilitated by specialized transporter proteins. These transporters are tightly regulated and differ in specificity and function:

Table 3. Schematic representation of Molecular Mechanisms and Metal Transporters

Transporter Family	Function	Examples
ZIP (ZRT/IRT-like proteins)	Uptake of Zn^{2+}, Fe^{2+}, Mn^{2+}, Cd^{2+}	Found in roots of *Lemna minor* and *Typha latifolia*
NRAMP (Natural Resistance-Associated Macrophage Proteins)	Metal transport across endosomal membranes	Involved in Cd and Fe uptake
HMA (Heavy Metal ATPases)	Efflux and detoxification of Cd^{2+}, Pb^{2+}, Zn^{2+}	*Eichhornia* and *Hydrilla* express HMA4-like genes
MTP (Metal Tolerance Proteins)	Vacuolar sequestration of metals	Prominent in *Phragmites australis*
ABC (ATP-Binding Cassette) transporters	Detoxification and efflux of conjugated metal complexes	Linked to phytochelatin-metal export

Williams & Mills (2005) noted that transporters such as HMA and MTP are essential for metal detoxification and tolerance in aquatic species, facilitating either sequestration into vacuoles or efflux from cells.

Chelation and Detoxification Compounds

In the cytoplasm, heavy metal ions are then bound through chelation by cysteine-rich peptides such as phytochelatins and metallothioneins. Phytochelatins (PCs): These are metal stress-responsive, enzymatically produced compounds that are produced under the peptidyl synthesis of glutathione by the enzyme phytochelatin synthase. They also form stable complexes especially with Cd, Pb, and Hg and the binding process is then transported into vacuoles to be sequestered (Cobbett and Goldsbrough, 2002; Grill et al., 1985). Metallothioneins (MTs): These are genes-encoded proteins that are characterized as being low-molecular weight and are highly binding to the divalent metal ions (Zn, Cu and Cd). They work in metal detoxification and homeostasis (Hassinen et al., 2009; Leszczyszyn et al., 2007). In *Lemna gibba*, Cd causes the up-regulation of PC synthase genes, suggesting an active process of detoxification through the synthesis of phytochelatins (Macek et al., 2008; Vogeli-Lange & Wagner, 1996).

Antioxidant Defence Systems: Heavy metals can cause oxidative stress since they produce oxidising free radicals, scavengers of reactive oxygen species (ROS), including hydrogen peroxide (H_2O_2), superoxide anion (O_2^{-1}), and hydroxyl radicals (OH^{-1}). In response to this, the hydrophytes trigger antioxidant mechanisms (enzymatic and those that are non-enzymatic). A) Enzyme antioxidants: These are superoxide dismutase (SOD): converts O_2^- to H_2O_2, catalase (CAT): H_2O_2 to water and oxygen), Ascorbate Peroxidase (APX): uses ascorbate to reduce H_2O_2, and, Glutathione reductase (GR): keeps glutathione (GSH) in its reduced form. In *Pistia stratiotes*, using Cr and Ni stress, the activity of antioxidant enzymes has increased significantly, which shows that it plays a big role in counteracting stress caused by oxidants (Rai, 2008; Singh et al., 2013; Sharma & Dietz, 2009). B) Non-enzymatic antioxidants: These are ascorbate (Vitamin C), glutathione (GSH), carotenoids, and flavonoids. They form direct ROS quenching molecules that are often induced by metal stress (Foyer & Noctor, 2011). There are especially flavonoids, which are not only antioxidants, but also metal chelators, which provide dual protection (Pourrut et al., 2011; Michalak, 2006).

Signal Transduction and Stress Gene Regulation

The exposure to heavy metal stimulates a cascade of MAPKs (Mitogen-Activated Protein Kinases) signal transduction pathways: These are found to play major roles in transducing signals of extracellular metal stress into corresponding intracellular responses, consisting of changes in gene expression (Yeh et al., 2007; Jonak et al., 2004). Calcium signaling (Ca^{2+} fluxes): responses to metal-induced oxidative stress is a universal secondary messenger, and it activates downstream protein kinases as well as transcription factors (Sanders et al., 2002; Singh et al., 2016). Hormonal control: Phytohormones, including abscisic acid (ABA), salicylic acid (SA), and ethylene (ET) exert control on antioxidant-related response, the expression of transporters and stress-related gene networks (Peleg & Blumwald, 2011; Vlot et al., 2009). Such signaling molecules control the synthesis of the genes associated with the process of metal detoxification, expression of transporters, and production of antioxidant enzymes. *Nelumbo nucifera* also presents downstream arsenic-induced genes that deal with detoxification, including stress-responsive transcription factors, such as WRKY and MYB (Zhou et al., 2022; Chen et al., 2021).

To cope with and detoxify heavy metals, hydrophytes use a complex defence system consisting of the preservation of the cell wall, complexation with heavy metals, vacuole sequestration, hormonal regulation of transporter actions, and the cellular antioxidant response (Verma & Dubey, 2003; Thapa et al., 2021). Such mutations not only enable them to survive in the waters of polluted environments but also to efficiently become a phytoremediation agent. Since molecular and physiological strategies are better understood, selection and engineering of high-performance species in constructed wetlands, wastewater treatment, and restoring aquatic ecosystems are key (Yan et al., 2020; Ali et al., 2013).

SOME IMPORTANT HYDROPHYTES WITH REMEDIATION POTENTIAL

Aquatic macrophytes have attracted much interest in the subject of water pollution control via phytoremediation. Out of a huge number of hydrophytic plants, there are quite a few that may attract the most attention because of the ability to grow very fast, accumulate biomass, tolerate heavy metals and produce bioactive compounds that help the plant in detoxification. This section gives an overview of four extensively researched hydrophytes i.e. *Eichhornia crassipes*, *Lemna minor*, *Pistia stratiotes*, and *Typha latifolia* on the basis of their phytoremediation mechanism, their metal specificity and potential use in real world applications.

1. *Eichhornia crassipes* (Water Hyacinth)

Eichhornia crassipes is an aggressive, fast-growing, free-floating aquatic plant that originated in South America. It has been given the tag of an invasive species, but nonetheless, it is commonly used in phytoremediation as a model plant because of its effectiveness in taking up metals and its ability to extensively produce biomass (Malik, 2007). It provides: quick growth and large surfaces to absorb metals, good capacity to absorb metals like Pb, Cd, Cr, Hg and As (Zhou et al., 2022; Rezania et al., 2016a), high phenolic acids, flavonoids and antioxidant enzymes production to counteract metal-induced oxidative stress (Wang et al., 2012). The processes here involved are rhizofiltration, phytoextraction and increased activities of the antioxidant enzymes such as SOD, CAT, and APX upon metal stress. Furthermore, the rhizosphere detoxification is achieved with the help of root exudates that bind the metals (Rai, 2008; Lu et al., 2010). The study of Rezania et al. (2016a) has shown that the maximum removal of cadmium (90%) and lead (85%) with the use of *E. crassipes* in the contaminated water occurred in 10 days. Along with that, Zhou et al. (2022) noted that it can be used in the reduction of arsenic concentrations in mine drainage, both by uptake and chemical transformation. It is already demonstrated in the technique of both constructed wetland and floating treatment wetlands (FTWs) and shows the potential of large-scale remediation of water (Zhang et al., 2014; Rezania et al., 2015) (**Table 4**).

2. *Lemna minor* (Duckweed)

Duckweed, scientific name *Lemna minor*, is a small, prolific, free-floating aquatic plant. Its ease of growth morphology, short lifespan, and fast vegetative growth rate make it suitable for batch and continuous flow systems of treating wastewater (Ziegler et al., 2016). It generates an abundance of glutathione, phytochelatins, and polyphenols and use them in the process of detoxification. It also stores heavy metals comprising Zn, Cu, Cd, Pb and As (Mkandawire & Dudel, 2005; Mishra et al., 2008). Nevertheless, it has a short life cycle and rapid reproduction that is fit to be harvested regularly. The processes implicated include: phytoextraction (translocation to fronds), metal chelation by intracellular thiol-rich compounds and an increase in glutathione-S-transferase (GST) activity during stress. In a study that was conducted by Mishra et al. (2008), *Lemna minor* had the capacity to accumulate a maximum of 150 mg/kg Cd in its tissues and caused a dramatic rise in the activities of antioxidant enzymes. Another study conducted by Mkandawire and Dudel (2005) reported that duckweed demonstrated treatment efficiency of 80% using synthetic waste water in the removal of arsenic. *L. minor* gets a common application as a polishing agent in sewage lagoons, industrial effluent polishing, and an effective bioassay model and the post-detoxified biomass can be fed to animals or used as organic fertiliser (Cheng & Stomp, 2009) (**Table 4**).

3. *Pistia stratiotes* (Water Lettuce)

The water lettuce (*Pistia stratiotes*) is a free-floating aquatic plant that has rosette-like leaves and fibrous roots. It is growing in shallow eutrophic waters and has a wide-range phytoremediation potential (Rezania et al., 2015). It concentrates Hg, Ni, Cr, Cu, Zn, Cd and it is effective in removing biochemical oxygen demand (BOD) and chemical oxygen demand (COD) (Kaur et al., 2012). Simplest processes were rhizofiltration and phytostabilization. Mucilages and polysaccharides because of root surface adsorption of metals. Increased metal exposure and peroxidase, and phenylalanine ammonia lyase (PAL). Rai, (2008) has showed that *P. stratiotes* had the potential to remove greater than 70 percent Cr and Cd in laboratory experiments of wastewater. Biochemical adjustment is related to the increased profile of secondary metabolites of the plant as alkaloids, flavonoids, and terpenoids, which increases with the metal's presence (Verma et al., 2020b). It is also applied in municipal wastewater treatment, techniques in agriculture run off and nutrient (N, P) elimination (Ali et al., 2013) (**Table 4**).

4. *Typha latifolia* (Common Cattail)

Typha latifolia is a wetland shrub that grows vertically, in tandem with extensive rhizome root structure and perennial plant-life. It is very common in established and natural wetlands and it has been rated most successful in sediment-stabilization and healing of metals (Salt et al., 1998). It is resistant to elevated amounts of Fe, Mn, Pb, Cu, Zn and As, and the extensive root system enables phytostabilisation and rhizosphere modification and helps in recruiting favourable metal-interconverting microorganisms within its rhizosphere. The simplest of these mechanisms is phytostabilization, a trapping of heavy metals by the roots; other mechanisms are redox potential and pH alteration of the rhizosphere to help precipitate the metals and upregulation of metallothionein and metal transporter genes. According to Salt et al. (1998), T. latifolia was able to suppress the mobility of the metals in the sediments by margin of 65%. Chandanshive et al. (2017) illustrated that more than 70 percent of Zn and Cu was retained in the root zone. It can be used in the treatment of mine drainage, sewerage in an urban area, shoreline protection, and the formation of habitats in large-scale wetland restoration (Zayed et al., 1998) (**Table 4**).

Table 4. Represents the different hydrophytes along with their mechanisms.

Species	Primary Mechanism	Notable Metals Removed	Application
Eichhornia crassipes	Phytoextraction, Rhizofiltration	Pb, Cd, Cr, As, Hg	Constructed wetlands, FTWs
Lemna minor	Phytoextraction, Chelation	Cd, Zn, Cu, Pb, As	Batch wastewater treatment
Pistia stratiotes	Rhizofiltration, Stabilization	Cr, Ni, Hg, Zn	Eutrophic lakes, drainage canals
Typha latifolia	Phytostabilization, Rhizosphere engineering	Zn, Cu, Pb, Mn, Fe	Wetlands, mine tailings, sediment remediation

Collectively, these species serve as cost-effective, low-maintenance, and eco-friendly tools in managing heavy metal pollution, supporting both environmental health and sustainable water resource management.

CURRENT INNOVATIONS AND FUTURE RESEARCH DIRECTIONS

Enhancing Phytoremediation: Synergistic Role of Microbes and Nanomaterials

The dynamics between rhizobacteria and fungi, as well as engineered materials, will enhance the efficiency of uptake. Although phytoremediation is environmentally friendly and economical, it will be constrained by low rates of metal uptake, low bioavailability of the contaminant, and plant stress in severe pollution sites (Hu et al., 2020). The effective and widely used method of volatilisation, stabilisation, degradation, or extraction of the pollutants by the vegetation and associated with them microorganisms are called phytoremediation (Luo et al., 2016; Zhou et al., 2019). Phytoremediation has been thoroughly explored in the last decades and seen as an effective measure to eliminate and destroy a lot of harmful contaminants, including antibiotics, HMs, landfill leachate, textile dyes, pesticides, hormones, petroleum, explosives, and even toxic gas (Markou et al., 2018; Pilon-Smits, 2005). The growing popularity of phytoremediation at government agencies and in industries is due to the low-cost and low available finance towards environmental governance (Vymazal, 2011).

Phytoremediation is a solar energy-driven process that can supply rehabilitated sites with ecological landscapes, which is also aesthetically appealing during the treatment of contaminants (Markou et al., 2018). It offers the sustainable way of nutrients recovery and the elimination of poisonous contaminants by LW. The species chosen to conduct phytoremediation must have faster growth rate, high biomass, long fiber root, it should be easy to manipulate, sensitive to pollutants, and it should be easy to grow and harvest. To evade biological invasion as a result of poor management, the plants are desirably indigenous. Moreover, carrying out phytoremediation does not demand as much labor or equipment compared to the common biological and physical-chemical ways of remediation as it is primarily a natural remediation mechanism. Plants have been endowed with a unique decontamination potential based on their unique metabolic and extractive ability with regard to decontamination potential (Pilon-Smits, 2005). In the meantime, the stems, roots, and leaves of the plants can offer homes to a range of microbes, which can intensify the treatment procedure through synchronous degrade pollutants (Pilon-Smits, 2005). In the same context, quite some investment into phytoremediation of livestock wastewater had been logged in over the last 20 years (Sooknah and Wilkie, 2004; Cheng and Stomp, 2009; Hu et al., 2019). The strategy will establish an effective cycle of circulation that takes full advantage of the residual value of the livestock wastewater, which will recover the nutrient collection, eliminate hazardous contaminants and reclaim waste water.

To address these issues, the combined growth of hydrophytes with beneficial microbes and nanoparticles has become a radical idea. Additions of plant growth-promoting rhizobacteria (PGPRs), mycorrhizal fungi and engineered nanomaterials considerably enhance the solubilization, uptake, stress resistance and the metals remediation efficiencies in an aqueous environment (Rajkumar et al., 2012; Khan et al., 2020).

1. Role of Rhizobacteria in Phytoremediation Enhancement

Plant Growth-Promoting Rhizobacteria (PGPRs) are good bacteria that parasitize the rhizosphere to increase plant growth by a number of biochemical processes. PGPRs colonizing hydrophytes have the capacity to solubilise heavy metals using organic acid secretion (e.g., citric acid; oxalic acid), produce siderophores to chelate iron and mobilise other metals, produce indole-3-acetic acid (IAA) and ACC deaminase to stimulate root growth and control ethylene induced stress via heavy metal exposure, increase antioxidant enzyme activity of host plants and form biofilms on roots surfaces, which mediate metal adsorption and stabilisation. The growth of *Pseudomonas putida* and *Bacillus subtilis* enhances the absorption of Cd and Zn in Lemna minor by increasing the development of roots and tolerability (Sinha & Paul, 2014). Such phytoremediation enhancement of As and Cr on *Eichhornia crassipes* that had been inoculated with PGPR was shown in increased biomass and fewer oxidative damages (Zhou et al., 2022). The advantages above all are that the metals are much more bioavailable and mobile, the roots have a larger surface area and ability to absorb them and also improve plant health in toxic environments.

2. Phytoremediation by Microalgae

Microalgae that may be tolerant to different types of water bodies were largely employed in effluent treatment (Luo et al., 2016). To satisfy its own growth requirements, microalgae-based systems can absorb water bodies organic matters and nutrients (Li et al., 2018a). In the interim, different high-value-added nutrients and chemical raw materials have been stored by microalgae, and they include proteins, lipids, polysaccharides, vitamins, and beta-carotene, which can be transformed into high-value products like biofuels and health products after processing (Li et al., 2020). Hence, the management of LW was given great care when dealing with the microalgae. To begin with, the microalgae *Chlorella vulgaris* studies eliminated the pollutants of cattle wastewater that have had promising outcomes and established that 62.30%, 81.16%, and 85.29% of COD, NH3-N and TP had been eliminated (Lv et al., 2018). Luo et al. (2016) indicated that in the cultivation of anaerobically treated SW and aerobically treated SW, *Coelastrella sp.* could remove greater than 90 percent of NH3-N and TP. Even better, nutrients in LW are primarily employed in growth and multiplication by microalgae once the nutrients are assimilated. These were in the nitrogen uptake of microalgae as part of the photosynthesis, and using synthetic glutamate in the cells of algae. In the meantime, phosphorus is a crucial macronutrient in the microalgae growth and metabolism, the primary forms of which exist in significant metabolites, including nucleotides, nucleic, and phospholipids.

Microalgae have divergent mechanisms of removing HMs compared to the antibiotics which include bioaccumulation and biosorption (Li et al., 2018a). First of all, biosorption is a non-metabolic process and thus this process is swift and does not consider the concept of a live cell (Markou et al., 2018). Meanwhile, biosorption can be reversed as well, which necessitates the need to take superior precautions that would prevent the desorption of HMs and preserve an augmented degree of removal (Kumar et al., 2014). Moreover, in contrast with biosorption, cells accumulate HMs with the help of active transmembrane transportation and diffusion, and then the HMs are bound to different cell compartments and organelles (Arguello et al., 2012). Notably, in the case of various HMs and microalgae, mechanisms might either be independent or concurrent.

3. Role of Mycorrhizal and Endophytic Fungi

Although phytoremediation using arbuscular mycorrhizal fungi (AMF) and endophytic fungi was historically evaluated in terrestrial systems, phytoremediation with AMF and endophytic fungi may play a role in phytoremediation of non-terrestrial plants that experience semi-submerged or emergent conditions (e.g. *Typha, Phragmites)*. The primary processes include extending the effective root zone by hyphal networks, sequestering metals in fungal biomass or converting metals to less toxic forms, inducing the synthesis of chelating agents and phytochelatins and enhancing the uptake of water and nutrients, particularly in nutrient-improverished polluted water. Li et al., 2021. In *Typha angustifolia, Rhizophagus irregularis* inoculation reduced the amount of damage to antioxidant defence because it increased Pb and Zn uptake (Li et al., 2021). Isolation of endophytic fungus *Penicillium funiculosum* in Pistia stratiotes resulted in an increased complement of Cd absorption and mitigation of growth inhibition by metals (Zhao et al., 2018). The most important advantages are the stronger metal translocation and plant-resistance and fungal symbiosis strengthens plant multi-metal stress tolerance.

4. Engineered Nanomaterials in Phytoremediation

Nanotechnology has also entered the phytoremediation process termed nanophytoremediation that applies nanomaterials (engineered nanomaterials) to increase metal availabilities, detoxification and plant growth. Nanomaterials are used in a variety of forms such as iron oxide nanoparticles (Fe_3O_4, nZVI) to perform redox, contaminant decreasing processes, carbon-based nanomaterials (graphene oxide, biochar nanoparticles) as sorption, carrier systems and nanoparticles based on silicon, zinc and titanium to induce antioxidants and metal mobilization. Its modes of activity were to chelate and solubilise the metal ions in water, root stimulation and photosynthesis by providing micronutrients, increased growth of ROS-scavenging enzymes to create microresidence to colonise and biotransform metals by microbes. Nanoparticles of Fe_3O_4 enhanced the bioaccumulation of Cd and Pb by the *Eichhornia crassipes* in constructed wetland microcosms (Ali et al., 2018). This enhanced the detoxification of arsenic in *Lemna minor* due to increased SOD activity and expression of glutathione because of the application of nano-silicon (Rastogi et al.,

2019). The nanoparticles at a low level of concentration enhance the activity of plants, but at a high level, they can result in phytotoxicity.

5. Microbe-Nanoparticle Synergy

Microbe-nanomaterial hybrids are a new direction of multifunctional phytoremediation platforms (Chaudhary et al., 2020; Singh et al., 2022). As an example, bio-functionalized nanoparticles produced by microbes can be arranged to be taken up by the plants with minimum toxicity (Rajput et al., 2018). *Hydrilla verticillata* that grew better in the presence of silver nanoparticles generated by PGPRs was observed to gain more Cr. *Typha latifolia* is an extractor of Zn and Cd in the composition with biochar and its bacteria, and does not decrease microbial activity and plant health (Kumar et al., 2022). Synergistic effects of the association of photophyta, microbes, and nanomaterials make phytoremediation significantly more effective in terms of metal bioavailability and solubilization, plant growth, antioxidant defense enrichment, maintenance of a healthy rhizospheric microbial community and, in some cases, an affordable alternative to chemical remediation (Sharma et al., 2020; Tripathi et al., 2021). With the development of research, integrated phytoremediation systems engineering nanomaterials and plant-associated microbes may be deemed as an innovative way of restoring polluted aquatic ecosystems (Pandey et al., 2023; Meena et al., 2020).

6. Applications in Constructed Wetlands and Natural Systems

Constructed wetland (CWs) and naturally occurring wetlands are nature-based systems that take into consideration the bioremediation capabilities of the hydrophytes, microbes and the substrates to treat the contaminated water bodies. As more heavy metals are polluting water resources because of industrial, mining, and agricultural runoff, such systems have played a key role in eco-sustainable wastewater treatment and ecosystem restoration (Vymazal, 2011; Zhang et al., 2020).

a. Design of Constructed Wetlands (CWs)

Constructed wetlands are engineered to replicate the structure and ecological functions of natural wetlands. Their design varies based on hydrology, flow type, plant species and intended pollutant removal objectives. Constructed wetlands (CWs) are engineered wetlands used for wastewater treatment, where natural or semi-natural processes involving flora, micro-organisms and substrate are used. They are mainly categorized by flow patterns and by vegetation. Surface flow (SF) wetlands, which mimic natural marshes by allowing water to flow over the soil surface, are generally vegetated by Typha and Phragmites, easy to operate, but require a large land area and can breed mosquitoes. Subsurface flow (SSF) wetlands, including horizontal subsurface flow (HSSF) and vertical flow (VF) systems, transport water below the surface level through a permeable medium, thus alleviating odour and surface algae growth, and improving pollutant reduction by increasing contact of the water with the plants roots and biofilms. Hybrid wetlands incorporate more than one wetland type, such as HSSF and VF, to maximize the removal of nutrients, heavy metals and pathogens, and are appropriate for high-strength or variable wastewater. Moreover, free water surface (FWS) wetlands include open water areas for treatment and wildlife habitat. Such versatility in configurations makes constructed wetlands adaptable to types of wastewater and treatment objectives (Vymazal, 2011; Wu et al., 2014; Yamashita et al., 2021; Zhang et al., 2020) (**Table 5**).

Table 5. Represents the different types of Constructed wetlands.

Type	Flow Regime	Plant Zone	Application
Surface Flow Wetlands (SF-CWs)	Water flows over soil	Emergent plants	Municipal/domestic wastewater
Subsurface Flow Wetlands (SS-CWs)	Water flows through substrate	Root zone (subsurface)	Industrial effluents, heavy metals
Hybrid Systems	Combination of SF and SS	Multiple plant zones	Complex wastewater (mixed contaminants)
Floating Treatment Wetlands (FTWs)	Water surface (free-floating plants)	Free-floating hydrophytes	Stormwater ponds, lakes, mining tailings

Substrates in CWs (e.g., gravel, sand, slag) provide surface area for microbial colonization and metal adsorption.

b. Role of Hydrophytes in Constructed Wetlands

The essence behind made wetland (CW) function is the hydrophytes. They also serve in the uptake of contaminants, rhizosphere oxidation, sediment stabilisation, as well as facilitating the microbial community's levels that help in the degradation of contaminants (Brix, 1997; Stottmeister et al., 2003; Vymazal, 2011). Typical hydrophytes applied in CWs come in the form of *Typha latifolia* (cattail), *Phragmites australis* (common reed), *Eichhornia crassipes* (water hyacinth), *Lemna minor* (duckweed) or *Pistia stratiotes* (water lettuce) (Reddy & Delaune, 2008; Akratos & Tsihrintzis, 2007). The species vary on their metal stress toleration, growth patterns and their repairing capabilities (Cheng et al., 2002; Rai, 2008).

As an example, *Typha* and *Phragmites* are commonly used in subsurface flow CWs by their strong root system and high flexibility (Brix et al., 2001; Calheiros et al., 2007) and *Eichhornia, Lemna* is more popular to be applied to floating treatment wetlands (FTWs) owing to their high expansion rate, convenient harvest, and high pollutants removal ability (Zimmels et al., 2006; Sirianuntapiboon et al., 2006).

c. Functioning of Wetland-Based Remediation Systems

Wetlands are characterized by the interactions of complex biogeochemical processes taking place among plants, microbes, substrates, and water (Mitsch & Gosselink, 2015; Kadlec & Wallace, 2009). The main processes they employ in removing the metals are: Phytoaccumulation, where metal is put into neurotrophy by roots and then stored in shoots or vacuoles, which have been widely reported in macrophytes like *Typha, Phragmites* and *Eichhornia* (Zhou et al., 2010; Marchand et al., 2010). Rhizofiltration refers to the adsorption and ion exchange of root surface, which is applied particularly well to the floating roots (Salt et al., 1998; Dhir, 2013). Sedimentation and Adsorption: These are the settlement of metals with particulate or the adsorption on the substrate material like sand, gravel, clay, or biochar mostly active in the subsurface flow CWs (Bolan et al., 2011; Reddy & D'Angelo, 1997). Microuber Transformations: Redox-driven reactions Precipitation, methylation or detoxification (e.g. Cr^{6+} Cr^{3+}) occur through metal-reducing or oxidizing bacteria (Gadd, 2000; Haferburg & Kothe, 2007). Biofilms, which are connected to roots, increase the rate of heavy metal transformations and fixation of heavy metals (Li et al., 2011). pH and Redox Alteration: The oxidation-reduction gradient in wetland areas permits the modulating effect, which tends towards spec suppressing heavy metals and making them more soluble (Batty & Younger, 2004). Plant roots that are emerging emit oxygen, which favors microaerophilic niches, which cause precipitation of iron, manganese, and arsenic (Brix et al., 2001). Metal Removal: Constructed wetlands have been shown to produce very high efficiencies in the removal of many metals including lead (Pb), cadmium (Cd), copper (Cu), zinc (Zn), chromium (Cr), mercury (Hg), and arsenic (As) in multiple field and laboratory studies (Fonder & Headley, 2013; Zhang et al., 2014; Wu et al., 2014) **(Table 6)**.

Table 6: Constructed wetlands showing efficiencies in the removal of different heavy metals.

System Type	Plant Species	Metal Removed	Removal Efficiency (%)	Reference
Surface Flow CW	*Typha latifolia*	Pb, Zn, Cu	65–85%	Mays & Edwards, 2001
Subsurface Flow CW	*Phragmites australis*	Cd, Pb, Cr	70–90%	Zhang et al., 2020
FTW (Floating wetland)	*Eichhornia crassipes*	Cd, As, Cr	80–95%	Rezania et al., 2016a
Hybrid System	*Typha + Lemna minor*	Multiple metals	>90% in industrial wastewater	Zhang et al., 2020

Case study: A pilot-scale subsurface CW planted with *Phragmites australis* and *Typha* near a mining site in China reduced Cd levels by 88% and Pb by 76% over six months (Zhang et al., 2020).

Applications in Natural Wetlands

The constructed system has the advantage of being designed controlled, and natural wetlands would offer significant mitigation services. The various hydrophytes found within such systems also assist in the buffering of agricultural runoff that contains heavy metals (Vymazal, 2011), buffering acid mine drainage (AMD) effects through precipitation of the metals (Sheoran and Sheoran, 2006) and bank stabilization of the rivers and the stream bed (Mitsch & Gosselink, 2015). They can, however, be implemented site-specifically, and they are dependent on hydrology, native vegetation and how ecologically sensitive an area is. As an example, iron, manganese and copper produced by industries in upstream sources are naturally removed by the native species *Cyperus papyrus* and *Typha capensis* in the Okavango Delta of Africa (Fisher and Acreman, 2004; Masamba and Mazvimavi, 2008). The substantial benefits associated with natural wetlands include the fact that they are economically friendly and energy cost-friendly (Kadlec and Wallace, 2009), enhance biodiversity and carbon capture (Zedler and Kercher, 2005) and can incorporate landscape restoration programs with little maintenance costs (Zedler and Kercher, 2005). Metal saturation of the material might limit long-term efficiency though (Kumar et al., 2022), the remediation rates can be slow (Verhoeven and Meuleman, 1999) and performance has been known to be affected by season (Iasur-Kruh et al., 2010).

7. FUTURE PROSPECTS

Projections entail incorporating nanomaterials and biochar in the form of substrates in constructed wetlands (CW) to increase the metal-sorption and removals capabilities of pollutants (Beesley et al., 2011; Lu et al., 2017). More so, metal tolerance and uptake are enhanced through synergistic plant microbe interactions when metal-tolerant microbes are inoculated in the hydrophytes (Glick, 2010; Ma et al., 2016). The use of the technologies based on remote sensing and IoT-based methods to monitor CW performance in real time to conduct data-driven adaptive management adds to the strength of these innovations (Zhao et al., 2020; Banerjee et al., 2022). The second future tendency is the design of the floating modular wetlands, which are, in turn, especially applicable to decentralized wastewater treatment in cities (Tanner and Headley, 2011; Dalahmeh et al., 2014).

Challenges and Limitations of Hydrophyte-Based Remediation

Among these problems there are such ones as biomass disposal, seasonal growth, metal re-release and ecological invasiveness. Hydrophyte-based remediation is also turning out to be a green, economical and sustainable intervention to remediate the levels of heavy metals in the waters. Nevertheless, biological, ecological and practical facets are limiting the realistic deployment of phytoremediation, especially in two forms, namely phytoremediation through the utilization of fish plants and phytoremediation by means of aquatic plants. Interpretation of these limitations is necessary to ensure that their performance is enhanced, they are sustainable and there are no adverse environmental effects of the system.

1. Biomass Disposal and Secondary Pollution

In hydrophyte-based remediation, the disposal of contaminated plant biomass following metal accumulation is one of the most burning problems. Plants grown during phytoremediation/treatment absorb a huge number of heavy metals in their tissues and are therefore susceptible to secondary pollution by their mishandling. The problems, such as the biomass that may have been harvested, may release the metals into the environment in the event that degradation is allowed or upon the above usage of the harvest without any treatment. Biomass containing potentially high metal concentrations can be costly to landfill or incinerate, leading to possible toxic emissions or leachates (Ali et al., 2013). The absence of any established safe pathways of using biomass, including through composting, production of bioenergy or biochar. Among measures put in place to remove them include thermal treatment (e.g. pyrolysis) to convert the biomass into biochar that immobilises the metals

(Beesley et al., 2011) and developing metal recovery methods to recover the metal using phytomining of plants high in biomass.

2. Seasonal Growth and Climate Dependency

The growth patterns of many hydrophytes are seasonal, which influences their biomass formation, metabolic rate and efficiency with which metals are absorbed (Brix & Schierup, 1989; Vymazal, 2007). This cyclical change may result in serious system performance during winter/dry periods, especially in temperate and arid weather (Bragato et al., 2006). Among other considerations are low photosynthetic activity during winter, which does not provide enough room to take up contaminants, and the dieback of aboveground mass, which can potentially cause the re-release of the built-up metals into the water column (Tanner, 2001; Calheiros et al., 2009). In addition, wetland systems are prone to changes in climate and extreme weather conditions (Ex: floods, droughts) that may disrupt the functioning of the system in the long-term (Erwin, 2009; Mitsch et al., 2010).

The solutions that were proposed to manage these difficulties include the plantation of multi-species which have complementary growth patterns in order to make the systems viable throughout the year. Besides, studies are being conducted on the breeding of climate-resistant species or genotypes of hydrophytes that can cope with wider ecological variability (Yamashita et al., 2021). The employment of controlled environments, including greenhouse wetlands (Molle et al., 2008; Vymazal & Kröpfelova, 2011) is another promising method that would help to maintain stable performance of remediation throughout all year.

3. Risk of Metal Re-Release and Remobilization

In systems dependent upon hydrophytes, the metals are stored on the plants themselves or within sediments and roots. But physical/chemical conditions are such that these metals can be mobilized into the water column at times, overriding remediation (Gadd, 2004; Weis and Weis, 2004). There are many environmental situations that may cause the re-release of metals such as plant decay under anaerobic circumstances, as well as alterations in pH, redox potential or microbial activity (e.g., sulfate-reducing bacteria) (Gilmour et al., 1992; Stumm and Morgan, 1996; Nancharaiah et al., 2016). Other reasons are ageing of the plant or root damages by pests, diseases or mechanical disruptions, which affect metal sequestration compared to senescence.

Among the more notable effects of these processes are the production of metal sulphides or methylated metal species (e.g. methylmercury) which can be more toxic and mobile than their parent forms and present, ultimately, a number of hazards associated with them including contamination of downstream aquatic eco-systems and bioaccumulation in aquatic food chains (Morel et al., 1998; Ullrich et al., 2001). Among the proposals to manage such risks, recurrent harvest before senescence, application of stabilization agents that hold the metals in contact and prevent the leaching of the metals because they are bound to the stabilisation agent (e.g. biochar or lime), and regular monitoring of the redox potential and pH of the root zones have all gained traction in potential strategies to address the risks (Kadlec and Wallace, 2009; Beesley et al., 2011; Mohan et al., 2014).

4. Ecological Invasiveness and Biodiversity Risk

Other hydrophytes which have been employed in remediation include *Eichhornia crassipes* (water hyacinth) and *Pistia stratiotes* (water lettuce), which is considered an invasive alien species in most parts (Lowe et al., 2000; Cilliers et al., 2003). Their swift reproduction, high prolificity and dominance in nutriment with the indigenous flora are some of the major hazards that they pose to the biodiversity of the ecosystem (Gopal, 1987). Such plants grow in dense mats which casts out sunlight, lower the concentration of dissolved oxygen and interferes with aquatic habitats (Villamagna and Murphy, 2010). Nutrient cycles are also distorted in their presence and growth of indigenous plant communities and microbes are suppressed (Shilla et al., 2006).

Moreover, the water supply in such areas might also be cut short when the evapotranspiration rates are high on the account of these species when these species occur in arid and semi-arid conditions (Patel, 2012). In Asia, Africa and the Americas, *Eichhornia crassipes* has taken over the supposedly wetlands and clogged waterways and impeded the nearby fisheries as well (Villamagna and Murphy, 2010; Ndimele et al., 2011). *Pistia stratiotes* has quenched indigenous water plants in lakes of India and Australia (Schooler et al., 2006; Reddy and DeBusk, 1984).

To counter these risks, through regulatory frameworks, local or non-invasive species, *Typha latifolia* and *Phragmites australis* are recommended to be used. Invasive species can be curbed by use of mechanical control, biological colonization (e.g. use of Neochetina spp on E. crassipes) or integrated weed management (Julien et al., 2001). There is a dire need to follow the rules of invasive species to the letter before one employs the plants to clean up by phytoremediation.

5. Limited Metal Specificity and Variable Uptake

Hydrophytes as single entities do not all have broad-spectrum remediation ability. Others are very selective or not efficient in biomass uptake of some metals, and this results in inconsistent performance among the contaminated sites (Zhou et al., 2007; Ali et al., 2013). As an example, some species, such as *Lemna minor*, can effectively concentrate zinc (Zn) but are unable to remove well lead (Pb) or mercury (Hg) (Mkandawire and Dudel, 2005; Zhao et al., 2015b). The performance greatly depends upon the water chemistry, nature of the substrate and co-contaminants such as the organic pollutants (Marchand et al., 2010; Zhang et al., 2020).

To eliminate these drawbacks, some of which are the limited effectiveness against varied contaminants, the proposed solutions are to use multi-species wetland systems to attack a wide variety of contaminants more easily (Fonder & Headley, 2013). Microbial augmentation or genetic engineering approaches are also being used to augment the mechanisms. The combination of these remedies with the addition of other treatment technologies (nanomaterials, biofilters) enables a better efficiency of removals and resilience of the system (Thakur et al., 2016; Rajasulochana and Preethy, 2016).

6. Knowledge Gaps and Technical Limitations

In spite of the developments, numerous scientific and technical challenges restrict the large-scale application of hydrophyte-based remediation. These are inadequate protocols that are not standardized ever implemented with regards to the plant selection, system design and monitoring, as well as the lack of comprehensive knowledge of the ecological effects in the long term (Bragato et al., 2006; Interstate Technology & Regulatory Council, 2008). The other hurdles are the issues of scaling up the laboratory conditions to field use and lack of policy support or incentive schemes, depending on the region (Luongo and Cutright, 2011; Zhang et al., 2014). Any future studies should be directed towards longitudinal field monitoring to observe any ecological succession or sustainability, which include, but are not limited to the development of decision-making tools of the best plant-metal combinations in the future (Marchand et al., 2010; Yadav et al., 2022). Broader adoption also hinges on the exploration of the economic viability including the cost-benefit analysis and life cycle (Dos Santos et al., 2016).

Although the use of hydrophytes as remediation agents in mitigating heavy metal pollution is an attractive aspect of nature, there are some issues associated with the actual practice of the remedy. They include disposal of contaminated biomass, seasonal curb variability, invasive species risks and secondary pollution (Schmidt et al., 2008; Ali et al., 2013). Hence, careful planning and monitoring, as well as technological innovations are necessary. Also, the combination of these systems with the concept of bioengineering, involving interested parties, and enhancing policy enlightenment will play an essential role in resolving the most possible benefits to the environment and society of the hydrophyte-based remediation activities (Tangahu et al., 2011; Vymazal and Březinová, 2015).

7. Practical and Biological Limitations

Hydrophyte representatives are promising, although the hydrophyte-based systems have undeniable practical and biological shortcomings at large scale.

Variability in Bioactive Compound Production

Natural variation of secondary metabolite production in hydrophytes including *Eichhornia crassipes*, *Lemna minor* and *Typha latifolia* is often species- and developmental-stage-dependent and can also be modified by nutrient and contaminant stress. Due to this variability, standardization and reproducibility of their bioactive profiles is difficult, which hampers their application in remediation or as a source of bioactives for pharmacological purpose (Saikia et al., 2023; Bayuo et al., 2024). In addition, genetic variations even among different genotypes of a species have been shown to modify enzyme expression associated with either metal chelation or antioxidant responses, resulting in an uneven non-uniform detoxification efficacy (Tao et al., 2020).

Scalability and Operational Challenges

Although laboratory and mesocosm investigations show elevated metal removal efficiencies, transposing these systems to realistic scenarios succeeds large challenges, including heterogeneous contaminant distribution, variable hydraulic loads and maintenance (Ntagia et al., 2023). Another challenge overcoming the large volumes of biomass associated with harvesting or decay is the risk of secondary pollution (Tan et al., 2023). Moreover, the population where the invasive tendencies of fast-growing hydrophytes such as water hyacinth have limited their safe large-scale deployment (Zhang et al., 2024).

Environmental and Climatic Influences

Temperature, light intensity, and seasonal variation of uptake kinetics and changes in enzymatic activity of hydrophytes are known to have a considerable impact on the uptake kinetics and enzymatic activity of hydrophytes. However, temperature rise affects metal lability and plant uptake, which may result in greater or lesser efficiency of remediation (Ma et al., 2020). Bioaccumulation patterns are modified further by shifts in pH, dissolved oxygen and/or competing ions, rendering performance site-specific and thus difficult to predict (Mao et al., 2024).

8. Planning, Monitoring and Technological Innovations- An Analytical Assessment

1. Genetic Engineering of Hydrophytes for Enhanced Metal Tolerance and Uptake

The genetic engineering of aquatic plants, a direction that holds a lot of promise, is to enhance their remediation efficiency. This incorporates the installation or excess of metallic transportation, sequestration and detoxification genes, in addition to stress-adaptive transcription factors and antioxidant enzymes (Cherian and Oliveira, 2005; Pilon-Smits, 2005). The genetic enhancements earmarked by this tendency are greater expression of metal transporters (e.g. ZIP, NRAMP, ABC proteins), more phytochelatins and metallothioneins, better root system architecture to encounter bigger pollutant surface areas and tolerance to oxidative stress and difficult environmental conditions (Clemens, 2006). To illustrate the point, *Arabidopsis thaliana* overexpression in AtPCS1 (phytochelatin synthase) showed enhanced cadmium (Cd) and arsenic (As) tolerance by increasing metal chelation and metal vacuolar sequestration (Gasic and Korban, 2007). In a further study, transgenic *Lemna minor* was genetically modified to produce more glutathione synthetase; it was observed that the level of uptake of mercury (Hg) and cadmium (Cd) was significantly higher, which was attributable to improved antioxidative and metal complexation ability (Liu et al., 2014). Though good, the use of genetically modified (GM) hydrophytes has biosafety, biosafety and ecological-related issues as mentioned in open architectural theories (Meagher, 2000). Other difficulties require consideration are the attitude of people, the existence of regulatory obstacles and the lack of species-specific transformation methods to aquatic plants, which are falling behind compared to terrestrial model species (Wang et al., 2017).

2. Multi-Species and Polyculture Wetland Systems

Single species phytoremediation systems do not necessarily need to handle multi metal contamination or mix pollutant contamination. Thus, in the future, the strategy of wetland design is moving to multi-species (poly-culture), where functional roles played by different hydrophytes are complementary (Ghosh and Singh, 2005; Brix et al., 2007). Among multiple benefits related to multi-species systems, functional complementarity, in which certain species will often have a higher ability to uptake metals and others that are more suited to structural stabilization or nutrient retrieval, as well as an increased resistance to both environmental stress and seasonal changes can be listed (Stottmeister et al., 2003). The systems also increase the biodiversity and ecological balance by raising the co-removal of various contaminants by removing heavy metals, nutrients (i.e., nitrogen and phosphorus) and organic compounds simultaneously (Wu et al., 2014; Zhang et al., 2019). As such, an example is the use of *Typha latifolia* (emergent) united with *Eichhornia crassipes* (floating) and *Lemna Minor* (free-floating) that was observed to be effective in the treatment of municipal effluents containing Pb, nitrogen, and pesticide residues due to their different uptake of mechanisms and growth habits (Zhou et al., 2022; Mishra and Tripathi, 2008). Future improvement demands optimization of plant combinations specific to the type of contaminant, further insight of interspecific interaction and competition of nutrients and integration of the native species to increase the compatibility and sustainability (Vymazal, 2011; Rai, 2008).

3. Integration with Emerging Green Technologies

Phytoremediation is also becoming hybrid biotechnological processes, with green and smart technologies being integrated, in a variety of forms, to enhance efficiency and applicability. a) Nanotechnology: Engineered nanomaterials (EM) (i.e. Fe_3O_4 [magnetite], nano zero-valent iron [nZVI] and biochar nanoparticle have been found have potential to increase metal solubility, bioavailability and plant uptake by altering the rhizosphere and facilitating redox reactions (Ali et al., 2018, Wang & Keller, 2008). It is also possible to control nutrients in the field, with the help of nano-sensors and nano-fertilizers (Kah et al., 2018). b) Microbial Inoculants: Plant growth-promoting rhizobacteria (PGPRs), mycorrhizae, endophytic bacteria, improve the heavy metal tolerance, nutrient acquisition, and biomass of plants, that also provide metal solubilization and detoxification (Khan et al., 2020; Glick, 2010). c) Biochar and Substrate Engineering: Synthesized biochar-based substrates (modified heave surfaces) and engineered biochars have the potential to enhance the immobilization capacity of metals and increase microbial colonization and also optimize the water retention and hydraulic scrunching capacity in consorted wetlands or phytoreactors (Beesley et al., 2011; Zhang et al., 2020). d) Digital Technologies: IoT, sensor networks, and remote-sensing technologies are suitable to measure the water quality, metal loads, plant growth, and wellbeing of the ecosystem in real-time to manage it as adaptive management (Ali et al., 2021; Pantelopoulos and Bourbakis, 2010). A possible future work is the introduction of constructed wetland systems using automated and AI-assisted operation, sensor-enabled systems, and bioelectrochemical constructed wetlands that combine the use of microbial fuel cells (MFCs) and phytoremediation to provide energy and degrade the pollutants at the same time (Yadav et al., 2012; Saba et al., 2021).

9. POLICY, GOVERNANCE AND MAINSTREAMING IN ENVIRONMENTAL MANAGEMENT

To achieve optimal application of phytoremediation, reinforcement by policies and incorporation in regulatory systems is necessary. The priority policy domains: a) Standardisation and Guidelines: Developing design, monitoring and harvesting standards for hydrophyte-based systems and standards involved in safe biomass disposal /reuse (Pivetz, 2001; Interstate Technology & Regulatory Council, 2009). b) Incentive Programs: Supporting the industries, municipalities and farmers shifting to the constructed wetlands with monetary or technical assistance (Vymazal, 2011). c) Invasive Species Regulation: Exclusion/exclusion and exclusion of hydrophyte species not native to an area to keep ecological disturbance at bay (Zedler and Kercher, 2004). d) Recognition of phytoremediation as an acceptable best management practice (BMP) as per law and water quality control and environmental impact assessments (EIs) (UNEP, 2018). e) Education and Community involvement: Community education to raise awareness about the cost-effectiveness of phytoremediation, the environmental benefits of phytoremediation and their potential in terms of a circular economy (Mitsch and Gosselink, 2015). Take the example of the EU water framework directive that promotes nature-based solutions such as wetlands to clean up water. The National Wetland Conservation Programme in India listed the use of the wetlands as wastewater treatment and restoring the habitat (MoEFCC, 2017).

Science, technology and policy- the convergence holds the future of hydrophyte-based remediation. New technologies like genetically enhanced aquatic flora, multi-tasking ecosystem and technology-integrated wetlands have a huge potential in overcoming the existing constraints (Salt et al., 1998; Aken et al., 2009). Yet, this advancement should be assisted by legal structures, the involvement of stakeholders and interdisciplinary work. The adoption of these trends will see phytoremediation take centre stage as far as the attainment of clean water, non-pollutive ecosystems, as well as sustainable growth and development are concerned.

CONCLUSION

In conclusion, hydrophyte-based phytoremediation stands at the intersection of ecological wisdom and technological innovation. With continued research, policy support and ecosystem-centred design, these humble aquatic plants have the potential to become green champions of environmental detoxification, fostering a healthier and more sustainable planet. The global water pollution, which is caused by heavy metals, can be solved by the use of a highly adaptable nature-based approach that uses hydrophyte-based remediation systems in the wetlands. Such methods, in the form of engineered constructed wetlands or simply with the help of existing natural systems, also align with the idea of sustainable development goals and ecological restoration principles. CWs can potentially form the backbone of eco-technological wastewater treatment in the urban and rural areas, as more efforts are still being made to innovate the design of the systems, integrate plant, microbe and material technologies.

Due to the increasing world burden of water contamination with heavy metals, there is a great desire to take hydrophyte-based remediation to the next level through scientific innovation, technology, and policy. Although traditional phytoremediation techniques have already demonstrated some success, the development of new strategies seeks to increase efficiency, flexibility and scalability via genetics, ecological engineering, synergistic technologies, as well as support by the institutions. Here, the authors will find out the fundamental research directions that will mark hydrophyte-based phytoremediation in the future. Top of Form

REFERENCES

Aken, B. V., Correa, P. A., & Schnoor, J. L. (2009). Phytoremediation of polychlorinated biphenyls: New trends and promises. *Environmental Science & Technology*, *43*(20), 7166–7176. PMID: 20384372

Akratos, C. S., & Tsihrintzis, V. A. (2007). Effect of temperature, HRT, vegetation and porous media on removal efficiency of pilot-scale horizontal subsurface flow constructed wetlands. *Ecological Engineering*, *29*(2), 173–191. DOI: 10.1016/j.ecoleng.2006.06.013

Alharthi, F., Althagafi, H. A., Jafri, I., Oyouni, A. A. A., Althaqafi, M. M., Al-Hijab, L. Y. A., Al-Hazmi, N. E., Elagib, S. M., & Naguib, D. M. (2024). Phytochemical composition and bioactivities of aquatic plants: Antioxidant, antiparasitic, antibacterial, and anticancer properties and mechanisms. *Plants*, *13*(15), 2148. DOI: 10.3390/plants13152148 PMID: 39124266

Ali, H., Khan, E., & Ilahi, I. (2019). Environmental chemistry and ecotoxicology of hazardous heavy metals: Environmental persistence, toxicity, and bioaccumulation. *Journal of Chemistry*, *2019*, 1–14. DOI: 10.1155/2019/6730305

Ali, H., Khan, E., & Ilahi, I. (2021). Green technologies for sustainable environmental remediation. *Environmental Chemistry Letters*, *19*, 1403–1421.

Ali, H., Khan, E., & Sajad, M. A. (2013). Phytoremediation of heavy metals—Concepts and applications. *Chemosphere*, *91*(7), 869–881. DOI: 10.1016/j.chemosphere.2013.01.075 PMID: 23466085

Ali, H., Khan, E., & Sajad, M. A. (2018). Nanotechnology and plant sciences: Nanoparticles and their impact on plant growth. In *Nanotechnology and plant sciences* (pp. 19–35). Springer., DOI: 10.1007/978-3-319-99651-4_2

Alloway, B. J. (2013). *Heavy metals in soils: Trace metals and metalloids in soils and their bioavailability*. Springer. DOI: 10.1007/978-94-007-4470-7

Arguello, J. M., Raimunda, D., & Gonzalez-Guerrero, M. (2012). Metal transport across biomembranes: Emerging models for a distinct chemistry. *The Journal of Biological Chemistry*, *287*(17), 13510–13517. DOI: 10.1074/jbc.R111.319343 PMID: 22389499

Bakrim, W. B., Ezzariai, A., Karouach, F., Sobeh, M., Kibret, M., Hafidi, M., Kouisni, L., & Yasri, A. (2022). *Eichhornia crassipes* (Mart.) Solms: A comprehensive review of its chemical composition, traditional use, and value-added products. *Frontiers in Pharmacology*, *13*, 842511. DOI: 10.3389/fphar.2022.842511 PMID: 35370709

Banerjee, S., Chatterjee, S., & Gupta, S. K. (2022). IoT-based smart constructed wetlands: A review on remote sensing and automation in wastewater treatment. *Journal of Environmental Management*, *317*, 115423.

Barakat, M. A. (2011). New trends in removing heavy metals from industrial wastewater. *Arabian Journal of Chemistry*, *4*(4), 361–377. DOI: 10.1016/j.arabjc.2010.07.019

Batty, L. C., & Younger, P. L. (2004). Seasonal variability of iron and manganese in surface water receiving mine drainage from abandoned coal mines. *Environmental Geochemistry and Health*, *26*(1), 33–42.

Bayuo, J., Karthik, C., & Bharagava, R. N. (2024). Recent and sustainable advances in phytoremediation of heavy metals from wastewater using aquatic plant species. *The Science of the Total Environment*, *925*, 171657. DOI: 10.1016/j.scitotenv.2024.171657 PMID: 39305882

Beesley, L., Moreno-Jiménez, E., & Gomez-Eyles, J. L. (2011). Effects of biochar and greenwaste compost amendments on mobility, bioavailability, and toxicity of inorganic and organic contaminants in a multi-element polluted soil. *Environmental Pollution*, *158*(6), 2282–2287. DOI: 10.1016/j.envpol.2010.02.003 PMID: 20219274

Bolan, N. S., Adriano, D. C., & Naidu, R. (2011). Role of phosphorus in (im)mobilization and bioavailability of heavy metals in the soil-plant system. *Reviews of Environmental Contamination and Toxicology*, *177*, 1–44. PMID: 12666817

Bragato, C., Brix, H., & Malagoli, M. (2006). Accumulation of nutrients and heavy metals in *Phragmites australis* and *Bolboschoenus maritimus* in a constructed wetland of the Venice Lagoon watershed. *Ecological Engineering*, *26*(3), 268–285. PMID: 16574288

Brix, H. (1997). Do macrophytes play a role in constructed treatment wetlands? *Water Science and Technology*, *35*(5), 11–17. DOI: 10.2166/wst.1997.0154

Brix, H., Arias, C. A., & Del Bubba, M. (2001). Media selection for sustainable phosphorus removal in subsurface flow constructed wetlands. *Water Science and Technology*, *44*(11–12), 47–54. DOI: 10.2166/wst.2001.0808 PMID: 11804137

Brix, H., Arias, C. A., & Johansen, N. H. (2007). Treatment of agricultural drainage water in a two-stage constructed wetland system. *Water Science and Technology*, *56*(3), 79–85. DOI: 10.2166/wst.2007.512 PMID: 17410843

Brix, H., & Schierup, H. H. (1989). The use of aquatic macrophytes in water pollution control. *Ambio*, *18*(2), 100–107.

Calheiros, C. S. C., Rangel, A. O. S. S., & Castro, P. M. L. (2007). Constructed wetland systems vegetated with different plants applied to the treatment of tannery wastewater. *Water Research, 41*(8), 1790–1798. DOI: 10.1016/j.watres.2007.01.012 PMID: 17320926

Calheiros, C. S. C., Rangel, A. O. S. S., & Castro, P. M. L. (2009). Treatment of industrial wastewater with two-stage constructed wetlands planted with *Typha latifolia* and *Phragmites australis*. *Bioresource Technology, 100*(13), 3205–3213. DOI: 10.1016/j.biortech.2009.02.017 PMID: 19289277

Carpenter, S. R., & Lodge, D. M. (1986). Effects of submerged macrophytes on ecosystem processes. *Aquatic Botany, 26*, 341–370. DOI: 10.1016/0304-3770(86)90031-8

Chambers, P. A., Lacoul, P., Murphy, K. J., & Thomaz, S. M. (2008). Global diversity of aquatic macrophytes in freshwater. *Hydrobiologia, 595*(1), 9–26. DOI: 10.1007/s10750-007-9154-6

Chandanshive, N. E., Kadam, S. K., Patil, S. M., & Jadhav, J. P. (2017). Integrated phytoremediation of heavy metals and decolorization of textile effluent using *Typha latifolia* in constructed wetland. *Ecotoxicology and Environmental Safety, 139*, 47–53.

Chaudhary, P., Shukla, P., & Sharma, A. K. (2020). Nanotechnology in wastewater management: A new avenue for environmental cleanup. *Environmental Technology & Innovation, 19*, 100830.

Chen, Y., Zhou, Q., Chen, Z., & Xu, J. (2021). Transcriptional responses and detoxification mechanisms of *Nelumbo nucifera* under arsenic stress. *Environmental and Experimental Botany, 186*, 104450.

Cheng, J. J., & Stomp, A. M. (2009). Growing duckweed to recover nutrients from wastewaters and for production of fuel ethanol and animal feed. *Clean (Weinheim), 37*(1), 17–26. DOI: 10.1002/clen.200800210

Cheng, S., Grosse, W., Karrenbrock, F., & Thoennessen, M. (2002). Efficiency of constructed wetlands in decontamination of water polluted by heavy metals. *Ecological Engineering, 18*(3), 317–325. DOI: 10.1016/S0925-8574(01)00091-X

Cherian, S., & Oliveira, M. M. (2005). Transgenic plants in phytoremediation: Recent advances and new possibilities. *Environmental Science & Technology, 39*(24), 9377–9390. DOI: 10.1021/es0511341 PMID: 16475312

Cilliers, C. J., Zeller, D. A., & Strydom, G. (2003). Short-term response of aquatic macrophyte communities following the removal of *Eichhornia crassipes* using herbicide and biological control. *African Journal of Aquatic Science, 28*(1), 73–76.

Clemens, S. (2006). Toxic metal accumulation, responses to exposure and mechanisms of tolerance in plants. *Biochimie, 88*(11), 1707–1719. DOI: 10.1016/j.biochi.2006.07.003 PMID: 16914250

Cobbett, C., & Goldsbrough, P. (2002). Phytochelatins and metallothioneins: Roles in heavy metal detoxification and homeostasis. *Annual Review of Plant Biology, 53*(1), 159–182. DOI: 10.1146/annurev.arplant.53.100301.135154 PMID: 12221971

Dalahmeh, S. S., Lalander, C. H., Pell, M., Vinnerås, B., & Jönsson, H. (2014). Floating treatment wetlands for domestic wastewater treatment: Effects of different plants and aeration strategies. *Water Science and Technology, 69*(4), 754–762.

Dhir, B. (2013). *Phytoremediation: Role of aquatic plants in environmental clean-up*. Springer. DOI: 10.1007/978-81-322-1307-9

Dos Santos, C. M., Vromman, V., & Bastien, C. (2016). Life cycle assessment of constructed wetlands for wastewater treatment: A review. *Environmental Progress & Sustainable Energy, 35*(6), 1680–1692.

Doyle, R. D. (2001). Effects of *Vallisneria americana* (L.) on community structure and ecosystem function in a shallow aquatic system. *Hydrobiologia, 448*(1), 139–146.

Dushenkov, V., Kumar, P. B. A. N., Motto, H., & Raskin, I. (1995). Rhizofiltration: The use of plants to remove heavy metals from aqueous streams. *Environmental Science & Technology, 29*(5), 1239–1245. DOI: 10.1021/es00005a015 PMID: 22192017

El Awady, F. R., Abbas, M. A., Abdelghany, A. M., El-Amier, Y. A., Darwish, H., & Fakhry, H. (2021). Silver modified hydrophytes for heavy metal removal from different water resources. *Biointerface Research in Applied Chemistry, 11*(2), 14555–14563. DOI: 10.33263/BRIAC116.1455514563

El-Liethy, M. A., Dakhil, M. A., El-Keblawy, A., Abdelaal, M., Halmy, M. W. A., Elgarhy, A. H., Kamika, I., El-Sherbeny, G. A., & Mwaheb, M. A. (2022). Temporal phytoremediation potential for heavy metals and bacterial abundance in drainage water. *Scientific Reports, 12*(1), 8223. DOI: 10.1038/s41598-022-11951-w PMID: 35581245

El-Sappah, A. H., Zhu, Y., Huang, Q., Chen, B., Soaud, S. A., Abd Elhamid, M. A., Yan, K., Li, J., & El-Tarabily, K. A. (2024). Plants' molecular behavior to heavy metals: From criticality to remediation. *Frontiers in Plant Science, 15*, 1423625. DOI: 10.3389/fpls.2024.1423625 PMID: 39280950

Erwin, K. L. (2009). Wetlands and global climate change: The role of wetland restoration in a changing world. *Wetlands Ecology and Management, 17*(1), 71–84. DOI: 10.1007/s11273-008-9119-1

European Commission. (2000). *Directive 2000/60/EC of the European Parliament and of the Council establishing a framework for the Community action in the field of water policy.*

Fisher, J., & Acreman, M. C. (2004). Wetland nutrient removal: A review of the evidence. *Hydrology and Earth System Sciences, 8*(4), 673–685. DOI: 10.5194/hess-8-673-2004

Fonder, N., & Headley, T. (2013). Systematic classification, structure, and size relationships in constructed wetlands for pollutant removal. *Ecological Engineering, 53*, 144–158.

Foyer, C. H., & Noctor, G. (2011). Ascorbate and glutathione: The heart of the redox hub. *Plant Physiology, 155*(1), 2–18. DOI: 10.1104/pp.110.167569 PMID: 21205630

Fu, F., & Wang, Q. (2011). Removal of heavy metal ions from wastewaters: A review. *Journal of Environmental Management, 92*(3), 407–418. DOI: 10.1016/j.jenvman.2010.11.011 PMID: 21138785

Gadd, G. M. (2000). Bioremedial potential of microbial mechanisms of metal mobilization and immobilization. *Current Opinion in Biotechnology, 11*(3), 271–279. DOI: 10.1016/S0958-1669(00)00095-1 PMID: 10851150

Gadd, G. M. (2004). Microbial influence on metal mobility and application for bioremediation. *Geoderma, 122*(2–4), 109–119. DOI: 10.1016/j.geoderma.2004.01.002

Gasic, K., & Korban, S. S. (2007). Expression of *Arabidopsis* phytochelatin synthase in Indian mustard (*Brassica juncea*) plants enhances tolerance for and accumulation of cadmium and arsenic. *Plant Molecular Biology, 65*, 519–529. DOI: 10.1007/s11103-007-9213-2

Gershenzon, J., & Dudareva, N. (2007). The function of terpene natural products in the natural world. *Nature Chemical Biology, 3*(7), 408–414. DOI: 10.1038/nchembio.2007.5 PMID: 17576428

Ghosh, M., & Singh, S. P. (2005). A review on phytoremediation of heavy metals and utilization of its byproducts. *Applied Ecology and Environmental Research, 3*(1), 1–18. DOI: 10.15666/aeer/0301_001018

Gilmour, C. C., Henry, E. A., & Mitchell, R. (1992). Sulfate stimulation of mercury methylation in freshwater sediments. *Environmental Science & Technology, 26*(11), 2281–2287. DOI: 10.1021/es00035a029

Glick, B. R. (2010). Using soil bacteria to facilitate phytoremediation. *Biotechnology Advances, 28*(3), 367–374. DOI: 10.1016/j.biotechadv.2010.02.001 PMID: 20149857

Gopal, B. (1987). *Water hyacinth*. Elsevier.

Grill, E., Winnacker, E. L., & Zenk, M. H. (1985). Phytochelatins: The principal heavy-metal complexing peptides of higher plants. *Science*, *230*(4726), 674–676. DOI: 10.1126/science.230.4726.674 PMID: 17797291

Guerinot, M. L. (2000). The ZIP family of metal transporters. *Biochimica et Biophysica Acta*, *1465*(1–2), 190–198. DOI: 10.1016/S0005-2736(00)00138-3 PMID: 10748254

Gustin, J. L., Loureiro, M. E., Kim, D., Na, G., Tikhonova, M., & Salt, D. E. (2009). MTP1-dependent Zn sequestration into shoot vacuoles suggests dual roles in Zn tolerance and accumulation in *Arabidopsis thaliana*. *The Plant Journal*, *57*(6), 1116–1127. DOI: 10.1111/j.1365-313X.2008.03754.x PMID: 19054361

Haferburg, G., & Kothe, E. (2007). Microbes and metals: Interactions in the environment. *Journal of Basic Microbiology*, *47*(6), 453–467. DOI: 10.1002/jobm.200700275 PMID: 18072246

Hall, J. L. (2002). Cellular mechanisms for heavy metal detoxification and tolerance. *Journal of Experimental Botany*, *53*(366), 1–11. DOI: 10.1093/jexbot/53.366.1 PMID: 11741035

Hasan, S. A., Fariduddin, Q., Ali, B., Hayat, S., & Ahmad, A. (2009). Cadmium: Toxicity and tolerance in plants. *Journal of Environmental Biology*, *30*(2), 165–174. PMID: 20121012

Haslam, E. (1996). Natural polyphenols (vegetable tannins) as drugs: Possible modes of action. *Journal of Natural Products*, *59*(2), 205–215. DOI: 10.1021/np960040+ PMID: 8991956

Hassinen, V. H., Tervahauta, A. I., Schat, H., & Kärenlampi, S. O. (2009). Plant metallothioneins – Metal chelators with ROS scavenging activity? *Plant Biology*, *11*(4), 418–426. DOI: 10.1111/j.1438-8677.2008.00158.x

Hu, H., Li, X., Wu, S., & Yang, C. (2020). Sustainable livestock wastewater treatment via phytoremediation: Current status and future perspectives. *Bioresource Technology*, *315*, 123809. DOI: 10.1016/j.biortech.2020.123809 PMID: 32682262

Hu, H., Zhou, Q., Li, X., Lou, W., Du, C., Teng, Q., Zhang, D. M., Liu, H. Y., Zhong, Y. Y., & Yang, C. P. (2019). Phytoremediation of anaerobically digested swine wastewater contaminated by oxytetracycline via *Lemna aequinoctialis*: Nutrient removal, growth characteristics and degradation pathways. *Bioresource Technology*, *291*, 121853. DOI: 10.1016/j.biortech.2019.121853 PMID: 31377510

Hussain, D., Haydon, M. J., Wang, Y., Wong, E., Sherson, S. M., Young, J., Camakaris, J., Harper, J. F., & Cobbett, C. S. (2004). P-type ATPase heavy metal transporters with roles in essential zinc homeostasis in *Arabidopsis*. *The Plant Cell, 16*(5), 1327–1339. DOI: 10.1105/tpc.020487 PMID: 15100400

Iasur-Kruh, L., Hadar, Y., & Minz, D. (2010). Seasonal dynamics of bacterial community composition in sediments of a constructed wetland. *Microbial Ecology, 60*(3), 492–503. DOI: 10.1007/s00248-010-9721-0

Interstate Technology & Regulatory Council. (2008). *Phytotechnologies technical and regulatory guidance and decision trees* (Revised). Phytotechnologies Team.

Interstate Technology & Regulatory Council. (2009). *Phytotechnologies: A technical and regulatory guidance document.*

Jaishankar, M., Tseten, T., Anbalagan, N., Mathew, B. B., & Beeregowda, K. N. (2014). Toxicity, mechanism and health effects of some heavy metals. *Interdisciplinary Toxicology, 7*(2), 60–72. DOI: 10.2478/intox-2014-0009 PMID: 26109881

Jeppesen, E.. (2012). Impacts of climate warming on lake fish community structure and potential effects on ecosystem function. *Hydrobiologia, 694*(1), 1–39. DOI: 10.1007/s10750-012-1182-1

Jonak, C., Nakagami, H., & Hirt, H. (2004). Heavy metal stress: Activation of distinct mitogen-activated protein kinase pathways by copper and cadmium. *Plant Physiology, 136*(2), 3276–3283. DOI: 10.1104/pp.104.045724 PMID: 15448198

Julien, M. H., Hill, M. P., Center, T. D., & Ding, J. (2001). *Biological and integrated control of water hyacinth (Eichhornia crassipes).* ACIAR Proceedings, 102.

Kadlec, R. H., & Wallace, S. D. (2009). *Treatment wetlands* (2nd ed.). CRC Press.

Kah, M., Kookana, R. S., Gogos, A., & Bucheli, T. D. (2018). A critical evaluation of nanopesticides and nanofertilizers against their conventional analogues. *Nature Nanotechnology, 13*(8), 677–684. DOI: 10.1038/s41565-018-0131-1 PMID: 29736032

Kaur, R., Mahajan, A., Kaur, R., & Sinha, D. (2012). Phytoremediation potential of water lettuce (*Pistia stratiotes*) for the removal of Pb and Cd. *African Journal of Environmental Science and Technology, 6*(6), 207–213.

Khan, M. I., Ahmad, A., Iqbal, M., & Hussain, M. (2020). Synergistic effects of microbes and nanoparticles in phytoremediation: Recent advances and challenges. *Environmental Research, 188*, 109765. DOI: 10.1016/j.envres.2020.109765

Khan, M. I. R., Asgher, M., & Khan, N. A. (2020). Phytoremediation of heavy metals and role of plant growth promoting microbes in the rhizosphere. In *Plant-microbe interactions in agro-ecological perspectives* (pp. 309–337). Springer., DOI: 10.1007/978-981-15-2696-6_13

Khan, M. I. R., Nazir, F., Asgher, M., Per, T. S., & Khan, N. A. (2009). Selenium and sulfur influence ethylene formation and alleviate cadmium-induced oxidative stress by improving proline and glutathione production in wheat. *Journal of Plant Physiology, 170*(2), 167–174. DOI: 10.1016/j.jplph.2012.08.018 PMID: 25462073

Khan, T., Shah, L., Khan, S., Wani, O. A., Sheikh, Z. N., Afroza, B., Rashid, R., & Baloch, F. S. (2025). Comprehensive review of multiomics applications and remediation of plant heavy metal toxicity. *Frontiers in Plant Science, 16*, 12440847. DOI: 10.3389/fpls.2025.12440847 PMID: 40956384

Krzesłowska, M. (2011). The cell wall in plant cell response to trace metals: Polysaccharide remodeling and its role in defense strategy. *Acta Physiologiae Plantarum, 33*(1), 35–51. DOI: 10.1007/s11738-010-0581-z

Kumar, K. S., Dahms, H. U., Lee, J. S., Kim, H. C., Lee, W. C., & Shin, K. H. (2014). Algal photosynthetic responses to toxic metals and herbicides assessed by chlorophyll *a* fluorescence. *Ecotoxicology and Environmental Safety, 104*, 51–71. DOI: 10.1016/j.ecoenv.2014.01.042 PMID: 24632123

Kumar, M., Ghosh, M., & Pandey, R. (2022). Microbial synthesized silver nanoparticles and their impact on phytoremediation of chromium by *Hydrilla verticillata. Environmental Science and Pollution Research International, 28*, 41635–41646. DOI: 10.1007/s11356-021-14292-5

Leszczyszyn, O. I., Imam, H. T., & Blindauer, C. A. (2007). Diversity and distribution of plant metallothioneins: A review of structure, properties and functions. *Metallomics, 1*(5), 416–430. PMID: 23694960

Li, H., Zhang, Q., Li, Y., & Shen, Z. (2021). Arbuscular mycorrhizal fungi enhanced phytoremediation potential of *Typha angustifolia* under metal stress. *The Science of the Total Environment, 764*, 142870. DOI: 10.1016/j.scitotenv.2020.142870

Li, X., Yang, C. P., Zeng, G. M., Wu, S. H., Lin, Y., Zhou, Q., Lou, W., Du, C., Nie, L. J., & Zhong, Y. Y. (2020). Nutrient removal from swine wastewater with growing microalgae at various zinc concentrations. *Algal Research, 46*, 101804. DOI: 10.1016/j.algal.2020.101804

Li, X., Yang, W. L., He, H. J., Wu, S. H., Zhou, Q., Yang, C. P., Zeng, G. M., Luo, L., & Lou, W. (2018). Responses of microalgae *Coelastrella* sp. to stress of cupric ions in treatment of anaerobically digested swine wastewater. *Bioresource Technology, 251*, 274–279. DOI: 10.1016/j.biortech.2017.12.058 PMID: 29288955

Li, Y., Shao, Y., Xu, J., Wei, Y., & Yu, Z. (2021). Biochar–microbe–plant interactions in heavy metal phytoremediation: A review. *Ecotoxicology and Environmental Safety, 208*, 111699.

Li, Y., Wang, C., & Zhang, W. (2011). Biofilms and their role in the removal of pollutants in constructed wetlands. *Environmental Science & Technology, 45*(3), 1158–1164. PMID: 21166456

Liu, D., Wang, W., & Liu, Y. (2014). Enhanced mercury and cadmium tolerance and accumulation in transgenic *Lemna minor* by overexpression of a glutathione synthetase gene. *Plant Cell, Tissue and Organ Culture, 119*(1), 127–137. DOI: 10.1007/s11240-014-0514-7

Liu, J., Zhu, Q., Zhang, Z., Xu, J., & Yang, J. (2014). Enhancement of phytoremediation potential of *Lemna minor* by genetic modification with glutathione synthetase gene. *Journal of Environmental Sciences (China), 26*(5), 962–970. DOI: 10.1016/S1001-0742(13)60488-8 PMID: 25079626

Lowe, S., Browne, M., Boudjelas, S., & De Poorter, M. (2000). *100 of the world's worst invasive alien species*. IUCN/SSC Invasive Species Specialist Group.

Lu, Q., He, Z. L., & Stoffella, P. J. (2017). Land application of biosolids in the USA: A review. *Applied and Environmental Soil Science, 2012*, 1–11. DOI: 10.1155/2012/201462

Lu, X., Kruatrachue, M., Pokethitiyook, P., & Homyok, K. (2010). Removal of cadmium and zinc by water hyacinth, *Eichhornia crassipes*. *ScienceAsia, 36*(3), 205–209. DOI: 10.2306/scienceasia1513-1874.2010.36.210

Luo, L., He, H. J., Yang, C. P., Wen, S., Zeng, G. M., Wu, M. J., Zhou, Z. L., & Lou, W. (2016). Nutrient removal and lipid production by *Coelastrella* sp. in anaerobically and aerobically treated swine wastewater. *Bioresource Technology, 216*, 135–141. DOI: 10.1016/j.biortech.2016.05.059 PMID: 27236400

Luongo, T., & Cutright, T. J. (2011). Characterization of plant root exudates and their impact on microbial communities. *International Journal of Phytoremediation, 13*(3), 211–224.

Lv, J. P., Liu, Y., Feng, J., Liu, Q., Nan, F. R., & Xie, S. L. (2018). Nutrient removal from undiluted cattle farm wastewater by the two-stage process of microalgae-based wastewater treatment. *Bioresource Technology, 264*, 311–318. DOI: 10.1016/j.biortech.2018.05.085 PMID: 29857286

Ma, T., Zhang, W., Fan, H., Huang, L., Xu, Q., Ma, G., Zhou, G., Meng, Y., Sun, J., Qiu, J., Gu, J., & Kang, X. (2020). Study on phytoremediation for heavy-metal-contaminated sediments by hydrophytes. *E3S Web of Conferences, 143*, 02020. https://doi.org/DOI: 10.1051/e3sconf/202014302020

Ma, Y., Rajkumar, M., & Freitas, H. (2011). Inoculation of plant growth promoting bacterium *Achromobacter xylosoxidans* strain Ax10 for the improvement of phytoremediation of metal-contaminated soils. *Chemosphere, 82*(10), 1383–1390. PMID: 21474162

Ma, Y., Rajkumar, M., Zhang, C., & Freitas, H. (2016). Beneficial role of bacterial endophytes in heavy metal phytoremediation. *Journal of Environmental Management, 174*, 14–25. DOI: 10.1016/j.jenvman.2016.02.047 PMID: 26989941

Macek, T., Macková, M., & Kas, J. (2008). Exploitation of plants for the removal of organics in environmental remediation. *Biotechnology Advances, 26*(5), 436–448. DOI: 10.1016/j.biotechadv.2008.05.007 PMID: 18565714

Madsen, J. D., Chambers, P. A., James, W. F., Koch, E. W., & Westlake, D. F. (2001). The interaction between water movement, sediment dynamics and submersed macrophytes. *Hydrobiologia, 444*(1–3), 71–84. DOI: 10.1023/A:1017520800568

Malik, A. (2007). Environmental challenge vis-à-vis opportunity: The case of water hyacinth. *Environment International, 33*(1), 122–138. DOI: 10.1016/j.envint.2006.08.004 PMID: 17010439

Mao, J., Zhang, X., Chen, Y., Liu, D., & Wang, L. (2024). Industrial wastewater treatment using floating wetlands: Performance and influencing factors. *Journal of Environmental Management, 352*, 120875.

Mao, L., Long, L., Qin, Q., Huang, Q., Yu, Y., Ding, L., Liu, G., Zhang, J., Cui, B., & Tan, X. (2025). *Nymphaea "Eldorado"* flower extract targets SERPINE1 to attenuate inflammatory and antioxidant crosstalk in zebrafish. *Frontiers in Pharmacology, 16*, 1612233. DOI: 10.3389/fphar.2025.1612233 PMID: 40717968

Marchand, L., Mench, M., Jacob, D. L., & Otte, M. L. (2010). Metal and metalloid removal in constructed wetlands, with emphasis on the importance of plants and standardized measurements: A review. *Environmental Pollution, 158*(12), 3447–3461. DOI: 10.1016/j.envpol.2010.08.018 PMID: 20869144

Markou, G., Wang, L., Ye, J. F., & Unc, A. (2018). Using agro-industrial wastes for the cultivation of microalgae and duckweeds: Contamination risks and biomass safety concerns. *Biotechnology Advances*, *36*(4), 1238–1254. DOI: 10.1016/j.biotechadv.2018.04.003 PMID: 29673973

Martinoia, E., Maeshima, M., & Neuhaus, H. E. (2007). Vacuolar transporters and their essential role in plant metabolism. *Journal of Experimental Botany*, *58*(1), 83–102. DOI: 10.1093/jxb/erl183 PMID: 17110589

Masamba, W. R. L., & Mazvimavi, D. (2008). The use of natural wetlands for wastewater treatment in developing countries: A case study of the Okavango Delta. *Physics and Chemistry of the Earth Parts A/B/C*, *33*(8–13), 684–688.

Mays, P. A., & Edwards, G. S. (2001). Comparison of heavy metal accumulation in a natural wetland and constructed wetlands receiving acid mine drainage. *Ecological Engineering*, *16*(4), 487–500. DOI: 10.1016/S0925-8574(00)00112-9

Meagher, R. B. (2000). Phytoremediation of toxic elemental and organic pollutants. *Current Opinion in Plant Biology*, *3*(2), 153–162. DOI: 10.1016/S1369-5266(99)00054-0 PMID: 10712958

Meena, R. K., Jatav, G. K., & Meena, H. (2020). Emerging role of nanotechnology in bioremediation: An innovative and eco-friendly approach. *Journal of Environmental Management*, *260*, 110023.

Michalak, A. (2006). Phenolic compounds and their antioxidant activity in plants growing under heavy metal stress. *Polish Journal of Environmental Studies*, *15*(4), 523–530.

Ministry of Environment. Forest and Climate Change (MoEFCC). (2017). *National plan for conservation of aquatic ecosystems (NPCA)*. Government of India.

Mishra, V. K., & Tripathi, B. D. (2008). Heavy metal pollution load in water and sediments of the Ganga River at Varanasi, India. *Environmental Monitoring and Assessment*, *142*(1–3), 395–402. DOI: 10.1007/s10661-007-9930-4

Mishra, V. K., Upadhyay, A. R., & Pandey, S. K. (2008). Heavy metal pollution induced due to coal mining effluent on surrounding aquatic ecosystem and its management through naturally occurring aquatic macrophytes. *Bioresource Technology*, *99*(5), 930–936. DOI: 10.1016/j.biortech.2007.03.010 PMID: 17475484

Mitsch, W. J., Bernal, B., & Hernandez, M. E. (2010). Ecosystem services of wetlands. *The International Journal of Biodiversity Science, Ecosystem Services & Management*, *6*(1–2), 1–4.

Mitsch, W. J., Bernal, B., & Hernandez, M. E. (2013). Ecosystem services of wetlands. *The International Journal of Biodiversity Science, Ecosystem Services & Management, 9*(1), 1–4. DOI: 10.1080/21513732.2015.1006250

Mitsch, W. J., & Gosselink, J. G. (2015). *Wetlands* (5th ed.). Wiley.

Mkandawire, M., & Dudel, E. G. (2005). Accumulation of arsenic in *Lemna gibba* L. (duckweed) in tailing waters of two abandoned uranium mining sites in Saxony, Germany. *The Science of the Total Environment, 336*(1–3), 81–89. DOI: 10.1016/j.scitotenv.2004.06.002 PMID: 15589251

Mohan, D., Sarswat, A., Ok, Y. S., & Pittman, C. U.Jr. (2014). Organic and inorganic contaminants removal from water with biochar, a renewable, low-cost and sustainable adsorbent – A critical review. *Bioresource Technology, 160*, 191–202. DOI: 10.1016/j.biortech.2014.01.120 PMID: 24636918

Molle, P., Liénard, A., Grasmick, A., Iwema, A., & Boucherie, H. (2008). A pilot-scale study of constructed wetlands for secondary and tertiary treatment of wastewater: Design and performance. *Ecological Engineering, 34*(2), 145–152.

Morel, F. M. M., Kraepiel, A. M. L., & Amyot, M. (1998). The chemical cycle and bioaccumulation of mercury. *Annual Review of Ecology and Systematics, 29*(1), 543–566. DOI: 10.1146/annurev.ecolsys.29.1.543

Nancharaiah, Y. V., Venkata Mohan, S., & Lens, P. N. L. (2016). Metals removal and recovery in bioreactors. *Environmental Science. Water Research & Technology, 2*(5), 800–817.

Ndimele, P. E., Kumolu-Johnson, C. A., & Anetekhai, M. A. (2011). The invasive aquatic macrophyte *Eichhornia crassipes* (Mart.) Solms: Problems and prospects. *Research Journal of Environmental Sciences, 5*(6), 509–520. DOI: 10.3923/rjes.2011.509.520

Nevo, Y., & Nelson, N. (2006). The NRAMP family of metal-ion transporters. *Biochimica et Biophysica Acta, 1763*(7), 609–620. DOI: 10.1016/j.bbamcr.2006.05.007 PMID: 16908340

Nriagu, J. O. (1996). A history of global metal pollution. *Science, 272*(5259), 223–224. DOI: 10.1126/science.272.5259.223

Ntagia, E.. (2023). Pilot and full-scale applications of floating treatment wetlands for urban and industrial effluents. *The Science of the Total Environment, 872*, 162053.

Pandey, A., Tripathi, N., & Tripathi, R. D. (2023). Engineered nano-bio interactions for enhanced phytoremediation: Current insights and future prospects. *The Science of the Total Environment*, *882*, 163600.

Pantelopoulos, A., & Bourbakis, N. G. (2010). A survey on wearable sensor-based systems for health monitoring and prognosis. *IEEE Transactions on Systems, Man, and Cybernetics. Part C, Applications and Reviews*, *40*(1), 1–12. DOI: 10.1109/TSMCC.2009.2032660

Patel, S. (2012). Threats, management and envisaged utilizations of aquatic weed *Eichhornia crassipes*: An overview. *Reviews in Environmental Science and Biotechnology*, *11*(3), 249–259. DOI: 10.1007/s11157-012-9289-4

Paudel, K. R., & Panth, N. (2015). Phytochemical profile and biological activity of *Nelumbo nucifera*. *Evidence-Based Complementary and Alternative Medicine : eCAM*, *2015*, 789124. Advance online publication. DOI: 10.1155/2015/789124 PMID: 27057194

Peleg, Z., & Blumwald, E. (2011). Hormone balance and abiotic stress tolerance in crop plants. *Current Opinion in Plant Biology*, *14*(3), 290–295. DOI: 10.1016/j.pbi.2011.02.001 PMID: 21377404

Pietta, P. G. (2000). Flavonoids as antioxidants. *Journal of Natural Products*, *63*(7), 1035–1042. DOI: 10.1021/np9904509 PMID: 10924197

Pilon-Smits, E. (2005). Phytoremediation. *Annual Review of Plant Biology*, *56*(1), 15–39. DOI: 10.1146/annurev.arplant.56.032604.144214 PMID: 15862088

Pivetz, B. E. (2001). *Phytoremediation of contaminated soil and groundwater at hazardous waste sites (EPA Ground Water Issue EPA/540/S-01/500)*. U.S. Environmental Protection Agency.

Polechońska, L., Klink, A., & Sokołowska, K. (2024). Rising temperature impacts the trace metal uptake and toxicity in aquatic plants – A case study of Ni and Co in *Elodea canadensis* Michx. *The Science of the Total Environment*, *955*, 177307. DOI: 10.1016/j.scitotenv.2024.177307 PMID: 39481559

Pourrut, B., Shahid, M., Dumat, C., Winterton, P., & Pinelli, E. (2011). Lead uptake, toxicity, and detoxification in plants. *Reviews of Environmental Contamination and Toxicology*, *213*, 113–136. PMID: 21541849

Prabha, P. S., Sadhana, S., Karthik, C., & Caroline, D. G. (2020). *In vitro* evaluation of *Hydrilla verticillata* for anti-adipogenesis activity on 3T3 L1 cell lines. *Pharmacognosy Magazine*, *16*(5), 498–505. DOI: 10.4103/pm.pm_234_20

Rai, P. K. (2008). Heavy metal pollution in aquatic ecosystems and its phytoremediation using wetland plants: An ecosustainable approach. *International Journal of Phytoremediation, 10*(2), 133–160. DOI: 10.1080/15226510801913918 PMID: 18709926

Rai, P. K. (2009). Heavy metal phytoremediation from aquatic ecosystems with special reference to macrophytes. *Critical Reviews in Environmental Science and Technology, 39*(9), 697–753. DOI: 10.1080/10643380801910058

Rai, P. K., Lee, S. S., Zhang, M., Tsang, Y. F., & Kim, K. H. (2022). Heavy metals in aquatic ecosystems and their phytoremediation using wetland plants: An eco-environmental perspective. *Environmental Geochemistry and Health, 44*, 1–28.

Rajasulochana, P., & Preethy, V. (2016). Comparison on efficiency of various techniques in treatment of waste and sewage water – A comprehensive review. *Resource-Efficient Technologies, 2*(4), 175–184. DOI: 10.1016/j.reffit.2016.09.004

Rajkumar, M., Ma, Y., & Freitas, H. (2012). Improvement of Ni phytostabilization by inoculation of Ni-resistant *Bacillus megaterium* SR28C. *Journal of Environmental Management, 95*, S231–S236. DOI: 10.1016/j.jenvman.2011.06.034 PMID: 23895909

Rajput, J. D., & Singh, S. P. (2023). Water chestnut (*Trapa natans* L.): Functional characteristics, nutritional properties and applications in food industry—A review. *Journal of Phytopharmacology, 12*(2), 209. DOI: 10.31254/phyto.2023.12209

Rajput, V. D., Minkina, T., Sushkova, S., Behal, A., & Mandzhieva, S. (2018). Interaction of nanoparticles with plants: Impacts on growth and phytotoxicity. *Nanomaterials (Basel, Switzerland), 8*(9), 745.

Rastogi, A., Zivcak, M., Sytar, O., Kalaji, H. M., He, X., Mbarki, S., & Brestic, M. (2019). Nanoparticles and plants: Interaction at the organismal level. *Plant Physiology and Biochemistry, 134*, 1–15. DOI: 10.1016/j.plaphy.2018.10.014

Raza, A., Tabassum, J., Zahid, Z., Charagh, S., Bashir, S., Barmukh, R., Khan, R. S. A., Barbosa, F.Jr, Zhang, C., Chen, H., Zhuang, W., & Varshney, R. K. (2022). Advances in 'Omics' approaches for improving toxic metal tolerance in plants. *Frontiers in Plant Science, 13*, 8764127. DOI: 10.3389/fpls.2021.794373

Reddy, K. R., & D'Angelo, E. M. (1997). Biogeochemical indicators to evaluate pollutant removal efficiency in constructed wetlands. *Water Science and Technology, 35*(5), 1–10. DOI: 10.2166/wst.1997.0152

Reddy, K. R., & DeBusk, T. A. (1984). Growth characteristics of aquatic macrophytes cultured in nutrient-enriched water: II. *Pistia stratiotes* L. *Journal of Environmental Quality*, *13*(1), 61–65.

Reddy, K. R., & Delaune, R. D. (2008). *Biogeochemistry of wetlands: Science and applications*. CRC Press. DOI: 10.1201/9780203491454

Rehman, A. U., Khan, A. U., Sohaib, M., & Rehman, H. (2024). Comparative analysis of nutritional properties, phytochemical profile, and antioxidant activities between red and green water chestnut (*Trapa natans*) fruits. *Foods*, *13*(12), 1883. DOI: 10.3390/foods13121883 PMID: 38928824

Rezania, S., Park, J., & Rupani, P. F.. (2016a). Phytoremediation potential and cadmium uptake by water hyacinth from wastewater: A laboratory to field study. *Environmental Science and Pollution Research International*, *23*, 13401–13411.

Rezania, S., Ponraj, M., Din, M. F. M., Taib, S. M., Song, S. T., & Sairan, F. M. (2015). The diverse applications of water hyacinth with main focus on sustainable energy and production for new era: An overview. *Renewable & Sustainable Energy Reviews*, *41*, 943–954. DOI: 10.1016/j.rser.2014.09.006

Rezania, S., Taib, S. M., Din, M. F. M., Dahalan, F. A., & Kamyab, H. (2016b). Comprehensive review on phytotechnology: Heavy metals removal by diverse aquatic plant species from wastewater. *Journal of Hazardous Materials*, *318*, 587–599. DOI: 10.1016/j.jhazmat.2016.07.053 PMID: 27474848

Ricachenevsky, F. K., Menguer, P. K., Sperotto, R. A., & Fett, J. P. (2013). Got to hide your Zn away: Molecular control of Zn accumulation and biotechnological applications. *Plant Science*, *211*, 1–12. PMID: 26025516

Saba, B., Batool, S., Ahmad, M. I., & Kalsoom, U. (2021). Bioelectrochemical wetlands: An emerging hybrid technology for wastewater treatment and energy generation. *Chemosphere*, *263*, 127951. DOI: 10.1016/j.chemosphere.2020.127951

Saikia, K., Dey, S., Hazarika, S. N., Handique, G. K., Thakur, D., & Handique, A. K. (2023). Chemical and biochemical characterization of *Ipomoea aquatica*: Genoprotective potential and inhibitory mechanism of its phytochemicals against α-amylase and α-glucosidase. *Frontiers in Nutrition*, *10*, 1304903. DOI: 10.3389/fnut.2023.1304903 PMID: 38192648

Salt, D. E., Blaylock, M., Kumar, N. P. B. A., Dushenkov, V., Ensley, B. D., Chet, I., & Raskin, I. (1995). Phytoremediation: A novel strategy for the removal of toxic metals from the environment using plants. *Nature Biotechnology*, *13*(5), 468–474. DOI: 10.1038/nbt0595-468 PMID: 9634787

Salt, D. E., Smith, R. D., & Raskin, I. (1998). Phytoremediation. *Annual Review of Plant Physiology and Plant Molecular Biology, 49*(1), 643–668. DOI: 10.1146/annurev.arplant.49.1.643 PMID: 15012249

Sanders, D., Pelloux, J., Brownlee, C., & Harper, J. F. (2002). Calcium at the crossroads of signaling. *The Plant Cell, 14*(Suppl. 1), S401–S417. DOI: 10.1105/tpc.002899 PMID: 12045291

Sarma, H., Sahariah, G., Bharadwaj, A., Sharma, D., Porasar, P., & Dutta, K. N. (2024). A systematic review on botanical background, phytochemical and pharmacological properties of *Nymphaea nouchali. Journal of Applied Pharmaceutical Research, 12*(6), 21–36. DOI: 10.69857/joapr.v12i6.714

Schmidt, R., Hrynkiewicz, K., & Baum, C. (2008). Management of plant biomass in metal-polluted areas: Challenges and opportunities. *Environmental Pollution, 153*(3), 567–576. PMID: 18571819

Schooler, S. S., Julien, M. H., & Walsh, G. C. (2006). Predicting the response of *Pistia stratiotes* to biological control agents using a population model. *Biological Control, 38*(3), 391–400.

Sekomo, C. B., Rousseau, D. P. L., Saleh, S. A., & Lens, P. N. L. (2012). Heavy metal removal in duckweed-based stabilization ponds as a polishing step for UASB treated domestic wastewater. *Ecological Engineering, 44*, 102–110. DOI: 10.1016/j.ecoleng.2012.03.003

Sharma, R. K., Tripathi, R. D., & Srivastava, S. (2020). Synergistic effects of PGPR and nanoparticles on phytoremediation of heavy metals. *Environmental Nanotechnology, Monitoring & Management, 13*, 100275.

Sharma, S. S., & Dietz, K. J. (2009). The relationship between metal toxicity and cellular redox imbalance. *Trends in Plant Science, 14*(1), 43–50. DOI: 10.1016/j.tplants.2008.10.007 PMID: 19070530

Sheoran, A. S., & Sheoran, V. (2006). Heavy metal removal mechanism of acid mine drainage in wetlands: A critical review. *Minerals Engineering, 19*(2), 105–116. DOI: 10.1016/j.mineng.2005.08.006

Shilla, D. A., Asaeda, T., & Fujino, T. (2006). Effects of floating macrophytes on the nutrient dynamics in a shallow lake. *Limnology, 7*(1), 27–34.

Singh, A., Shahid, M., Hussain, S., & Tripathi, R. D. (2016). Signaling in plants under heavy metal stress. In *Plant Metal Interaction* (pp. 299–309). Elsevier.

Singh, H., & Chandra, R. (2014). Performance of *Pistia stratiotes* and *Eichhornia crassipes* for phytoremediation of pulp and paper mill wastewater in field conditions. *International Journal of Phytoremediation, 16*(12), 1181–1191.

Singh, H., Kumar, A., & Sharma, A. (2020). Secondary metabolites of plants in phytoremediation: Mechanisms and significance. *Plant Physiology Reports, 25*, 710–725. DOI: 10.1007/s40502-020-00512-3

Singh, H. P., Mahajan, P., Kaur, S., Batish, D. R., & Kohli, R. K. (2013). Chromium toxicity and tolerance in plants. *Environmental Chemistry Letters, 11*(3), 229–254. DOI: 10.1007/s10311-013-0407-5

Singh, J., Lee, B. K., & Kumar, V. (2022). Role of nanomaterials in phytoremediation: Recent advances and future outlook. *Chemosphere, 287*, 131996. DOI: 10.1016/j.chemosphere.2021.131996 PMID: 34455120

Singh, M., Muller, G., & Singh, I. B. (2005). Heavy metals in freshly deposited sediments of the Gomati River (a tributary of the Ganga River): Effects of human activities. *Environmental Geology, 48*(3), 410–420. DOI: 10.1007/s00254-005-1297-9

Singh, S., Parihar, P., Singh, R., Singh, V. P., & Prasad, S. M. (2015). Heavy metal tolerance in plants: Role of transcriptomics, proteomics, metabolomics, and ionomics. *Frontiers in Plant Science, 6*, 1143. DOI: 10.3389/fpls.2015.01143 PMID: 26904030

Sinha, S., & Paul, D. (2014). Enhanced phytoremediation of heavy metals by *Lemna minor* and associated microbes from industrial effluent. *International Journal of Phytoremediation, 16*(1), 1–12. DOI: 10.1080/15226514.2012.669801 PMID: 24912211

Sirianuntapiboon, S., Chularat, P., & Sansak, J. (2006). Removal of nitrogen and phosphorus from wastewater using vertical surface-flow constructed wetland planted with narrow leaf cattails. *ScienceAsia, 32*(3), 319–326. DOI: 10.2306/scienceasia1513-1874.2006.32.319

Sooknah, R. D., & Wilkie, A. C. (2004). Nutrient removal by floating aquatic macrophytes cultured in anaerobically digested flushed dairy manure wastewater. *Ecological Engineering, 22*(1), 27–42. DOI: 10.1016/j.ecoleng.2004.01.004

Sosa, D., Alves, F. M., Prieto, M. A., Pedrosa, M. C., Heleno, S. A., Barros, L., Feliciano, M., & Carocho, M. (2024). *Lemna minor*: Unlocking the value of this duckweed for the food and feed industry. *Foods, 13*(10), 1435. DOI: 10.3390/foods13101435 PMID: 38790736

Stottmeister, U., Wießner, A., Kuschk, P., Kappelmeyer, U., Kästner, M., Bederski, O., Müller, R. A., & Moormann, H. (2003). Effects of plants and microorganisms in constructed wetlands for wastewater treatment. *Biotechnology Advances*, *22*(1–2), 93–117. DOI: 10.1016/j.biotechadv.2003.08.010 PMID: 14623046

Stumm, W., & Morgan, J. J. (1996). *Aquatic chemistry: Chemical equilibria and rates in natural waters* (3rd ed.). Wiley-Interscience.

Tan, H. W.. (2023). Phytoremediation of heavy-metal-contaminated waters: Challenges including secondary pollution and biomass handling. *The Science of the Total Environment*, *872*, 162411. DOI: 10.1016/j.scitotenv.2023.162411

Tangahu, B. V., Abdullah, S. R. S., Basri, H., Idris, M., Anuar, N., & Mukhlisin, M. (2011). A review on heavy metals (As, Pb, and Hg) uptake by plants through phytoremediation. *International Journal of Chemical Engineering*, *2011*(1), 1–31. DOI: 10.1155/2011/939161

Tanner, C. C. (2001). Plants as ecosystem engineers in subsurface-flow treatment wetlands. *Water Science and Technology*, *44*(11–12), 9–17. DOI: 10.2166/wst.2001.0804 PMID: 11804163

Tanner, C. C., & Headley, T. R. (2011). Components of floating emergent macrophyte treatment wetlands influencing removal of stormwater pollutants. *Ecological Engineering*, *37*(3), 474–486. DOI: 10.1016/j.ecoleng.2010.12.012

Tao, M., Zhang, W., Fan, H., Huang, L., & Xu, Q. (2020). *E3S Web of Conferences, 143,* 02020.

Tchounwou, P. B., Yedjou, C. G., Patlolla, A. K., & Sutton, D. J. (2012). Heavy metal toxicity and the environment. In *Molecular, Clinical and Environmental Toxicology* (Vol. 101, pp. 133–164). Springer. https://doi.org/DOI: 10.1007/978-3-7643-8340-4_6

Thakur, S., Dangi, P., Pandey, A., & Shukla, R. (2016). Engineered nanomaterials for water treatment and remediation: Challenges and opportunities. *Environmental Technology & Innovation*, *5*, 161–173.

Thapa, G., Sadhukhan, A., Panda, S. K., & Sahoo, L. (2021). Molecular mechanistic model of heavy metal tolerance in plants: Interactions between signaling pathways. *Plant Signaling & Behavior*, *16*(4), 1870063.

Thomaz, S. M., & Cunha, E. R. (2010). The role of macrophytes in habitat structuring in aquatic ecosystems: Methods of measurement, causes and consequences on animal assemblages' composition and biodiversity. *Acta Limnologica Brasiliensia*, *22*(2), 218–236. DOI: 10.4322/actalb.02202011

Tripathi, D. K., Singh, S., Singh, S., Mishra, S., Chauhan, D. K., & Dubey, N. K. (2021). Nanomaterials in the environment: Fate, transport, toxicity and interactions with plants. *Environmental Technology & Innovation*, *21*, 101200.

Ullrich, S. M., Tanton, T. W., & Abdrashitova, S. A. (2001). Mercury in the aquatic environment: A review of factors affecting methylation. *Critical Reviews in Environmental Science and Technology*, *31*(3), 241–293. DOI: 10.1080/20016491089226

United Nations Environment Programme (UNEP). (2018). *Nature-based solutions for water*. UNEP.

Varol, M. (2011). Assessment of heavy metal contamination in sediments of the Tigris River (Turkey) using pollution indices and multivariate statistical techniques. *Journal of Hazardous Materials*, *195*, 355–364. DOI: 10.1016/j.jhazmat.2011.08.051 PMID: 21890271

Verhoeven, J. T. A., & Meuleman, A. F. M. (1999). Wetlands for wastewater treatment: Opportunities and limitations. *Ecological Engineering*, *12*(1–2), 5–12. DOI: 10.1016/S0925-8574(98)00050-0

Verma, P. K., Sinha, A., & Singh, V. K. (2020a). Role of phytochemicals in phytoremediation of heavy metals: Mechanistic insights and future perspectives. *Environmental Technology & Innovation*, *18*, 100684. DOI: 10.1016/j.eti.2020.100684

Verma, R., Sinha, S., & Chauhan, P. (2020b). Recent trends in phytoremediation: A critical review. *Environmental Science and Pollution Research International*, *27*, 13855–13868. DOI: 10.1007/s11356-020-07904-3

Verma, S., & Dubey, R. S. (2003). Lead toxicity induces lipid peroxidation and alters the activities of antioxidant enzymes in growing rice plants. *Plant Science*, *164*(4), 645–655. DOI: 10.1016/S0168-9452(03)00022-0

Verma, S., Gupta, R., & Singh, N. (2020c). Mechanisms of heavy metal removal by aquatic plants and their future prospects. *Environmental Science and Pollution Research International*, *27*(25), 31725–31737. DOI: 10.1007/s11356-020-09512-9

Villamagna, A. M., & Murphy, B. R. (2010). Ecological and socio-economic impacts of invasive water hyacinth (*Eichhornia crassipes*): A review. *Freshwater Biology*, *55*(2), 282–298. DOI: 10.1111/j.1365-2427.2009.02294.x

Vlot, A. C., Dempsey, D. A., & Klessig, D. F. (2009). Salicylic acid, a multifaceted hormone to combat disease. *Annual Review of Phytopathology*, *47*(1), 177–206. DOI: 10.1146/annurev.phyto.050908.135202 PMID: 19400653

Vögeli-Lange, R., & Wagner, G. J. (1996). Relationship between cadmium-induced phytochelatin accumulation and the response of protein synthesis to cadmium in suspension-cultured plant cells. *Plant Science, 114*(2), 147–155.

Vögeli-Lange, R., & Wagner, G. J. (1996). Relationship between cadmium, glutathione and cadmium-binding peptides (phytochelatins) in leaves of intact tobacco seedlings. *Plant Science, 114*(1), 11–18. DOI: 10.1016/0168-9452(95)04299-7

Vymazal, J. (2007). Removal of nutrients in various types of constructed wetlands. *The Science of the Total Environment, 380*(1–3), 48–65. DOI: 10.1016/j.scitotenv.2006.09.014 PMID: 17078997

Vymazal, J. (2011). Constructed wetlands for wastewater treatment: Five decades of experience. *Environmental Science & Technology, 45*(1), 61–69. DOI: 10.1021/es101403q PMID: 20795704

Vymazal, J. (2014). Constructed wetlands for treatment of industrial wastewaters: A review. *Ecological Engineering, 73*, 724–751. DOI: 10.1016/j.ecoleng.2014.09.034

Vymazal, J., & Březinová, T. (2015). The use of constructed wetlands for removal of pesticides from agricultural runoff and drainage: A review. *Environment International, 75*, 11–20. DOI: 10.1016/j.envint.2014.10.026 PMID: 25461411

Vymazal, J., & Kröpfelová, L. (2011). *A three-stage constructed wetland for wastewater treatment: Performance studies*. Springer.

Wang, C., & Keller, A. A. (2008). Natural and engineered nano-scale iron for environmental remediation. *Trends in Analytical Chemistry, 27*(2), 152–158. DOI: 10.1016/j.trac.2007.10.010

Wang, J., Zhang, C., Dai, H., & Zhang, Z. (2017). Advances in the genetic engineering of aquatic plants for phytoremediation. *Environmental Science and Pollution Research International, 24*, 17118–17131. DOI: 10.1007/s11356-017-9351-6

Wang, W., Wang, Y., Zhang, Y., & Gao, F. (2012). Bioaccumulation and antioxidative responses of *Eichhornia crassipes* to heavy metals. *Ecotoxicology (London, England), 21*(3), 681–689. DOI: 10.1007/s10646-012-0852-4 PMID: 22113456

Wang, W., Wang, Y., Zhang, Y., & Gao, F. (2012). Bioaccumulation and antioxidative responses of *Eichhornia crassipes* to heavy metals. *Ecotoxicology (London, England), 21*(3), 681–689. DOI: 10.1007/s10646-012-0852-4 PMID: 22113456

Weis, J. S., & Weis, P. (2004). Metal uptake, transport and release by wetland plants: Implications for phytoremediation and restoration. *Environment International, 30*(5), 685–700. DOI: 10.1016/j.envint.2003.11.002 PMID: 15051245

Wetzel, R. G. (2001). *Limnology: Lake and river ecosystems* (3rd ed.). Academic Press.

Williams, L. E., & Mills, R. F. (2005). P1B-ATPases—An ancient family of transition metal pumps with diverse functions in plants. *Trends in Plant Science*, *10*(10), 491–502. DOI: 10.1016/j.tplants.2005.08.008 PMID: 16154798

Wolverton, B. C., & McDonald, R. C. (1981). The water hyacinth: From prolific pest to potential provider. *Ambio*, *10*(2–3), 82–87.

Wu, S., Kuschk, P., Brix, H., Vymazal, J., & Dong, R. (2014). Development of constructed wetlands in performance intensifications for wastewater treatment: A nitrogen and organic matter targeted review. *Water Research*, *57*, 40–55. DOI: 10.1016/j.watres.2014.03.020 PMID: 24704903

Wu, S., Wallace, S., Brix, H., Kuschk, P., Kirui, W. K., Masi, F., & Dong, R. (2014). Treatment of industrial effluents in constructed wetlands: Challenges, operational strategies and overall performance. *Environmental Pollution*, *192*, 245–261. DOI: 10.1016/j.envpol.2014.05.007 PMID: 25792030

Yadav, A. K., Dash, P., Mohanty, A., Abbassi, R., & Mishra, B. K. (2012). Performance assessment of innovative constructed wetland-microbial fuel cell for electricity production and dye removal. *Ecological Engineering*, *47*, 126–131. DOI: 10.1016/j.ecoleng.2012.06.029

Yadav, R. K., Saini, R., & Sharma, A. (2022). Decision support tools for sustainable phytoremediation: A review of frameworks and practical applications. *Ecotoxicology and Environmental Safety*, *238*, 113597. DOI: 10.1016/j.ecoenv.2022.113597

Yadav, S. K. (2010). Heavy metals toxicity in plants: An overview on the role of glutathione and phytochelatins in heavy metal stress tolerance of plants. *South African Journal of Botany*, *76*(2), 167–179. DOI: 10.1016/j.sajb.2009.10.007

Yamashita, T., Ishikawa, M., & Fujiwara, T. (2021). The effects of temperature and seasonal changes on pollutant removal in constructed wetlands: A review. *Journal of Environmental Chemical Engineering*, *9*(1), 104680. DOI: 10.1016/j.jece.2020.104680

Yan, A., Wang, Y., Tan, S. N., Mohd Yusof, M. L., Ghosh, S., & Chen, Z. (2020). Phytoremediation: A promising approach for revegetation of heavy metal-polluted land. *Frontiers in Plant Science*, *11*, 359. DOI: 10.3389/fpls.2020.00359 PMID: 32425957

Yeh, C. M., Chang, I. F., Lu, C. A., Wu, Y. M., Lu, M. Y., & Shih, M. C. (2007). A novel family of transcription factors involved in plant defense response. *Plant Molecular Biology*, *64*(6), 771–787. DOI: 10.1007/s11103-007-9227-4

Zayed, A., Gowthaman, S., & Terry, N. (1998). Phytoaccumulation of trace elements by wetland plants: I. Duckweed. *Journal of Environmental Quality*, *27*(3), 715–721. DOI: 10.2134/jeq1998.00472425002700030032x

Zedler, J. B., & Kercher, S. (2004). Causes and consequences of invasive plants in wetlands: Opportunities, opportunists, and outcomes. *Critical Reviews in Plant Sciences*, *23*(5), 431–452. DOI: 10.1080/07352680490514673

Zedler, J. B., & Kercher, S. (2005). Wetland resources: Status, trends, ecosystem services, and restorability. *Annual Review of Environment and Resources*, *30*(1), 39–74. DOI: 10.1146/annurev.energy.30.050504.144248

Zhang, D., Gersberg, R. M., Ng, W. J., & Tan, S. K. (2014). Removal of pharmaceuticals and personal care products in aquatic plant-based systems: A review. *Environmental Pollution*, *184*, 620–639. DOI: 10.1016/j.envpol.2013.09.009 PMID: 24080393

Zhang, D. Q., Jinadasa, K. B. S. N., Gersberg, R. M., Liu, Y., Ng, W. J., & Tan, S. K. (2020). Application of constructed wetlands for wastewater treatment in developing countries – A review of recent developments (2010–2020). *Journal of Environmental Management*, *293*, 112862. DOI: 10.1016/j.jenvman.2021.112862

Zhang, D. Q., Jinadasa, K. B. S. N., Gersberg, R. M., Liu, Y., Tan, S. K., & Ng, W. J. (2014). Application of constructed wetlands for wastewater treatment in developing countries – A review of recent developments (2000–2013). *Journal of Environmental Management*, *141*, 116–131. DOI: 10.1016/j.jenvman.2014.03.015 PMID: 24784754

Zhang, X., Xia, H., Li, Z., & Zhuang, P. (2014). Potential of four ornamental plants in remediation of contaminated soils with heavy metals. *Ecological Engineering*, *70*, 244–252. DOI: 10.1016/j.ecoleng.2014.05.003

Zhang, X., Xia, H., Li, Z., Zhuang, P., Gao, B., & Gao, Y. (2014). Removal efficiency of heavy metals in a pilot-scale constructed wetland system and accumulation in plant tissues. *Ecological Engineering*, *64*, 356–362. DOI: 10.1016/j.ecoleng.2013.12.044

Zhang, Y., Li, Z., & Wang, L. (2024). Environmental problems of emerging toxic metals and remediation techniques: Role of phytoremediation and microbes. *Environmental Pollution*, *334*, 122479. DOI: 10.1016/j.envpol.2024.122479

Zhang, Y., Shao, Y., Tian, S., & He, Z. (2020). Influence of co-contaminants on heavy metal uptake and accumulation in plants: A review. *Environmental Pollution*, *257*, 113613. DOI: 10.1016/j.envpol.2019.113613 PMID: 31838392

Zhang, Y., Zhang, M., Yao, Y., & Hu, R. (2020). Biochar for environmental management: A meta-analysis of its effects on soil nitrogen transformation. *The Science of the Total Environment*, *731*, 139105. DOI: 10.1016/j.scitotenv.2020.139105

Zhao, F. J., Ma, J. F., & Meharg, A. A. (2015a). Arsenic uptake, metabolism, and toxicity in plants. *The New Phytologist*, *206*(4), 1063–1078. DOI: 10.1111/nph.13281

Zhao, F. J., Ma, Y., Zhu, Y. G., Tang, Z., & McGrath, S. P. (2015b). Soil contamination in China: Current status and mitigation strategies. *Environmental Science & Technology*, *49*(2), 750–759. DOI: 10.1021/es5047099 PMID: 25514502

Zhao, Y., Liu, J., & Zhang, J. (2020). Remote sensing technologies for real-time monitoring and optimization of constructed wetlands: A review. *The Science of the Total Environment*, *749*, 141438. DOI: 10.1016/j.scitotenv.2020.141438

Zhao, Y., Wang, Z., Zhang, Q., & Wang, C. (2018). Enhancing Cd phytoremediation efficiency in *Pistia stratiotes* through endophytic fungi colonization. *Environmental Science and Pollution Research International*, *25*(16), 15408–15416. DOI: 10.1007/s11356-018-1747-9

Zhou, C., Xu, Y., Song, X., & Yang, Y. (2022). Phytoremediation of heavy metal-contaminated water and sediments using aquatic plants: A review. *Ecological Engineering*, *179*, 106608. DOI: 10.1016/j.ecoleng.2022.106608

Zhou, Q.. (2022). Phytoremediation of heavy metals using aquatic macrophytes: Mechanisms and applications. *Environmental Pollution*, *298*, 118800. DOI: 10.1016/j.envpol.2021.118800

Zhou, Q., Chen, Y., Gao, Y., & Xu, J. (2022). Arsenic-induced oxidative stress and the regulatory role of WRKY transcription factors in *Nelumbo nucifera*. *Environmental Pollution*, *294*, 118646. DOI: 10.1016/j.envpol.2021.118646

Zhou, Q., Li, X., Lin, Y., Yang, C. P., Tang, W. C., Wu, S. H., Li, D. H., & Lou, W. (2019). Effects of copper ions on removal of nutrients from swine wastewater and on release of dissolved organic matter in duckweed systems. *Water Research*, *158*, 171–181. DOI: 10.1016/j.watres.2019.04.036 PMID: 31035194

Zhou, Q., Sun, T., Wang, X., & Shan, Y. (2010). Mechanisms of heavy metal uptake and accumulation by aquatic macrophytes. *Journal of Environmental Sciences (China)*, *22*(5), 716–723. DOI: 10.1016/S1001-0742(09)60185-5 PMID: 20608508

Zhou, Q., Zhang, J., Fu, J., Shi, J., & Jiang, G. (2007). Biomonitoring: An appealing tool for assessment of metal pollution in the aquatic ecosystem. *Analytica Chimica Acta*, *606*(2), 135–150. DOI: 10.1016/j.aca.2007.11.018 PMID: 18082645

Ziegler, P., Adelmann, K., Zimmer, S., Schmidt, C., & Appenroth, K. J. (2016). Relative in vitro growth rates of duckweeds (Lemnaceae) – The most rapidly growing higher plants. *Plant Biology*, *17*(s1, S1), 33–41. DOI: 10.1111/plb.12184 PMID: 24803032

Zimmels, Y., Kirzhner, F., & Malkovskaja, A. (2006). Application of *Eichhornia crassipes* and *Pistia stratiotes* for treatment of urban sewage in Israel. *Journal of Environmental Management*, *81*(4), 420–428. DOI: 10.1016/j.jenvman.2005.11.014 PMID: 16597484

Chapter 5
Hydrophytes as a Sustainable Tool for Bioremediation of Heavy Metals in Wetland Ecosystems

Richa Arora
JIS University, India

Ishita Mitra
JIS University, India

Suparna Ghosh
https://orcid.org/0000-0002-9200-1291
JIS University, India

Dibyarupa Pal
https://orcid.org/0000-0002-8564-1485
JIS University, India

ABSTRACT

Wetlands are essential for preserving biodiversity, ecological balance, and water quality. Increasing human activity has resulted in serious heavy metal contamination from several avenues. They are hazardous, persistent, and bioaccumulative, pose major threats to the environment and human. Bypassing expensive conventional remediation approaches, natural hydrophyte-based phytoremediation absorb, stabilize, or transform pollutants, provides a sustainable and economical substitute. Their

DOI: 10.4018/979-8-3373-4037-1.ch005

heavy metal uptake and detoxification capabilities are done by root systems, high biomass, and symbiotic relationships with rhizospheric microorganisms. Challenges, include seasonal growth fluctuations, invasive potential, metal saturation, and safe biomass disposal, despite benefits like cheap operating costs, biodiversity support, and biomass valorization. In accordance with sustainable development objectives, hydrophyte-based remediation provides an effective strategy to restore polluted wetlands, fostering ecological resilience.

INTRODUCTION

Wetland habitats are dynamic, critical and versatile ecosystems that exist at the land-water interface, comprising a wide range of ecological territory encompassing marshes, swamps, bogs, and fens (Mitsch et al., 2015). Often referred to as the "kidneys of the earth," these areas remain saturated throughout the year, generating unique lives that sustain both water and terrestrial habitats. Wetlands are far from being wastelands, as they provide enormous ecological benefit. They act as natural water filters, trapping pollutants and sediments, as well as essential sponges that absorb excess rainfall, reducing flooding, nurturing biodiversity and replenishing groundwater resources (Alikhani et al., 2021; Wu et al., 2023). Furthermore, they are among the world's most biologically productive ecosystems, providing critical breeding and rearing grounds for several species of fish, birds, amphibians, and invertebrates. Despite their critical roles in biodiversity, water purification, and climate regulation, wetland habitats are among the most vulnerable in the world, inflicted with diverge contaminants, especially heavy metals. Escalating anthropogenic activities like mining, urbanization, agricultural runoff, and industrial discharges release Heavy metals (HM) like Lead (Pb), Cadmium (Cd), Mercury (Hg), Arsenic (As), and Chromium (Cr) in the system (Li et al., 2022).

Heavy metals do not disintegrate over a stretch, in contrast to organic pollutants; instead, they carry on in the different water sources like surface or ground water, thereafter accumulating in in the sediments that sustain aquatic life. HM impose severe toxicity, reproductive failure, and mortality to the existing flora and fauna population of wetland inclusion of human lives. Interestingly, these metals have the ability to get biomagnified as they move high in the food chain, and place top predators like fish-eating birds and mammals at risk. The most upsetting part is that the wetland's natural filtering capacity is burdened with entrapped organics, heavy metals, creating a long-term pool of contaminants (Ali & Khan, 2019).

Numerous efficient traditional techniques are applied for heavy metal removal which includes chemical precipitation, ion exchange, activated carbon induced adsorption and electrochemical methods. Traditional methods are efficacious, but they produce poisonous sludge and wasted filters that generates disposal problem. The processes are energy-intensive and expensive. For instance, chemical precipitation turns dissolved metals into insoluble forms producing secondary contaminants, which has to be disposed with caution (Sharma et al., 2019). Regardless of their efficacy, membrane filtration methods like reverse osmosis requires huge energy sources and are costly to maintain in laboratory or industrial scale settings. Hence more bio-based process are desired for their sustainability and cost effectiveness.

Natural processes involving plants and/or microbes offer green solutions which are beneficial to the environment and provides an easy-go solution to the rising problems of waste treatment by chemical means. One of the most sought after approach is remediation by plants or called as phytoremediation, which involves mainly the elaborate roots of the plant to remove, or detoxify heavy metals. Among the usage of plants, aquatic plants or hydrophytes have gained popularity in bioremediation of heavy metals (Pang et al., 2023; Pal et al., 2025).

Hydrophytes, are extremely thriving and adaptible plant species that can withstand any stressful environment. Several aquatic species have their own niche for existence, like the submerged plants (*Hydrilla verticillata* and *Ceratophyllum demersum*), emergent plants (*Typha latifolia* grass) and (*Phragmites australis* reeds), and the floating plants (*Nymphaea sp.* and *Eichhornia crassipes*). They have unique physiological as well as genomic flexibility like supple and hollow stems, specialized sumptuous roots, invasiveness and distinctive metabolic processes, to endure the presence of heavy metals in aquatic habitat. Usually these plant species are widely recognized growing in the swampy areas where they can exert multiple pathways to detoxify entrapped heavy metals. Their enormous root systems, high biomass production, and ability to tolerate metal pollutants make them ideal candidates for heavy metal bioremediation (Zhao et al., 2018).

To remove heavy metals from the environment, hydrophytes use a variety of bioremediation strategies. Phytoextraction draws contaminants out of the water column and uptake metals from sediments or water by roots, which then are translocated to the aerial part of the tissues. Plant roots adsorb or precipitate heavy metals on their root surface by Rhizofiltration (Chan et al., 2004; Vangronsveld et al., 2009). By immobilizing heavy metals in their root zones, plants stop them from migrating into water or the food chain (Ghosh & Singh, 2005). This process is known as phytostabilization. Certain metals can be successfully removed from the water or soil by the process of phytovolatilization, in which particular plants change the metals into volatile forms that are emitted into the atmosphere. Certain species can volatilize elements such as mercury, transforming them into less hazardous gaseous forms (Lasat, 2002).

Multiple traits of hydrophytes enable them to tolerate elevated levels of heavy metals. The surface area available for metal sorption and absorption is increased by extensive and fine root systems. Metals are bound by the organic ligands such as phytochelatins and metallothioneins, which lessen the toxicity of metals in plant tissues (Zhou & Lai, 2018). Increased antioxidant levels aid plants in reducing oxidative stress driven by metal-induced reactive oxygen species. In order to maintain cellular equilibrium, active transporters can remove excess metals from cells.

The rhizosphere, usually considered as the bio-geochemical factory, is the thin region of soil encircling the plant roots, plays a distinguished role in decontamination of heavy metals. This dynamic root zone is the kernel for microbial activity and interactions amongst themselves and with plant roots. The microbial dynamism exists as the plant roots release variety of plant exudates, which incorporates sugars, hormones, acids, and enzymes which are mainly utilized as nutrition by the rhizosphere microbial population. These exudates have the potential to change soil chemistry which can make the entrapped soil metals mobile by making them more soluble (Li et al., 2019; Kabeer et al., 2022).

Hydrophyte-based remediation can be performed on large-settings and in areas where land resources are constrained. Their physiological specialization like elaborate aerenchyma tissue have the ability to draw up heavy metals through water column and store, stabilize or degrade in their vacuoles. After the uptake of HMs, the entire plant biomass can be sacrificed for easy metal recovery, without generating any secondary wastes. Thus, phytoremediation offers hassle free and cost sustainable remediation but definitely with certain limitations like slower rate of detoxification, varied metal tolerance of plant species, and the problem of invasiveness. Research is ongoing to identify and engineer hyperaccumulator species with enhanced metal uptake capacities, improve microbial-plant partnerships, and develop integrated treatment systems combining phytoremediation with other methods. Constructed wetlands are engineered systems that mimic natural wetlands, often planted with hydrophytes to treat industrial or municipal wastewater. In one case study performed by Kumari and Tripathy (2015), at different wetland Ghats of Varanasi, Kanpur, India, using *Typha latifolia* and *Phragmites australis*, different heavy metals was removed. The wetland received urban sewage and industrial effluents brought by Ganga River and after 14 days of treatment with *P. australis* and *T. latifolia* alone and in conjunction showed *P. australis* as better performer than *T. latifolia* for Cr, Cu, Cd, Fe, Ni, Zn and Pb removal, while synergistic action displayed better heavy metal removal capacity from wastewater samples.

Furthermore, maintaining safety and environmental integrity depends on knowing how metals behave in plants and how they could spread across food webs (Meagher, 2000). Technological developments in ecological engineering, genetic engineering, and nanotechnology have the potential to improve hydrophyte-based remediation techniques and overcome present obstacles. In wetland environments, hydrophytes offer a flexible and sustainable way to reduce heavy metal contamination. They can effectively tolerate, acquire, and convert heavy metals owing to their distinct physiological characteristics and intricate rhizosphere interactions. Utilizing aquatic plants' inherent remediation capabilities presents a viable option to restore contaminated wetlands, preserve biodiversity, and safeguard public health as environmental issues worsen.

1. HEAVY METAL CONTAMINATION IN WETLANDS

There are several ways that heavy metals, including Cd, Pb, Hg, As, and Cr, can infiltrate wetland ecosystems (Figure 1), including:

- Industrial effluents, such as those from mining industries, the tanning process, burning coal, steel manufacture, and electroplating.
- Acid mine drainage and the mobilization of metals like Pb and Cd into water and soils are frequent outcomes of mining and smelting operations.
- Metals from phosphate fertilizers, herbicides, and fertilizer additives are carried into wetland systems through agricultural runoff (Li et al., 2022).
- Sewage, wastewater from industries, and stormwater from urban areas.

Figure 1. Different sources of heavy metal contamination in wetlands

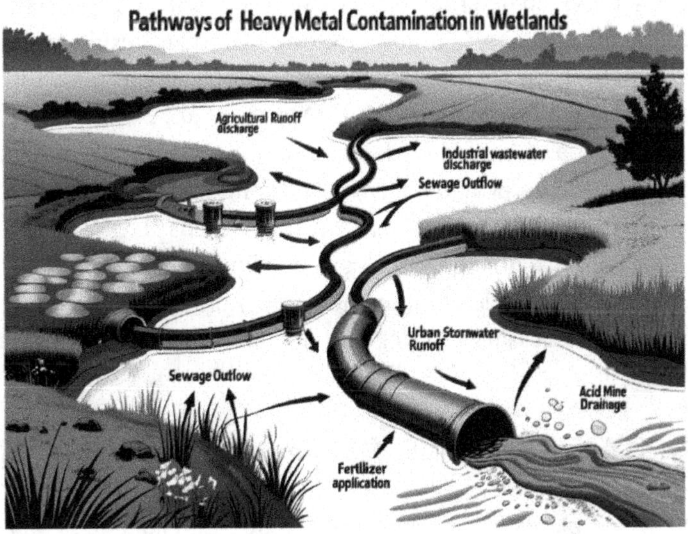

Heavy metals do not break down over time like organic contaminants do; instead, they accumulate in sediments and enter food webs through bioaccumulation and biomagnification, especially in aquatic creatures and waterfowl (Jamil Emon et al., 2023; Leaphart et al., 2022).

Ecosystems and human health are seriously threatened by heavy metals. A persistent and extremely harmful heavy metal, cadmium (Cd) binds to the sediments of wetlands, hinders plant growth, interferes with photosynthesis, and changes the microbial communities in the soil (Genchi et al., 2020). Aquatic plants and animals easily bioaccumulate it, which then spreads to humans through food chains (Nnaji et al., 2023). Long-term exposure results in osteoporosis and irreparable kidney damage. In addition, oxidative stress, cardiovascular disease, lung, prostate, and breast malignancies are associated with cadmium. While extended exposure affects several organ systems and poses major ecological and public health problems, acute inhalation can cause flu-like symptoms, respiratory distress, and pulmonary edema (Faroon et al., 2012).

A persistent neurotoxic, lead (Pb), slows down plant development, impairs enzyme activity, and affects biodiversity as it builds up in sediments and biota. It can bioaccumulate throughout aquatic food chains and harm the human neural system, especially in children, leading to cognitive impairments, anaemia, and kidney malfunction (Collin et al., 2022).

In aquatic systems, mercury (Hg), a neurotoxin, transforms into methylmercury, which enters food webs and magnifies in aquatic organisms. In wildlife, it hinders growth, behaviour, and reproduction; in humans, it results in neurological impairments, cardiovascular problems, and developmental difficulties (Kumar et al., 2024).

A hazardous metalloid that lingers in sediments, arsenic (As) alters plant metabolism and lowers biodiversity. Chronic exposure in humans results in neurotoxicity, cardiovascular disease, cancer, and skin blemishes. It endangers the health of wildlife by bioaccumulating in aquatic food chains (Marisa et al., 2013)

Trivalent Cr (III) and hazardous hexavalent Cr (VI), which is mutagenic and carcinogenic, are the two primary forms of chromium (Cr). It hinders microbial diversity, photosynthesis, and plant growth in wetlands. Exposure to Cr (VI) in humans results in organ damage, skin ulcers, and lung cancer (Sharma et al., 2022).

Considering the magnitude of these effects, economical and sustainable remediation techniques are necessary to safeguard human health and ecological integrity.

2. HYDROPHYTES: CLASSIFICATION AND CHARACTERISTICS

Hydrophytes are plants that have evolved to thrive in aquatic or wet conditions where the root zone's oxygen supply is constrained by water saturation. They have physiological, anatomical, and morphological adaptations that enable them to thrive, reproduce, and perform effectively in these environments. Hydrophytes are often categorized into the following groups according to their ecological niche and growth habit:

3.1 Emergent Hydrophytes: These plants are rooted in the anoxic, contaminated sediments and aerial parts radiate above the waterline, making a biological bridge between the polluted substrate and the atmosphere. Their main adaption is abundant aerenchyma tissue, which oxygenates the buried roots and transports metal ions from the sediment and water. These plants use phytoextraction to uptake metals like Pb, Zn and Cd into their root systems and transfer them to stems and leaves, where they are stored in vacuoles. Examples: *Typha latifolia* (cattail), *Phragmites australis* (common reed).

3.2 Submerged Hydrophytes: They spend their entire lives underwater and absorb heavy metals directly from the water column rather than sediments. Hydrophyte roots play rapid role in absorbing impurities by rhizofiltration. The leaves' bear thin cuticle with high surface-to-volume ratio for direct uptake of heavy metals and the entire plant body undergo photosynthesis in poor light conditions. Examples: *Hydrilla verticillata, Vallisneria spiralis* and *Ceratophyllum demersum.*

3.3 Free-floating Hydrophytes: These species float freely on the water's surface with their roots directly into the water column, unlike submerged or emergent plants. They are suitable for rhizofiltration because their huge dangling fibrous root surfaces collect metal ions from the water like Pb, Cd, and As. Due to their quick growth, rapid adaptation capability, these invasive species generate huge turnover in short time period which could be beneficial for heavy metal uptake and easy harvest later on. Examples: *Eichhornia crassipes* (water hyacinth), *Lemna minor* (duckweed) and *Pistia stratiotes* (water lettuce).

3.4 Floating-leaved Hydrophytes: Although the root of these plants lie in the sediments at the bottom of the pool, their big, buoyant leaves float just above the surface. Because of their atypical habitat, they can clean up heavy metals in two main ways. Firstly, roots and rhizomes draw metals straight from the sediments by phytoextraction, secondly, pollutants from the water column are absorbed by their petioles, stalks and leaves. Example: *Nymphaea* spp. (water lily) and *Nelumbo* spp. (lotus).

Hydrophytes are perfect for bioremediation in wetlands because they have unique adaptations which let them survive in their natural environment. They can improve water quality, absorb surplus nutrients, and sequester heavy metals by their specialized aerenchyma tissue. Their large root surface area optimizes the absorption of nutrients and pollutants. Decreased lignification, along with a thin cuticle, helps in the transport of solutes and water. Antioxidant enzyme systems, such as peroxidase, catalase, and superoxide dismutase, help reduce oxidative stress caused by contaminants.

3. MECHANISMS OF HEAVY METAL BIOREMEDIATION BY HYDROPHYTES

Hydrophytes contribute to bioremediation through several interrelated mechanisms (Figure 2)

Figure 2. Different methods of phytoremediation

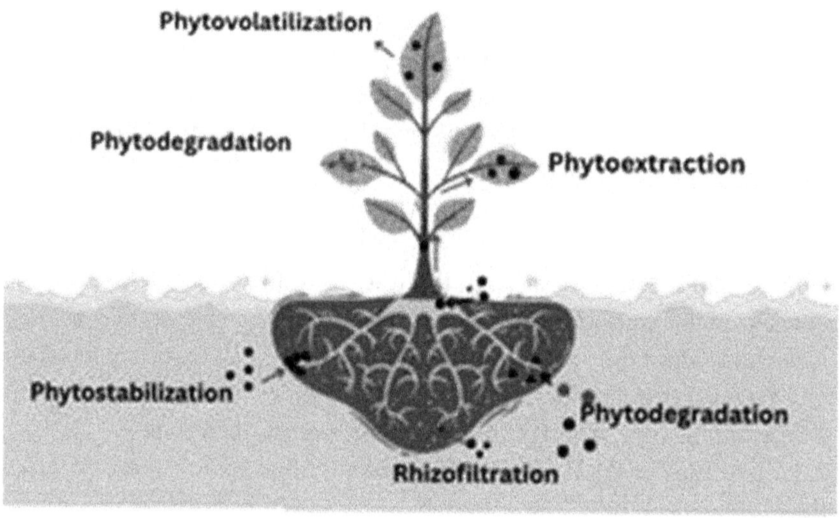

3.1 Phytoextraction

Phytoextraction is a key mechanism of phytoremediation, wherein plants absorb heavy metals (HMs) or pollutants from the soil or water and accumulate them in their harvestable tissues, such as roots, stems, and leaves, which can later be processed or disposed of safely. Certain plants, such as species of *Scirpus sp.*, exhibit excellent phytoextraction capacity due to their inherent metallo tolerance, high adaptability to polluted environments, and easily harvestable biomass. These plants are classified as natural hyperaccumulators, capable of thriving in heavily contaminated sites while producing substantial biomass, making them effective candidates for metal pollution mitigation (Ali et al., 2019). *Eichhornia crassipes* accumulates significant amounts of Cd, Pb, Zn, and Mn (Zhou et al., 2007)

An interesting example of phytoextraction using high-biomass non-accumulator plants was observed in an industrial region near Plovdiv, Bulgaria, where *Gossypium* species (cotton plants) demonstrated the ability to reduce accumulated levels of zinc (Zn), cadmium (Cd), copper (Cu), and lead (Pb) in their fibres. Angelova et al. (2004) observed that boiling water lowered the metal concentration in the fibers to levels similar to plants from uncontaminated areas. Moderate quantities of heavy metals were also accumulated by other regularly cultivated medicinal plants, including peppermint (*Mentha piperita*) and lavender (*Lavandula angustifolia*), indicating that phytoextraction can be used to a large range of plant species. Alpine pennygrass (*Noccaea caerulescens*) is another such hyperaccumulator species that can absorb heavy metals and store in its aerial parts. It was found that *N. caerulescens* decreased 50% cadmium contamination levels from soils after two or three harvest cycles (Kumar et al., 1995). Species like Mustard (*Brassica juncea*) have also demonstrated Cd, Cr, Cu, Ni, Pb, and Zn metal-removing capabilities from soil and hydroponic environments (Salt et al., 1995). The free-floating aquatic plant *Eichhornia crassipes*, possess spread-out fibrous root system, with quick propagation rate of the plant, and huge biomass, all of which make it an effective pollutant remover (Zhou et al., 2007). Genetically modified (GM) plants have been developed to improve removal of specific heavy metal. For instance, in the transgenic *Brassica juncea,* YCF1 gene when introduced deliberated increased tolerance towards Cd and Pb and significantly improved metal accretion in the shoot tissues (Song et al., 2003).

A specific class of proteins called heavy metal transporters (HMTPs) allows hyperaccumulators to efficiently translocate metals from their roots across cellular membranes. Proteins on the plasma and tonoplast membranes allow metals to be transported and stored in vacuoles, protecting cells from their harmful effects. Enabling effective accumulation of metals, this pathway is vital for cellular homeostasis maintenance.

3.2 Phytostabilization

Phytostabilization is a method of remediation that makes use of plant species that have fibrous root systems that can absorb, disperse, and immobilize heavy metal contaminants in the soil. By changing contaminants into forms that are less soluble or bioavailable, these plants lessen their environmental impact and make them less mobile. The perfect phytostabilizer would be able to bind to roots rather than shoots, have low translocation factor (TF), and be very resistant to changes in pH, salt, and moisture.

One plant that could justify the bioremediation is red canary grass (*Phalaris arundinacea*). According to an earlier research group working on European aquatic bodies, *P. arundinacea* stores trace amounts of Cd, Pb, Co, and Zn metals from polluted sediments. Korzeniowska and Stanislawska-Glubiak (2019) particularly emphasized on concentrations of Co and Zn found in the grass with high potential for biomonitoring and phytostabilization processes. Researchers have found that metal transfer to shoots is limited by root vacuole sequestration mechanisms (e.g., phytochelatins) and tissue-specific expression of Cd transport proteins (HMA3) (Henrique et al., 2021). The genetic regulation highlighted the plant's ability to stabilize soils contaminated with Cd and Co since they exhibit low shoot mobility.

Earlier research on *P. arundinacea* mainly concentrates on sequestering Cd and Zn, but it's potential towards stabilizing other metals like Ni are also being studied keeping in mind the resilience, adaptability and vigorous biomass capacity of the plant (Korzeniowska & Stanislawska-Glubiak, 2019).

3.3 Phytodegradation

Phytodegradation uses plants to enzymatically break down or alter xenobiotics, pollutants that are not normally present in the environment. Plant tissues or the surrounding rhizosphere are usually the site of this process.

Mechanisms of Phytodegradation can include intracellular metabolism, extracellular enzyme activity, or rhizosphere co-metabolism. In intracellular metabolism, plants absorb contaminants through their roots and metabolize them internally using enzymes like dehalogenases, peroxidases, and laccases. These enzymes facilitate the breakdown of complex pollutants into less harmful substances (Pivetz, 2001). Some plants secrete enzymes into the soil, aiding in the degradation of pollutants externally. This extracellular degradation is particularly effective for breaking down persistent organic pollutants. Plant roots exude organic compounds that serve as carbon sources for rhizospheric microorganisms. These microbes, in turn, enhance the degradation of contaminants through co-metabolic processes.

Several notable plant species can effectively perform Phytodegradation. Wheatgrass, *Triticum aestivum* has also shown potential in extracting and metabolizing various contaminants, including aristolochic acids and heavy metals, from both water and soil. Its rapid growth and adaptability make it a suitable candidate for phytoremediation (Kwok & Chan, 2022). Wetland plants, such as *Typha domingensis* and *Pontederia rotundifolia*, have been found to accumulate heavy metals, Cr, Ni, Zn and Pb from polluted waters (Maine et al., 2022). *Eichhornia crassipes,* a quite popular wetland plant removes heavy metals from contaminated soil sediments and due to its large biomass can also be processed later with ease for rare metal recovery (Sompura et al., 2024). Contaminated sludge and soil generated from untreated industrial effluent, radioactive waste and petroleum contamination, are the best sites for phytodegradation process. It offers more ecological and worthwhile alternative to traditional clean-up methods involving microbial communities also.

3.4 Rhizofiltration

Rhizofiltration engrosses plant roots to remove contaminants like HMs and radioactive substances in an eco-friendly manner from aquatic environments. Mostly the HMs are absorbed and concentrated by the roots, which function as natural filters. Later on after the absorption is over, the plant roots are collected and disposed, which removes the contaminants also. Rhizofiltration works best for treating surface wastewater and groundwater (Alsafran et al., 2022; Biswal, 2025). The plants involved in this procedure are usually grown in water instead of soil, a technique known as hydroponic cultivation. This allows their large root networks to grow directly in the polluted water and since plants grow in presence of natural sunlight, the process becomes less expensive.

The process of metallic contaminant uptake through rhizofiltration involves several crucial stages. Due to concentration gradients and water absorption, pollutants first move toward the root surface. Then metal ions attach to negatively charged sites on the root epidermis, restricting their mobility. Metal ions are transported into the root cells across cell membranes with the help of certain transport proteins. Depending on the plant species and metal type, metals can either be immobilized after entering the root cells or transferred to the shoots via the xylem (Sabreena et al., 2022).

For a plant species to be effective in rhizofiltration, it should possess some special characteristics. It should be able to withstand elevated concentrations of toxic metals. It should possess an extensive root system to enhance contaminant uptake. The species with rapid growth rates and high biomass facilitates quicker remediation and easier harvesting.

Several plant species have demonstrated efficacy in rhizofiltration. A wetland plant, *Carex pendula* has shown significant potential in removing lead (Pb) from contaminated water, primarily accumulating the metal in its root biomass (Yadav et al., 2011). Research has shown that sunflowers (*Helianthus annuus*), effectively remove up to 95% of Uranium (U) from water-based solutions in just one day (Cooney, 1996). Water hyacinths (*Eichhornia crassipes*), are famous for their ability to rhizofilter heavy metals like Cd, Cr, and Pb out of water sources. Surface water treatment (removal of contaminants from waterways impacted by farming runoff or urban pollution), groundwater remediation (cleaning up damaged aquifers), and industrial effluents are all potential applications of the technique.

Thus, this technique offers an effective and cheap solution to remove the problem of heavy metal contamination from water sources.

Phytovolatilization

In this process, the plants absorb pollutants, metabolize them into volatile components which are subsequently released into the atmosphere mainly through leaves. Elements like selenium (Se), mercury (Hg), and arsenic (As) as well as organic chemicals like trichloroethylene, benzene, nitrobenzene, phenol, and atrazine have been remedied using this procedure (Lone et al., 2008).

The widespread pollution and toxicity of arsenic make it a major environmental concern. Arsenic (As) hyperaccumulators are a group of plant species that are able to withstand and even accumulate very high As concentrations. The first known plant to hyperaccumulate arsenic was *Pteris vittata*, more often known as the Chinese brake fern, which was reported by Ma et al. (2001). Researchers have found that phytovolatilization can shorten the time it takes to remediate soils contaminated with arsenate by reducing the amount of arsenate-laden plant material that is produced. Arsenic is more bioavailable when it dissolves in soil thanks to organic acids released by the plant's roots, such as oxalic acid. After being absorbed, arsenate [As(V)] is converted to arsenite [As (III)], a form that is more mobile, inside the plant. Up to 95% of the absorbed arsenic accumulates in the above-ground tissues, but a considerable amount of As (III) is translocated to the fronds (Ma et al., 2001).

In another example of *P. vittata,* the plant volatilizes 90% of the arsenic from the contaminated soil and release into the atmosphere from their frond in a greenhouse set up. But again air emission of arsenic compounds possess a threat to the environment (Sakakibara et al., 2010).

Moreover, *P. vittata*'s ability to hyperaccumulate arsenic is influenced by factors such as availability of phosphorus in the soil. It has been established that high phosphorus levels in the plant roots contribute to the plant's arsenic uptake and tolerance. The fern's rapid growth, unique adaptations, makes it further a prime candidate for phytoremediation efforts in arsenic-contaminated sites.

4. KEY HYDROPHYTES IN HEAVY METAL BIOREMEDIATION

4.1 *Eichhornia crassipes*

The water hyacinth, or *Eichhornia crassipes*, is an infamous aquatic invasive plant that has its roots in South America. The invasiveness of this plant poses a threat to various water bodies and ponds because it limits boat traffic, affects aquaculture, and creates a myriad of other problems. Despite the negativities, this plant offers unique phytoremediating potential for eliminating Heavy metals, hazardous industrial effluents, and chemical pollutants. With its large root system, and quick growth rate, water hyacinth absorb heavy metals like Lead, cadmium, mercury, and arsenic and store in its fibrous roots resulting in better water quality and less eutrophication.

The increased rise in mining and agricultural activities has sumptuous release of HMs into water bodies, contributing ominously to water pollution. *Eichhornia crassipes* being invasive can accumulate huge amounts of cadmium (Cd), lead (Pb), zinc (Zn), magnesium (Mg), and arsenic (As) into its system. Even, biochar derived from water hyacinth biomass has emerged as an encouraging tool in the field of phytoremediation as it can absorb more HMs with incraesed surface area (Mitra et al., 2025). In a study conducted by Nyamunda et al. (2019) the effectiveness of magnetic biochar derived from the water hyacinth was explored in the phytoremediation of zinc (Zn^{2+}) and copper (Cu^{2+}) from industrial wastewater. The magnetic biochar validated a high adsorption capacity, enhanced porous structure, and more availability of surface functional groups. Thus, water hyacinth can be used in wastewater phytoremediation in both ways, through biomass and with the help of its biochar (Nyamunda et al., 2019; Mitra et al., 2025).

Lu et al., 2018 and Sompura et al., 2024, investigated the phytoremediation efficiency of *E. crassipes* in treating HMs like lead (Pb) and chromium (Cr) whereas Eid, 2020 reported their high tolerance towards Cd, Ni and Pb. In another study by Singh et al., 2022, HMs from glass factory were successfully remediated by the water hyacinth. Removal from 89.66% to 93.55% was achieved in cases of heavy metals like Pb and Cu, whereas Fe, Cd and Mn were also considerably high. Thus *E. crassipes* offers an inexplicable approach towards HM remediation which could offer promising results and applicability especially in economically poor regions with limited access to advanced treatment technologies.

4.2 *Lemna gibba*

Lemna gibba is visible in aquatic habitats as free-floating small aquatic plant of 3 to 5 mm in diameter. It belongs to Lemnaceae, the duckweed family and is widely acclaimed as ecological indicator. It grows profusely without any cultivation hassles and has unique ability to absorb pollutants from water through the process of accumulation, biodegradation, and phytoextraction.

The plant has shown great promise to absorb HMs like cadmium (Cd), lead (Pb), copper (Cu), arsenic (As), uranium (U) and zinc (Zn). Its roots and lower fronds tend to accumulate the most metals, often supported by root-associated microbial communities with metal clearance rate of 70 to 95%. Since *L. gibba* remain immersed in water most of the times, they are able to withstand high concentrations of hexavalent chromium, arsenic, and copper ions, which are omnipresent in wastewater (Aslanzadeh, et al., 2024; Rai and Nongtri, 2024; Al-Baldawi, et al., 2022). According to Sabó et al. (2023), *L. gibba* can also accumulate gadolinium, a rare earth metal from contaminated water. In a separate study, Turker and Baran (2018) demonstrated that *Lemna gibba*, when used in combination with chitosan, significantly enhances the removal of boron from drinking water. Boron is a compound that is used for many medicinal purposes, in agriculture, in fire retardants, in soaps and detergents, but despite being a helpful chemical compound in many aspects, it becomes harmful if the boron concentration increases in drinking water. It affects reproductive health and can cause many other health problems related to the liver, kidney, stomach, intestines, and brain, and can ultimately lead to many serious issues. This synergistic approach not only improved pollutant uptake but also provided a cost-effective and eco-friendly solution for boron remediation in potable water systems. The study highlighted the potential of integrating plant-based phytoremediation with biopolymer technologies for improved water treatment outcomes.

Furthermore, Chaudhary and Sharma (2019) investigated the plant's ability to remediate Cr and Cd, two highly toxic heavy metals, and observed not only high uptake efficiency but also ultrastructural deformations in plant tissues like chloroplast damage under metal stress. *L. gibba*'s bioaccumulation and sensitive indicator of environmental metal toxicity functions are highlighted by these results, which provide light on the organism's remediation capacities and stress responses. This aquatic plant proposes numerous benefits that make it highly suitable for phytoremediation applications especially for Cr, As and Cu. This plant is easy to harvest due to its floating nature and diminutive size, which allows for regular clearance of accumulated pollutants. Moreover, it can survive in both aerobic and anaerobic water settings, demonstrating its remarkable adaptability in both small-scale laboratory systems and large-scale field applications. As a model organism for environmental monitoring, it is extensively utilized in ecotoxicological testing supported by the standards, Organisation for Economic Co-operation and Development (OECD guidelines).

4.3 *Pistia stratiotes*

Pistia stratiotes, often called water lettuce or water cabbage, is a type of aquatic plant. Originating in the Nile Valley close to Lake Victoria, this plant is now known to be pantropical because it is native to both the Eastern and Western Hemispheres. Another name for this plant is Nile cabbage. Although it is cultivated for its aesthetic value, this plant is invasive in nature preferred in wetland systems.

The plant's rosettes-shaped leaves are meaty and float on airy fabric. It has the ability to grow quickly and create thick mats; its roots dangle freely in water, allowing it to efficiently contact pollutants. *Pistia* plants propagate quickly and absorb HMs in different experimental settings, which has been repeatedly proved by many investigators working on this forefront. According to earlier studies, the plant can effectively remove 90-95% of heavy metals like Pb, Cd, Cu and Zn from wastewater. In addition, Zahari et al., 2021 elaborated the role of microbial communities in *Pistia* rhizosphere for rhizodegradation of Pb. Chan et al., 2022 demonstrated the use of electro-assisted phytoremediation for efficient removal of Pb and Fe from wastewater and soil environment by *P. stratiotes*. Accumulation of these heavy metals was observed more at roots based on Bioaccumulation Factor (BCF) and Translocation (TF) values. These results verify the fact that the plant is resilient and efficient in collecting and detoxifying heavy metals of wide range.

Pistia stratiotes can be considered a good phytoremediating plant especially for removal of Pb, Cr, Ni, Co and Zn, more so because of its high biomass production rate, low maintenance cost, and high tolerance to metal pollutants.

4.4 *Ipomea aquatica*

Ipomoea aquatica is commonly known as water spinach. It belongs to the Convolvulaceae (Morning glory) family and the same genus as the sweet potato (*Ipomoea batatas*). It is a semi-aquatic, herbaceous plant that is edible and also has medicinal and pharmacological properties. It is widely found in Southeast Asia, South Asia, and East Asia. It features hollow, milky-sapped stems that can trail along moist soil or float in water, rooting at each node and often reaching lengths of 2–3 m or even up to 70 ft in optimal conditions. The leaves are simple, alternate, and typically 5–15 cm long, with arrowhead-shaped (sagittate) to lanceolate blades. Its showy, funnel or trumpet-shaped flowers span about 3–5 cm across and are usually white to pale lilac with a mauve center, borne singly or in small clusters in the leaf axils. Propagation occurs readily by both seeds and vegetative stem cuttings, which root easily when in contact with water or moist soil.

Ipomoea aquatica (water spinach) has demonstrated strong effectiveness in removing heavy metals from water and soil through mechanisms such as rhizofiltration and phytoextraction. In hydroponic setups, it accumulates lead primarily in roots and lower stems, with bioconcentration factors exceeding 1, indicating substantial uptake, though translocation to aerial parts is limited; its stems can even act as metal reservoirs. It also shows impressive cadmium absorption, with root BCF values ranging from 375 to over 2,200 L/kg and shoot values up to 144 L/kg in wastewater conditions. Additionally, in industrial wastewater treatment, studies report removal of over 1,500 mg/kg of aluminum and iron within ten days, evidencing rapid uptake in both root and stem tissues (Hanafiah et al., 2020). Field trials from automotive and industrial sites further confirm the plant's capacity to extract Co,

Mn, Zn and Cr from polluted soils. Its rapid propagation ensures quick recycling of heavy metals from polluted sites but challenges like proper disposal should be also taken care off to prevent re-release of toxins. In another study by Hisam et al. 2022, the water spinach, removed noteworthy concentration of Zn, Cu and Cd from wastewater. The plants exhibited remarkable uptake when exposed to 5–15 ppm of these metals for 20 days. Among them noteworthy is the Cd accumulation which reached 14 mg/kg in biomass, while 82% removal was observed in case of Co and Zn from the water contaminants.

Ipomoea aquatica as a powerful, phytoremediator suitable for removing heavy metals like Cd, Cu, and Zn, from shallow, aquatic surroundings. Its rapid growth and ease of propagation makes it supreme for controlled cleanup systems and could be used as a sustainable remediation strategy.

4.5 *Azolla pinnata*

This is a tiny, free-floating aquatic fern, also refereed as water velvet of Salviniaceae family. They form dense mats with 1-2 mm long velvety leaves on calm water surfaces, doubling its biomass within two days. The fern is typically used as biofertilizer and often observed in environments like rice paddies, ditches, and ponds. *A. pinnata* is a very efficient, fast-growing, nitrogen-fixing phytoremediating plant proven to be effective across a spectrum of wastewater types, particularly for heavy-metal contamination from industrial effluents. In a study by Hanafy et al., 2018, *Azolla* was exposed to increasing concentration of several heavy metals like lead, copper and cobalt, and it was observed the fern accumulated till 45 ppm/liter for all metals. In another study by Zeid et al. (2024), concluded *Azolla* could effectively remediate Nickel (Ni) contaminated water and observed that that uptill 190 ppm was the tolerable concentration.

Aquaculture is a major food sector today, contributing to a large part of the food protein for human consumption. Still, this form of farming also works as a source of heavy metal pollution, which contributes to persistent problems in food chain and environment. Mercury, cadmium and lead gains easy access to human food chain and cause significant health risk. Mitigation by plants through process of adsorption, guided by microbial assisted remediation is widely employed by scientists (Emenike et al., 2022). In 2022, Adabembe et al. reported that *A. pinnata* in constructed wetland setups removed 3–82% of heavy metals from aquaculture wastewater within five days, yielding water suitable for irrigation according to FAO/WHO standards.

Heavy metals one of the biggest concern for pollution remediating researchers, because these pollutants are most abundant and are very difficult to degrade and cause disturbance both in natural ecosystem and human health, but among many plants which can phytoremediate heavy metals through various techniques such as biodegradation, bioaccumulation, phytoextraction and phytostabilization and few other techniques of phytoremediation *Azolla pinnata* can also phytoremediate heavy metals from drainage water. Earlier research has also confirmed its capacity to hyperaccumulate metals like lead and zinc from sewage and industrial effluents over 42 days. The results show that *A. pinnata* can be used for two purposes: first, as a cheap way to clean water that are high in pollutants; second, as a biomass producer, which could lead to sustainable wastewater management.

5. CASE STUDIES AND APPLICATIONS

Advantages of Hydrophyte-Based Bioremediation

- **Economically sustainable:** They are nature's inherent filtering units which operates on their own or requires very low maintenance when performed in field conditions.
- Low **operational costs:** Presence of natural nutrients encourages rhizosphere microbes without any special nutrient supplement.
- **Reduced carbon footprint:** No toxic waste gases liberated and no energy or power input required to initiate phytoremediation.

5.1 Eco-Friendly & Sustainable

- **Reclamation of Resources:** Hydrophytes after HM absorption can be sacrificed to retrieve rare and important resources.
- **Composting:** The plant biomasses can be composted for agricultural purposes after retrieving HMs.
- **Bioenergy manufacture:** These plants possess high biomass which can be delignified and used for fuel generation.

5.2 Aesthetic & Biodiversity Benefits

- **Improving Ecosystem:** Man-made wetlands are aesthetically pleasing than other treatment facilities. They infuse an ecological purpose beyond merely filtering water.
- **Auxiliary Flora & Fauna:** Fish, birds, microbes, aquatic insects, and even endangered or rare species of flora and fauna find a home in these wetlands, which boosts biodiversity in the area. Hydrophytes are common in wetland ecosystems because they inherently prefer wet areas where they can grow and thrive.

- **Cultural, Recreational & Educational Value:** People use these areas for leisure, field trips, and environmental appreciation; wetlands are frequently incorporated into ecotourism or local education, which improves cultural participation. It is generally recognized that exposure to green and blue areas, such as wetlands, improves human health and well-being by lowering stress and enhancing mental and physical health.

5.3 Services for Regulation and Support

- **Water and Nutrient Cycle:** Hydrophytes assist broader hydrologic activities, including flood reduction, groundwater recharge, and stormwater management, as well as regulating nutrient levels (N, P) and facilitating sediment retention.
- **Maintaining Ecosystem Services:** Water purification, recreation, and other ecosystem services are supported by habitat construction and nutrient cycling.

In cities, constructed wetlands contribute to stormwater absorption, flood delay, and water reuse.

6. CHALLENGES AND LIMITATIONS

6.1 Seasonal Growth Fluctuations & Plant Performance

Since hydrophytes, including *Eichhornia, Typha, Phragmites,* and *Lemna*, are sensitive to light, temperature, and photoperiod, they exhibit significant seasonal variation in growth.

Because water hyacinth development is constrained in colder climates and frequently stops in the winter, bioremediation efforts are reduced during those times, making operations season-dependent. Furthermore, some species' slower development rates may limit their ability to be used widely and effectively throughout the year.

6.2 Ecological Risks and Invasiveness

Water hyacinth species, such as *Eichhornia crassipes*, are known for their rapid growth and ecological disturbance. Under warm, nutrient-rich conditions, it can quickly double biomass and cover water bodies in a matter of weeks, changing oxygen levels, obstructing light, and displacing native biodiversity. Secondary contamination, especially oxygen depletion and deterioration of water quality, can result from the post-harvest biomass's ability to break down, degrade, and sink. Combining mechanical removal, biological agents (such weevils), and nutrient control, all of which have drawbacks and compromises, is necessary to manage the spread of invasive species.

6.3 Metal Saturation and Phytotoxicity

Metals are accumulated by hydrophytes, but exceeding thresholds might damage plant health or prevent uptake.

The capacity of constructed wetlands is limited. Saturation could happen, preventing long-term buildup, unless the system media is changed or the biomass is routinely removed and replenished.

6.4 Biomass Disposal & Secondary Contamination

To avoid environmental hazards, accumulated metal-laden plant material must be handled carefully. The difficulties in disposing of biomass are well known; it needs to be gathered, processed, or valorized, yet improper disposal could return pollutants to the environment.

Emerging alternatives include pyrolysis-based biochar manufacturing, which stabilizes toxins and produces a sorptive product for energy, carbon sequestration, or soil amendment. However, there are still hazards of residual pollutants, inconsistent feedstock availability, and the requirement for reliable logistics.

As an alternative, water hyacinth compost or additives have potential; however, there are questions over their agronomic safety, nutrient release patterns, heavy metal concentration, and widespread acceptance.

6.5 Site-Specific Efficacy & Context Sensitivity

The environmental and chemical context is crucial to the efficacy of bioremediation. Effectiveness varies across water chemistry, pollutant types and loads, flow regimes, pH, salinity, and redox conditions. Only metals and pollutants within root zones are remediated; deeper pollution is unaffected by the rooting depth, which restricts the reach of contaminants.

7. ENHANCING BIOREMEDIATION EFFICIENCY

7.1 Genetic Engineering

Phytoremediation properties are being continuously improved by genetic alteration and sophisticated breeding methods. Plants have been modified to express proteins that detoxify or bind to metals. Plants that combine high accumulation and robust growth have been produced by crossing hyperaccumulators with high-biomass species or by employing chemical mutagens. Engineered plants exhibit improved metal absorption, translocation, and sequestration by overexpressing genes encoding metal transporters (e.g., ZIP, MTP, HMA) or metal chelators (phytochelatins, metallothioneins). This makes it possible to modify rapidly developing hydrophytes with hyperaccumulation and heavy metal tolerance. But the intricacy of characteristics, environmental hazards, legal issues, and the unproven extensive field implementation continue to be major barriers.

7.2 Microbial Synergy (Plant–Microbe Partnerships)

The efficacy of phytoremediation is significantly increased when hydrophytes are paired with favourable microbes. Through siderophore generation, organic acid secretion, phosphate solubilization, ethylene mitigation, and root architecture enhancement, Plant Growth-Promoting Rhizobacteria (PGPR) improve biomass and pollutant uptake while also increasing metal bioavailability and plant stress tolerance (Yan et al., 2020). Arbuscular mycorrhizae (AMF) increase the surface area of roots, enhance the uptake of nutrients and metals, and lessen toxicity. In soils under metal stress, mycorrhizal helper bacteria (MHBs) promote fungal colonization and increase resistance. Without requiring genetic alteration, these synergistic relationships improve plant health and absorption capacity, providing a strong, environmentally beneficial improvement.

7.3 Sequential Planting (Species Mixing)

Using combinations of hydrophytes, sequential or hybrid systems can target multiple metals more effectively. By combining vertical-flow and horizontal-flow constructed wetlands (CWs) with diverse plant species, hybrid CWs maximize the removal of different contaminants by utilizing their unique characteristics (e.g., rooting depth, absorption methods).

7.4 Periodic Harvesting

Regular biomass removal maintains system effectiveness and prevents pollutant re-release. It reduces crowding, accelerates growth, and sustains high biomass production and pollutant uptake efficacy.

8. FUTURE DIRECTIONS

Omics technology are advancing hydrophytes as a sustainable bioremediation method in wetland habitats. Hydrophytes can be genetically modified to improve metal uptake, tolerance, and accumulation for more efficient site clean-up. Combining hydrophytes with metal-resistant bacteria and mycorrhizal fungus may boost phytoremediation efficacy. Advanced omics like genome decrypting, transcript decoding, protein translation, and metabolite analysis explore the cryptic dynamic mechanism essential for heavy metal bioremediation from wetland environments. Finding important genes for heavy metal sequestering and studying their proteins, metabolites, and detoxification sites will illuminate their metabolic and genetic pathways. By combining data from several omics studies, correlation between plants and microbes interaction and how they respond to metal stress, will help us restore wetlands more effectively.

9. CONCLUSION

An efficient, environmentally conscious, and sustainable method of controlling heavy metal contamination in wetland environments is hydrophyte-based bioremediation. While public awareness campaigns might encourage acceptance and involvement, incorporating these technologies into urban planning can improve water management. Frequent monitoring guarantees effectiveness and safety, and explicit rules for the reuse or disposal of biomass aid in avoiding secondary pollution. Future developments in omics technologies (genomics, proteomics, and metabolomics) can enhance understanding and optimize metal uptake, and climate-resilient designs will guarantee performance throughout the year. Additionally, hybrid systems that combine phytoremediation with complementary technologies, such as solar-powered aeration or the application of biochar, may increase the effectiveness of remediation. Through mechanisms like phytoaccumulation, phytostabilization, and phytovolatilization, hydrophytes, including *Eichhornia crassipes, Typha,* and *Phragmites*, have an inherent capacity to absorb, stabilize, and even transform harmful metals. Frequently, they do this in conjunction with helpful microorganisms. Hydrophyte-based wetlands can improve biodiversity, repair damaged ecosystems, and treat wastewater. When used cautiously, they can offer an inexpensive, low-maintenance, and visually pleasing way to combat pollution, promoting the objectives of sustainable development and ecological health.

REFERENCES

Al-Baldawi, I. A., Yasin, S. R., Jasim, S. S., Abdullah, S. R. S., Almansoory, A. F., & Ismail, N. I. (2022). Removal of copper by Azolla filiculoides and Lemna minor: Phytoremediation potential, adsorption kinetics and isotherms. *Heliyon*, *8*(11), e11456. DOI: 10.1016/j.heliyon.2022.e11456 PMID: 36406685

Ali, H., & Khan, E. (2019). Trophic transfer, bioaccumulation, and biomagnification of non-essential hazardous heavy metals and metalloids in food chains/webs—Concepts and implications for wildlife and human health. *Human and Ecological Risk Assessment*, *25*(6), 1353–1376. DOI: 10.1080/10807039.2018.1469398

Alikhani, S., Nummi, P., & Ojala, A. (2021). Urban wetlands: A review on ecological and cultural values. *Water (Basel)*, *13*(22), 3301. DOI: 10.3390/w13223301

Alsafran, M., Usman, K., Ahmed, B., Rizwan, M., Saleem, M. H., & Al Jabri, H. (2022). Understanding the Phytoremediation Mechanisms of Potentially Toxic Elements: A Proteomic Overview of Recent Advances. *Frontiers in Plant Science*, *13*, 881242. DOI: 10.3389/fpls.2022.881242 PMID: 35646026

Angelova, V., Ivanova, R., & Todorov, T. (2004). Use of *Gossypium hirsutum* for phytoextraction of heavy metals from contaminated soils. *Journal of Environmental Protection and Ecology*, *5*(4), 886–891.

Aslanzadeh, M., Saboora, A., & Moradlou, O. (2024). Phytoremediation potential of duckweed (Lemna minor L.) for hexavalent chromium removal in synthetic wastewater: Unveiling physiological response and defense mechanisms against excessive heavy metal uptake. *International Journal of Environmental Science and Technology*, *21*(16), 10155–10174. DOI: 10.1007/s13762-024-05721-6

Biswal, T. (2025). Rhizofiltration: A Sustainable Green Technology for Remediation of Heavy Metals from Aquatic Systems. In *Recent Advances in Bioremediation and Phytoremediation* (pp. 1–21). Springer Nature Switzerland. DOI: 10.1007/978-3-031-77884-1_1

Chan, M. Y., Tee, C. S., Chai, T. T., Sim, Y. L., & Beh, W. L. (2022). Evaluation of electro-assisted phytoremediation (EAPR) system for heavy metal removal from synthetic leachate using Pistia stratiotes. *International Journal of Phytoremediation*, *24*(13), 1376–1384. DOI: 10.1080/15226514.2022.2031863 PMID: 35191343

Chan, Y. C., Brown, P., & Chan, M. (2004). Phytoremediation of metal-contaminated soils. *Environmental Pollution*, *133*(3), 393–401. DOI: 10.1016/S0269-7491(01)00286-X PMID: 12009137

Chaudhary, E., & Sharma, P. (2019). Chromium and cadmium removal from wastewater using duckweed-Lemna gibba L. and ultrastructural deformation due to metal toxicity. *International Journal of Phytoremediation*, *21*(3), 279–286. DOI: 10.1080/15226514.2018.1522614 PMID: 30698020

Collin, M. S., Venkatraman, S. K., Vijayakumar, N., Kanimozhi, V., Arbaaz, S. M., Stacey, R. S., & Swamiappan, S. (2022). Bioaccumulation of lead (Pb) and its effects on human: A review. *Journal of Hazardous Materials Advances*, *7*, 100094. DOI: 10.1016/j.hazadv.2022.100094

Cooney, C. M. (1996). Sunflowers remove radionuclides from water in ongoing phytoremediation field tests. *Environmental Science & Technology*, *30*(5), 194A. DOI: 10.1021/es962219c

Eid, E. M., Galal, T. M., Sewelam, N. A., Talha, N. I., & Abdallah, S. M. (2020). Phytoremediation of heavy metals by four aquatic macrophytes and their potential use as contamination indicators: A comparative assessment. *Environmental Science and Pollution Research International*, *27*(11), 12138–12151. DOI: 10.1007/s11356-020-07839-9 PMID: 31984462

Emenike, E. C., Iwuozor, K. O., & Anidiobi, S. U. (2022). Heavy Metal Pollution in Aquaculture: Sources, Impacts and Mitigation Techniques. *Biological Trace Element Research*, *200*(10), 4476–4492. DOI: 10.1007/s12011-021-03037-x PMID: 34813030

Faroon, O., Ashizawa, A., Wright, S., . . . (2012). *Toxicological Profile for Cadmium*. Atlanta (GA): Agency for Toxic Substances and Disease Registry (US); Chapter: POTENTIAL FOR HUMAN EXPOSURE.

Genchi, G., Sinicropi, M. S., Lauria, G., Carocci, A., & Catalano, A. (2020, May 26). The Effects of Cadmium Toxicity. *International Journal of Environmental Research and Public Health*, *17*(11), 3782. DOI: 10.3390/ijerph17113782 PMID: 32466586

Ghosh, M., & Singh, S. P. (2005). A review on phytoremediation of heavy metals and utilization of it's by products. *Asian Journal of Energy and Environment*, *6*(4), 18.

Hanafiah, M. M., Zainuddin, M. F., Mohd Nizam, N. U., Halim, A. A., & Rasool, A. (2020). Phytoremediation of Aluminum and Iron from Industrial Wastewater Using *Ipomoea aquatica* and *Centella asiatica*. *Applied Sciences (Basel, Switzerland)*, *10*(9), 3064. DOI: 10.3390/app10093064

Hanafy, R. R., Eweda, W. E., Zayed, M. S., & Khalil, H. (2018). Potentiality of using a. pinnata to bioremediate different heavy metals from polluted draining water. *Magallat Itihad al-Gami'at al-'Arabiyyat Lil-Dirasat Wa-al-Buhut al-Zira'iyyat*, *26*(1), 359–372. DOI: 10.21608/ajs.2018.14022

Hisam, N. I. B., Zakaria, M. Z., Azid, A., Bakar, M. F. A., & Samsudin, M. S. (2022). Phytoremediation process of water spinach (Ipomoea aquatica) in absorbing heavy metal concentration in wastewater. *Journal Of Agrobiotechnology, 13*(1S), 131–144. DOI: 10.37231/jab.2022.13.1S.322

Jamil Emon, F., Rohani, M. F., Sumaiya, N., Tuj Jannat, M. F., Akter, Y., Shahjahan, M., Abdul Kari, Z., Tahiluddin, A. B., & Goh, K. W. (2023, June 6). Bioaccumulation and Bioremediation of Heavy Metals in Fishes — A Review. *Toxics, 11*(6), 510. DOI: 10.3390/toxics11060510 PMID: 37368610

Kabeer, R., VP, S., CS, P. K., AP, T., V, S., EK, R., & KR, B. (. (2022). Role of heavy metal tolerant rhizosphere bacteria in the phytoremediation of Cu and Pb using Eichhornia crassipes (Mart.) Solms. *International Journal of Phytoremediation, 24*(11), 1120–1132. DOI: 10.1080/15226514.2021.2007215 PMID: 34846266

Korzeniowska, J., & Stanislawska-Glubiak, E. (2019). Phytoremediation potential of Phalaris arundinacea, Salix viminalis and Zea mays for nickel-contaminated soils. *International Journal of Environmental Science and Technology, 16*(4), 1999–2008. DOI: 10.1007/s13762-018-1823-7

Kumar, A., Kumar, V., Chawla, M., Thakur, M., Bhardwaj, R., Wang, J., & Rinklebe, J. (2024). Bioremediation of mercury contaminated soil and water: A review. *Land Degradation & Development, 35*(4), 1261–1283. DOI: 10.1002/ldr.4989

Kumari, M., & Tripathi, B. D. (2015). Efficiency of Phragmites australis and Typha latifolia for heavy metal removal from wastewater. *Ecotoxicology and Environmental Safety, 112*, 80–86. DOI: 10.1016/j.ecoenv.2014.10.034 PMID: 25463857

Kumari, M., & Tripathi, B. D. (2015). Efficiency of Phragmites australis and Typha latifolia for heavy metal removal from wastewater. *Ecotoxicology and Environmental Safety, 112*, 80–86. DOI: 10.1016/j.ecoenv.2014.10.034 PMID: 25463857

Kwok, Hong-Ching, & Chan, Wan (2022). Wheatgrass (Triticum aestivum) as an Efficient Phytoremediation Plant for Aristolochic Acid-Contaminated Water and Arable Soil. *ACS Publications.* doi:. 2c00057.s001.DOI: 10.1021/acsagscitech

Lasat, M. M. (2002). Phytoremediation of metals. *Brazilian Journal of Plant Physiology, 14*(3), 109–119.

Leaphart, J. C., Abercrombie, S. A., Borchert, E. J., Bryan, A. L.Jr, & Beasley, J. C. (2022). Bioaccumulation of Mercury and Radiocesium in Waterfowl Introduced to a Site with Legacy Contamination. *Environmental Toxicology and Chemistry, 41*(10), 2479–2487. DOI: 10.1002/etc.5444 PMID: 35866467

Li, C., Wang, H., Liao, X., Xiao, R., Liu, K., Bai, J., Li, B., He, Q. (2022). Heavy metal pollution in coastal wetlands: A systematic review of studies globally over the past three decades. *Journal of Hazardous Materials*, 424(Pt A), 127312. .DOI: 10.1016/j.jhazmat.2021.127312

Li, W., Wang, Z., & Li, J. (2019). Microbial reduction of hexavalent chromium: Mechanisms and applications. *Environmental Science & Technology*, *53*(1), 14–34.

Lone, M. I., He, Z. L., Stoffella, P. J., Yang, X. E. (2008 Mar). Phytoremediation of heavy metal polluted soils and water: Progresses and perspectives. *Journal of Zhejiang University Science B*, 9(3), 210-220. doi:. B0710633.DOI: 10.1631/jzus

Lu, B., Xu, Z., Li, J., & Chai, X. (2018). Removal of water nutrients by different aquatic plant species: An alternative way to remediate polluted rural rivers. *Ecological Engineering*, *110*, 18–26. DOI: 10.1016/j.ecoleng.2017.09.016

Ma, L. Q., Komar, K. M., Tu, U. C., Zhang, W., Cai, Y., & Kennelley, E. D. (2001). A fern that hyperaccumulates arsenic. *Nature*, *409*(6820), 579. DOI: 10.1038/35054664 PMID: 11214308

Maine, M. A., Hadad, H. R., Camaño Silvestrini, N. E., Nocetti, E., Sanchez, G. C., & Campagnoli, M. A. (2022). Cr, Ni, and Zn removal from landfill leachate using vertical flow wetlands planted with Typha domingensis and Canna indica. *International Journal of Phytoremediation*, *24*(1), 66–75. DOI: 10.1080/15226514.2021.1926909 PMID: 34077330

Marisa, L., de Reyniès, A., Duval, A., Selves, J., Gaub, M. P., Vescovo, L., Etienne-Grimaldi, M.-C., Schiappa, R., Guenot, D., Ayadi, M., Kirzin, S., Chazal, M., Fléjou, J.-F., Benchimol, D., Berger, A., Lagarde, A., Pencreach, E., Piard, F., Elias, D., & Boige, V. (2013). Gene expression classification of colon cancer into molecular subtypes: Characterization, validation, and prognostic value. *PLoS Medicine*, *10*(5), e1001453. DOI: 10.1371/journal.pmed.1001453 PMID: 23700391

Meagher, R. B. (2000). Phytoremediation of toxic metals. *Bioscience*, *50*(6), 469–478.

Mitra, I., De, K., & Pal, D. (2025). Biochar: Insinuations in Agriculture and Bioremediation. *Food and Scientific Reports*, *6*(1), 62–66.

Mitsch, W. J., Bernal, B., & Hernandez, M. E. (2015). Ecosystem services of wetlands. *The International Journal of Biodiversity Science, Ecosystem Services & Management*, *11*(1), 1–4. DOI: 10.1080/21513732.2015.1006250

Nnaji, N. D., Onyeaka, H., Miri, T., & Ugwa, C. (2023). Bioaccumulation for heavy metal removal: A review. *SN Applied Sciences*, *5*(5), 125. DOI: 10.1007/s42452-023-05351-6

Nyamunda, B. C., Chivhanga, T., Guyo, U., & Chigondo, F. (2019). Removal of Zn (II) and Cu (II) ions from industrial wastewaters using magnetic biochar derived from water hyacinth. *Journal of Engineering, 2019*, 1–11. Advance online publication. DOI: 10.1155/2019/5656983

Pal, D., Bandyopadhyay, P., Bhowmick, S., & Maji, P. (2025). Heavy Metal Detection and Antibacterial Activity of Phytosynthesized Copper oxide (CuO) Nanoparticles using Leaf Extract of Colocasia esculenta. *International Journal of Ecology and Environmental Sciences, 51*(1), 161–166. DOI: 10.55863/ijees.2025.0524

Pang, Y. L., Quek, Y. Y., Lim, S., & Shuit, S. H. (2023). Review on phytoremediation potential of floating aquatic plants for heavy metals: A promising approach. *Sustainability (Basel), 15*(2), 1290. DOI: 10.3390/su15021290

Pivetz, B. E. (2001). *Phytoremediation of contaminated soil and ground water at hazardous waste sites*. US Environmental Protection Agency, Office of Research and Development, Office of Solid Waste and Emergency Response.

Rai, P. K., & Nongtri, E. S. (2024). Heavy metals/-metalloids (As) phytoremediation with Landoltia punctata and Lemna sp.(duckweeds): Coupling with biorefinery prospects for sustainable phytotechnologies. *Environmental Science and Pollution Research International, 31*(11), 16216–16240. DOI: 10.1007/s11356-024-32177-5 PMID: 38334920

Sabreena, H. S., Hassan, S., Bhat, S. A., Kumar, V., Ganai, B. A., & Ameen, F. (2022). Phytoremediation of Heavy Metals: An Indispensable Contrivance in Green Remediation Technology. *Plants, 11*(9), 1255. DOI: 10.3390/plants11091255 PMID: 35567256

Sakakibara, M., Watanabe, A., Inoue, M., Sano, S., & Kaise, T. (2010). Phytoextraction and phytovolatilization of arsenic from As-contaminated soils by Pteris vittata. In *Proceedings of the annual international conference on soils, sediments, water and energy*. 12(1) 26.

Sharma, P., Singh, S. P., Parakh, S. K., & Tong, Y. W. (2022). Health hazards of hexavalent chromium (Cr (VI)) and its microbial reduction. *Bioengineered, 13*(3), 4923–4938. DOI: 10.1080/21655979.2022.2037273 PMID: 35164635

Sharma, S., Singh, S., & Goyal, P. (2019). Heavy metal removal from wastewater using chemical and biological methods. *Water Science and Technology, 80*(12), 2155–2169.

Singh, J., Kumar, V., Kumar, P., & Kumar, P. (2022). Kinetics and prediction modeling of heavy metal phytoremediation from glass industry effluent by water hyacinth (Eichhornia crassipes). *International Journal of Environmental Science and Technology*, *19*(6), 5481–5492. DOI: 10.1007/s13762-021-03433-9

Sompura, Y., Bhardwaj, S., Selwal, G., Soni, V., & Ashokkumar, K. (2024). Unrevealing the potential of aquatic macrophytes for phytoremediation in heavy metal-polluted wastewater. *Journal of Current Opinion in Crop Science*, *5*(1), 48–61. DOI: 10.62773/jcocs.v5i1.233

Song, W. Y., Sohn, E. J., Martinoia, E., & Lee, Y. J. (2003). Engineering tolerance and accumulation of lead and cadmium in transgenic plants. *Nature Biotechnology*, *21*(8), 914–919. DOI: 10.1038/nbt850 PMID: 12872132

Vangronsveld, J.. (2009). Phytoremediation of contaminated soils and sediments: A review. *Environmental Science and Pollution Research International*, *16*(6), 715–731.

Wu, Y., Sun, J., Hu, B., Zhang, G., & Rousseau, A. N. (2023). Wetland-based solutions against extreme flood and severe drought: Efficiency evaluation of risk mitigation. *Climate Risk Management*, *40*, 100505. DOI: 10.1016/j.crm.2023.100505

Yadav, B. K., Siebel, M. A., & van Bruggen, J. J. (2011). Rhizofiltration of a heavy metal (lead) containing wastewater using the wetland plant Carex pendula. *Clean (Weinheim)*, *39*(5), 467–474. DOI: 10.1002/clen.201000385

Yan, A., Wang, Y., Tan, S. N., Mohd Yusof, M. L., Ghosh, S., & Chen, Z. (2020, April 30). Phytoremediation: A Promising Approach for Revegetation of Heavy Metal-Polluted Land. *Frontiers in Plant Science*, *11*, 359. DOI: 10.3389/fpls.2020.00359 PMID: 32425957

Zahari, N. Z., Tuah, P. M., & Rahim, S. A. (2021, September). Inoculation of Bacillus cereus enhance phytoremediation efficiency of Pistia stratiotes and Eichhornia crassipes in removing heavy metal Pb. []. IOP Publishing.]. *IOP Conference Series. Earth and Environmental Science*, *847*(1), 012012. DOI: 10.1088/1755-1315/847/1/012012

Zeid, I., Ghaly, E. K., & Shedeed, Z. A. (2024). Azolla pinnata as a phytoremediator: Improves germination, growth and yield of maize irrigated with Ni-polluted water. *Scientific Reports*, *14*(1), 22284. DOI: 10.1038/s41598-024-72651-1 PMID: 39333677

Zhao, F. J.. (2018). Phytoremediation of heavy metals: A review. *Environmental Pollution*, *234*, 785–794.

Zhou, H., & Lai, C. (2018). Chelation strategies in phytoremediation of heavy metals. *Frontiers in Plant Science*, *9*, 1290.

Zhou, Q., Zhang, J., Fu, J., Shi, J., & Jiang, G. (2007). Biomonitoring: An appealing tool for assessment of metal pollution in the aquatic ecosystem. *Analytica Chimica Acta, 606*(2), 135–150. DOI: 10.1016/j.aca.2007.11.018 PMID: 18082645

KEY TERMS AND DEFINITIONS

Hydrophytes: They are the plants that have adaptations to thrive in aquatic environments or in soils where the moisture content is abundant and the oxygen concentration is generally less.

Wetlands: Wetlands are unique ecosystems where the land is persistently or periodically saturated with water, creating conditions that sustain distinct, water-adapted plants and wildlife.

Heavy metals: They are a class of dense elements characterized by high atomic weight and density, which is almost five times that of water, many of which exhibit toxic properties even at low concentrations.

Phytoremediation: Phytoremediation is an eco-friendly approach that harnesses the ability of green plants to remove pollutants from the environment.

Chapter 6
Hydrophytes for Sustainable Agriculture: Achieving SDG3 by Soil Reclamation

Neha Minj
https://orcid.org/0009-0008-4233-4865
Amity University, Mumbai, India

Shashi Bala Singh
https://orcid.org/0000-0003-2448-2393
Amity University, Mumbai, India

Sagar Barage
Amity University, Mumbai, India

ABSTRACT

Hydrophytes, or aquatic plants, play a significant role in sustainable agriculture by supporting the achievement of Sustainable Development Goals (SDGs), particularly SDG 3 (Good Health and Well-being) through bioremediation and soil reclamation, water purification, and enhanced carbon storage. Land reclamation practice is the process of converting heavy metal contaminated water bodies and soil into usable land for various human activities like agriculture, urbanization, and industrial use. Phytoremediation is a cost-effective and environmentally friendly method in the revitalization of the environment by utilizing plants to restore polluted water and land sources. Some hydrophytes such as water cabbage (Pistiastratiotes), duckweed (Lemnagibba), and water hyacinth (Eichhorniacrassipes) are floating hydrophytes

DOI: 10.4018/979-8-3373-4037-1.ch006

famous for their pollution absorption capability mostly for metals in industrial waste water. This chapter discuss about reclamation of soil based on re-vegetation with hydrophytes.

1. INTRODUCTION

The United Nations Sustainable Development Goals (SDGs), particularly SDG 3: Good Health and Well-Being, which prioritizes healthy lives and well-being for all, is dependent on sustainable agriculture (FAO, 2020). Public health and agricultural practices are closely related due to the fact that food production requires the growing of food on healthy soil that is free from harmful toxins and residues. Nevertheless, soil degradation caused by pollution, nutrient depletion, salinization, waterlogging, and heavy metal contamination (i.e., arsenic, cadmium, lead, etc.) complicates food security and agriculture across large regions of the world.

Soil reclamation has become an important approach in attempting to restore soil health and supporting sustainable agricultural systems in order to address these problems. Hydrophytes, which are plants typically found in flooded or waterlogged conditions, represent a potential benign approach to remediation (Brix, 1997).Species of hydrophytes like *Typhalatifolia, Phragmitesaustralis, Eichhorniacrassipes,* and *Cyperusalternifolius*possess physiological and morphological features that allow them to thrive under highly saline, contaminated, or hypoxic conditions (Colmer& Flowers, 2008). The physiological and morphological properties enable hydrophytes to facilitate soil restoration through a variety of abiotic and biotic mechanisms such as promoting microbial activity, immobilizing or extracting metals, reducing salinity, enhancing organic matter content, and phytoremediation of both organic and inorganic contaminants (Shahid et al., 2018).Hydrophytes are also a source of other ecosystem services, including water quality control, reducing eutrophication, improving soil-water interactions, and creating habitats that indirectly facilitate the development of crops (Vymazal, 2011).

There are three objectives of this chapter:

i) To provide an overview of the ecological, physiological, and remedial mechanisms of hydrophytes in soil reclamation.
ii) To evaluate their potential role in improving public health and agriculture sustainability in the context of SDG3.
iii) To promote the growth of hydrophyte-based soil reclamation as a scalable, sustainable, health-centered intervention by identifying gaps in the current research and outlining research goals moving forward.

By completing these objectives, this chapter depicts hydrophytes as a green technology that can provide human health and food security while addressing food resilience and environmental conservation.

2. HYDROPHYTES ROLE IN ECOLOGICAL BALANCE

Hydrophytes, or aquatic plants, are key players in freshwater and wetland systems and help to ensure ecological stability and balance. As specialized plants that can include macroalgae, mosses, ferns, and angiosperms, hydrophytes have adapted to living in water or waterlogged soil and are often referred to as "ecosystem engineers" due to their strong impact on their surrounding environment (Wetzel, 2001). Hydrophytes contribute ecologically to primary productivity, habitat, water purification, and sediment stabilization, among other functions.

2.1 Basis of the Aquatic Food Web and Habitat Hydrophytes as primary producers are the ones who form the foundation of the aquatic food chain by the process of turning solar energy in the form of photons into biomass through the chemical reaction called photosynthesis. This biomass serves as a food source for small organisms, like zooplankton, aquatic insects, fish, and mammals and even birds (New Hampshire Department of Environmental Services, 2019). In addition to direct consumption, their decay supports the detrital food web and thereby the microbial activity, which is simultaneously crucial for nutrient cycling. Hydrophytes are a major source of structural habitat and refuge. The submerged plants, such as *Hydrilla* and *Vallisneria*, create "underwater forests", positioning and drawing many fish, amphibians, invertebrates, etc. (Beckett et al., 1992). Moreover, the emergent plants, such as cattails (*Typha*) and bulrushes (*Scirpus*), not only provide the birds and small mammals with hiding places and nesting sites, respectively, but also the wide variety of flora and fauna becomes a life-supporting system of the ecosystem (New Hampshire Department of Environmental Services, 2019).

2.2 Water Quality Regulation and Bioremediation Hydrophytes are utilized as filters of nature. They capture and remove nutrients from the water directly and also actively contribute to nutrient sequestration and phytoremediation. The process through which they remove the excess nutrients from the water and the soil is called phytoremediation, and the main nutrients that they absorb are nitrates and phosphates, which are then used for their growth (Sondergaard& Moss, 1998). The controlling of eutrophication, which leads to the formation of damaging algal blooms, is thus significant, and hydrophytes play a major role in preventing it by being the source of the constructing wetland wastewater treatment, which is sustainable and low-cost. There are certain hydrophytes that can take up heavy metals and other contaminants from the ground through their roots; this process is known as phytoextraction. These plants can then hold the metals in their tissues. The invasive water hyacinth (*Eichhorniacrassipes*) and broadleaf cattail (*Typhalatifolia*) are two plant species that have been studied for their ability to uptake and remove such metals as zinc (Zn) and copper (Cu) from water; hence, they are being considered for both ecological restoration and environmental engineering (Farid et al., 2023).

2.3 Sediment Stabilization and Erosion Control Rooted hydrophytes in particular are the most important players in the process of sediment stabilisation and shoreline protection. The extensive root and rhizome systems of hydrophytes create a natural mesh that binds the bottom substrate, thus preventing it from being re-suspended into the water column. Water turbidity as a result of sediment binding is reduced, which leads to greater light penetration, thus facilitating higher rates of photosynthesis and improved overall water clarity (Houwing et al., 2002).This process is very important for keeping the aquatic ecosystem productive and healthy.

2.4 Bioindicators and Climate Change Mitigation The composition and health of hydrophyte communities give an excellent indication of the ecological status of an aquatic system. Hydrophytes, being stationary, allow the presence, absence, or dominance of certain species to reflect the environmental conditions over a long period of time, e.g., nutrient loading, pollution, or hydrological changes ((Farid et al., 2023; De et al., 2019).For instance, a gradual shift towards invasive and fast-growing species usually signifies that nutrient pollution is high. Like all plants, hydrophytes also play a role in mitigating climate change by taking in atmospheric carbon dioxide (CO_2) during the process of photosynthesis. Wetlands, in this case, are the world's largest carbon sinks, as they bury in their highly vegetated soils around 20-30% of the global carbon pool; thus, the conservation of hydrophyte communities can be seen as an investment in the health of the planet (Lal, 2008).

Table 1. Hydrophytes Ecological Roles in Preserving Ecosystem Balance

Ecological Role	Contribution to Ecosystem Balance	Example Plants	References
Primary Production & Food Web Support	Forms the base of the aquatic food chain; provides energy for zooplankton, invertebrates, fish, and waterfowl.	*Hydrilla*, *Vallisneria* (submerged); *Lemna* (duckweed), *Pistia* (floating)	(New Hampshire Department of Environmental Services, 2019; De et al., 2019)
Habitat & Refuge Provision	Creates complex structure, offering shelter, spawning, and nursery grounds for fish and macro-invertebrates.	Submerged: *Myriophyllum*; Emergent: *Typha* (cattail)	(Beckett et al., 1992; De et al., 2019)
Water Oxygenation	Releases dissolved oxygen into the water column during photosynthesis, vital for aerobic aquatic life.	Submerged species (e.g., *Ceratophyllum*)	(New Hampshire Department of Environmental Services, 2019)
Nutrient Sequestration (Eutrophication Control)	Absorbs excess nitrogen and phosphorus from the water and sediment, limiting the growth of harmful algae.	*Phragmites* (emergent); *Eichhorniacrassipes* (floating)	(Søndergaard& Moss, 1998; Farid et al., 2023)
Sediment Stabilization	Root and rhizome systems bind the bottom substrate, reducing erosion, re-suspension, and turbidity.	Rooted emergent and submerged plants (e.g., *Scirpus*, *Potamogeton*)	(Houwing et al., 2002)
Bioremediation & Water Purification	Accumulates and sequesters heavy metals and pollutants from industrial wastewater (phytoextraction).	*Typhalatifolia*, *Eichhorniacrassipes* (water hyacinth)	(Farid et al., 2023)
Bioindication	Community structure and presence/absence reflect the trophic status and overall water quality of the ecosystem.	Certain species of *Ceratophyllum* or invasive macrophytes	(De et al., 2019)
Carbon Sequestration	Stores atmospheric carbon dioxide (CO_2) in biomass and sediment, contributing to climatechangemitigation in wetlands.	Wetland plants and deep-rooted macrophytes	(Lal, 2008; Wetzel, 2001)

3. HEAVY METAL CONTAMINATION OF SOIL AND WATER BODIES

The presence of heavy metals in the soil and water sources is a major global problem concerning environmental quality, food security, and human health that in a direct way hinders the realization of Sustainable Development Goal 3 (SDG 3): Good Health and Well-being (UN, 2015).

3.1 Sources and Characteristics of Heavy Metal Contamination

The term 'heavy metal (HM) pollutants' includes metallic elements and metalloids, and in general, they indeed have higher density (Tchounwou et al., 2012). More significantly, they are typically non-biodegradable and very persistent in the environment, which is one feature that contributes to their hazardous tendency to bioaccumulate in trophic levels and in the related food chain (Jaishankar et al., 2014). The most important and toxic HMs of relevance to the environment are arsenic (As), cadmium (Cd), chromium (Cr), lead (Pb), and mercury (Hg). These features indicate the seriousness of their presence as an environmental pollutant.

3.1(a) Sources from Human Activity

Human activity is the primary contributor to elevated heavy metals (HM) in agricultural systems.

i) **Industry Effluent and Mining:** Activities like metal smelting, leather tanning, refining, and mining have deliberately or inadvertently released massive amounts of HMs into the adjacent land and water systems (Raj, 2021).
ii) **Agriculture:** Contaminants enter via agricultural practices. The use of inorganic phosphate fertilizers makes substantial contributions of Cd, Pb, Cr, and other metals, as those metals are not naturally occurring in the phosphate rock but nevertheless are common impurities. Moreover, the application of certain sludges and pesticides from sewage and using polluted water for irrigation have all contributed to the accumulation of HMs in farmland (Rashid et al., 2023).
iii) **Waste disposal:** Poorly managed industrial and municipal solid waste disposal and atmospheric deposition from burning fossil fuels (Priya et al., 2023) further increase environmental hazards.

3.1(b) Persistence and Mobility in the Environment

Once introduced to the environment, HMs remains present for extended durations. Bioavailability (the portion that can be absorbed by a living organism) is primarily influenced by the soil's pH, organic matter concentration, and the particular chemical structure (or speciation) of the metal (Jachym et al., 2018). HMs can also easily move from contaminated soils (the pedosphere) to surface water and groundwater (the hydrosphere) and thus expand their effect in ecological systems.

3.2 Effect on Human Health and SDG 3

Heavy metal contamination is a direct impact on human health and is consistent with SDG Target 3.9, which states to substantially reduce the number of deaths and illnesses from hazardous chemicals and air, water, and soil pollution and contamination by 2030 (UN, 2015).

3.2(a) Toxicity and Bioaccumulation

HMs are systemic toxins and can enter the human body as a result of contaminated food and/or water ingestion, inhalation, or through the dermal route (Jaishankar et al., 2014). An additional risk of HMs is that they are non-degradable, so they can bioaccumulate in a biological system and, even at low and prolonged exposure, can pose significant health risks (Tchounwou et al., 2012).

3.2(b) Health Outcomes

The exposure of toxic HMs may cause serious disorders:

i) **Neurological Damage:** Pb and Hg are known neurotoxins that have detrimental effects on children, which affect their cognitive development, leading to neurological impairments (Jaishankar et al., 2014).
ii) **Carcinogenicity:** As and Cr are identified as carcinogens and are associated with increased risk of different cancers.
iii) **Organ Damage:** Cd can lead to kidney and liver dysfunction, and high exposures to HMs over time can also cause cardiovascular complications (Jaishankar et al., 2014).

Moreover, HMs may contaminate and threaten the safety of the agricultural food chain, as it is an inevitable consequence of changing soil and water back into a healthy condition necessary for public health and SDG 3.

3.3 Hydrophytes for Soil and Water Remediation

Typical methods for the remediation of heavy metals (e.g., chemical precipitation, membrane filtration) can be expensive, energy intensive, and produce toxic sludge as a secondary waste (Priya et al., 2023). In contrast, phytoremediation, the use of plants to remediate contaminated environments, is an ecologically benign, sustainable, and cost-effective alternative to soil and water remediation (Kumar et al., 2024; Jachym et al., 2018).

3.3(a) Function of Hydrophytes

Aquatic macrophytes, or hydrophytes, are highly effective in phytoremediating contaminated water and wet/waterlogged soils due to their high quantity of biomass and ability to take up contaminants directly from the water through roots or from the surface of the whole plant (Saha et al., 2022).

The two main mechanisms (Jachym et al., 2018) used by hydrophytes for phytoremediation include:

i) **Phytoextraction (or Phytoaccumulation):** This mechanism involves the absorption of HMs through the root of a plant, followed by the translocation and ultimately the accumulation of HMs in the aboveground biomass or shoots and/or leaves of a plant.
ii) **Rhizofiltration:** This process involves the adsorption or precipitation of contaminants onto the root surface of aquatic plants. This process is primarily applied to clean water bodies that have become contaminated.

3.3(b) Fields of Hydrophytes

Many emergent and free-floating species of hydrophytes could function as hyper-accumulators (i.e., species that bioaccumulate HMs at concentrations 100 times greater than species that do not):

i) **Water Hyacinth (*Eichhorniacrassipes*):** Water Hyacinth has been extensively researched for its bioaccumulation of metals (e.g., Cd, Cu, Zn and Pb) into its biomass (especially in its roots) from contaminated water (Muramoto& Oki, 1983; Agunbiade et al., 2009) .
ii) **Water Lettuce (*Pistiastratiotes*):** Water lettuce can accumulate Pb and Zn and has in some cases exhibited high bioconcentration factors (Farid et al., 2023; Saha et al., 2022).
iii) **Duckweed (*Lemna* spp. and *Spirodela* spp.):** Duckweed is often recognised for a rapid growth rate which correlates with high removal efficiencies for HMs (e.g., Cd and Zn) (Agunbiade et al., 2009).
iv) **Cattail (*Typhalatifolia*):** An emergent macrophyte that has been studied for the removal of heavy metals (e.g., Cu, Zn) from a variety of environments (Farid et al., 2023).

Figure 1. **Mechanism of Heavy Metal Contamination:** *The visual emphasizes how anthropogenic activities lead to contamination that threatens ecological balance and human well-being.*

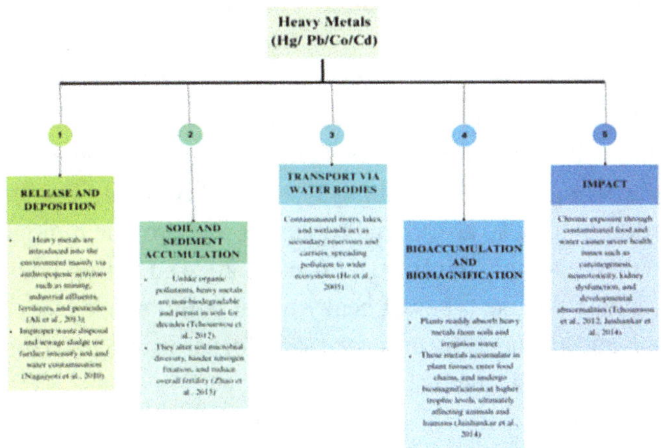

4. ROLE OF HYDROPHYTES FOR SOIL RECLAMATION

Hydrophytes provide an affordable and environmentally friendly means for phytoremediation of soils and sediments contaminated by heavy metals and other contaminants. *Typha*, *Phragmites*, *Lemna*, and *Eichhornia* species, rhizofiltration, phytoextraction, and phytostabilization reduce metals and nutrients in constructed and natural water-bodies (Buta et al., 2014). Depending on the selection criteria of the species chosen for hydrophytes to be successful includes resistance to local stressors, root structure, and rate of heavy metal absorption, and hydrophytes can be conditioned to the environment (Inelova et al., 2023; Priya et al., 2023).

Hydrophytes play an interconnected, specialized role in phytoremediation as a result of their unique rhizosphere and large surface area, leading to conditions ideal for microbial consortia. This microbiota plays critical roles in the degradation of organic matter and heavy metal sequestration/conversion (Rai, 2008; Nzengung& Jeffers, 2001). Despite their operational effectiveness, several limitations impede the broad implementation of this technology, including the disposition of contaminant-laden biomass. Cropped plant tissues can leach considerable contaminant load. In addition, sporadic seasonal growth patterns may result in variability of success (Rai, 2008; Rai, 2009). Scientists have begun to address these limitations by investigating downstream applications of harvested biomass, i.e., bioenergy production, to create value while inhibiting waste generation (Rai, 2008). Aquatic plants are extending beyond their historical use in metal remediation. Emerging studies are demonstrating that aquatic plants can also absorb drug residues, organic contaminants, and now emerging contaminants such as microplastics, therefore expanding their applications in contemporary environmental remediation (Kim et al., 2024).

Additionally, species-level tolerance mechanisms regulate the process of remediation too; for example, some species are able to actively accumulate metals and store the contaminant in vacuoles or the cell wall (phytostabilizartion), while other plant species have fast growth rates and accumulate biomass pollutants that can be harvested using multiple cropping systems (Priya et al., 2023). Studies have shown that hydrophytes also absorb microplastics and provide biological filtration, thus adding to their ecological utility (Kim et al., 2024). Hydrophytes have general applicability against diverse types of heavy metal pollutants, dyes, pesticides, and pharmaceuticals and are therefore universal tools for use in integrated bioremediation systems (Buta et al., 2014; Inelova et al., 2023).

Table 2 collates these different strategies by listing the repair processesfollowed by hydrophytes and giving examples of typical species.

Table 2. Remediation mechanisms of hydrophytes in soil reclamation

Mechanism of Remediation	Description	Examples (Species)	References
Phytoextraction	Uptake and accumulation of metals and pollutants in plant tissues, which are subsequently eliminated by harvesting.	Typha, Lemma, Eichhornia	(Rai, 2008; Buta et al., 2014; Priya et al., 2023; Rai, 2009)
Phytostabilization	Root sorption and decreased bioavailability cause metals in soils to become immobile.	Phragmites, Typha	(Rai, 2008; Priya et al., 2023)
Rhizofiltration	Direct removal of pollutants from a water/ soil solution by means of root absorption and adsorption.	Lemna, Eichhornia	(Rai, 2008; Rai, 2009)
Removal of Organics	Enzymatic breakdown and absorption of organic contaminants.	Lemna, Typha	(Rai, 2008; Nzengung& Jeffers, 2001; Tront et al., 2007)
Microplastics	Adsorption, filtration, and root entrapment of microplastics in sediments and water.	Iris, Lythrum	Kim et al., 2024

Soil recovery is facilitated by hydrophytes through a range of interlinked processes such as the uptake, conversion, and stabilisation of contaminants. The schematic image in Figure 2 consolidates the process of soil reclamation mediated by hydrophytes.

Figure 2. Mechanism of soil reclamation by hydrophytes: The progressive procedure by which hydrophytes aid in the cleanup of polluted soils is depicted in this flowchart.

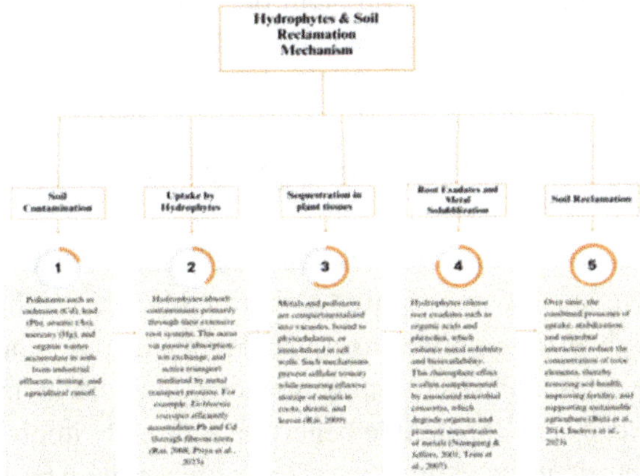

5. ROLE OF HYDROPHYTES IN SUSTAINABLE AGRICULTURE

Hydrophytes are crucial for connecting aquatic ecosystems and sustainable agriculture. They boost soil fertility, assist nutrient cycling, improve water management, and directly contribute to the production of food and feed in addition to their capacity to eliminate contaminants. In the globe, For example, rice (*Oryzasativa*) is a staple crop, and Azolla is utilised as a high-protein feed for cattle and poultry as well as a biofertilizer (Uddin et al., 2021). According to (Ekoko et al., 2021), duckweed (*Lemna* spp.) offers high-protein biomass for animal feed and may be used in human nutrition. In agricultural settings, coontail (*Ceratophyllumdemersum*) and water hyacinth (*Eichhorniacrassipes*) provide organic materials and animal feed. Hydrophytes promote agro-ecological methods and sustainable development by serving as bio-filters and bio-resources (Inelova et al., 2023).

5.1 Hydrophytes as Food and Feed Resources

In agriculture, some hydrophytes are directly utilized as food crops and fodder. Rice (*Oryzasativa*) is the most widely cultivated hydrophyte in the world, serving as a food source for billions of individuals. Azolla is a small floating fern that serves as a high-protein feed for cattle and poultry feed as well as a biofertilizer (Uddin et al., 2021). Because they have high biomass as well as nutritional value, some hydrophytes such as *Hydrillaverticilata, Spirodela*polyrhiza (greater duckweed),

Ceratophyllumdemersum (Coontail) and *Eichhorniacrassipes* (water hyacinth) are employed as green manure or animal fodder. Also, *Lemna* species, commonly referred to as duckweed, yield high-protein biomass suitable for animal feed and could potentially find applications in human nutrition due to their beneficial amino acid balance (Ekoko et al., 2021).

5.2 Benefits of Soil Fertility and Agro-ecology

Hydrophyte's root system facilitates soil aeration, aggregate stability, and nutrient conservation, leading to improved soil structure and fertility (Inelova et al., 2023). Their growth also supportsdecreased erosion and loss of nutrients, reducing the negative impacts of intensive farming methods. In addition, most hydrophytes, like*Lemna*(duckweed), are the double benefit of indirectly improving soil fertility by re-circulating nutrients in addition to offering high-protein biomass to cattle and potentially to human consumption (Ekoko et al., 2021).

5.3 Climate Adaptation and Water Regulation

Hydrophytes play a crucial role in both climate change adaption and hydrological management in agricultural areas. Plant that enhances water entry and water holding, such as *Typha* (cattail) and *Phragmitesaustralis*(commom reed), reduce the risk of flooding and preserve soil moisture during times of drought. While *Saccharunspontaneum* and *Scripus*species stabilise riverbanks and reduce soil erosion in flood areas. *Eichhorniacrassipes* (water hyacinth) and *Lemna* (duckweed) helps to control water levels in canals and ponds (Inelova et al., 2023). Hydrophytes stabilize soil, regulate water tables, and reduce salt effects, making them vital components of climate-resilient agriculture systems (Ekoko et al., 2021).

5.4 Phytoremediation of Agricultural Run-Off

Aqua systems are often contaminated with metals, fertilisers, and pesticides due to agricultural development at the expense of aquatic ecosystems. Hydrophytes counteract such pollutants by phytoremediation through phytosorption and entrapment of pollutants in plant tissues, transformation of toxic chemicals into less toxic equivalents, and stabilisation of contaminants in sediments or rhizospheres. For instance, the cattail (*Typha*), and the common reed (*Phragmitesaustralis*), are excellent at absorbingphosphate and nitrogen from nutrient-enriched runoff, this reduces eutrophication in downstream water bodies (Rai, 2008; Inelova et al., 2023). The water hyacinth (*Eichhorniacrassipes*) and the duckweed *Lemna*effectively absorb heavy metals including copper, lead, and cadmium while encouraging microbial breakdown of remaining pesticides in the water column. As natural biofilters, species such as *Ceratophyllumdemersum* and *Hydrillaverticilata* also immobilise metals and trap sediments in their root zones. Hydrophytes are vital parts of sustainable agricultural landscapes because they absorb nutrients, agrochemicals, and sediments, protecting freshwater biodiversity and enhancing water quality (Uddin et al., 2021).

5.5 Integrated Agro-Ecological Systems

Hydrophytes many functions point to their importance in integrated agricultural systems. When Azolla, Lemna, or fish farming is combined with rice paddies, multi-trophic systems are produced that maximise water, nutrient, and biomass fluxes while concurrently supplying food, fertiliser, and ecological services (Ekoko et al., 2021).

Table 3. Role of hydrophytes in sustainable agriculture

Agricultural role	Contribution of Hydrophyte	Example (Plants)	References
Crop production	Food, forage	Rice, Lemna	Inelova et al., 2023; Ekoko et al., 2021
Soil fertility	Nutrient cycling, green manure	Azolla	Inelova et al., 2023; Uddin et al., 2021
Runoff filtration	Remediation of nutrients/ metals	Lemna, Typha	Rai, 2008; Uddin et al., 2021
Soil structure	Erosion control, waterlogging	Rice, Typha	Spence, 1967; Ekoko et al., 2021

6. HEALTH IMPLICATIONS AND ACHIEVING SDG3

The United Nations' Sustainable Development Goal 3 (SDG3) focuses on ensuring healthy lives and promoting well-being for all. Environmental pollution is a major obstacle to reaching this objective, especially heavy metal poisoning in soil and water, which poses serious health concerns to the general people. Long-term exposure to lead, arsenic, mercury, and cadmium has been linked to developmental delays, kidney and liver damage, neurological abnormalities, and cancerous consequences (Shahid et al., 2018).

Hydrophytes offer an economical and sustainablemeans of reducing such risks because they are natural phytoremediators. They remove pollutants from aquatic bodies and sediments, reducing the accumulation of harmful substances in the food chain and thus enhancing the health and well-being of humans (Zhakypbek et al., 2024).

6.1 Heavy Metals and Human Health

Heavy metals accumulate in aquatic and agricultural produce, hence bioaccumulate and biomagnify in food webs (Priya et al., 2023). Children and pregnant women are part of the vulnerable populations who can be affected by the toxic exposures that are derived from this. For example, cadmium is associated with renal disease and osteoporosis, while exposure to lead causes cognitive impairment and learning disabilities (Shahid et al., 2018). Populations dependent on fish-based foods are threatened with neurological complications as a result of mercury bioaccumulation in fish, whereas arsenic exposure results in dermal lesions and a high risk of cancer.

6.2 Role of Hydrophytes in Reducing Exposure

Toxic metals such as cadmium, lead, mercury, and arsenic, can be taken up and immobilized by hydrophytes like *Typha latifolia* and *Eichhornia crassipes* (Sharma et al., 2023; Priya et al., 2023). Hydrophytes filter the aquatic ecosystem and irrigation water from harmful contaminants and thus enhance agricultural and fishery safety. Besides reducing the incidence of waterborne diseases, some plants, including*Phragmitesaustralis*, are also useful in minimizing nitrate and pathogen contamination (Karwal et al., 2024). Duckweed, Lemna minor, effectively eliminates many heavy metals from polluted waterways, avoiding the buildup of toxins in crops and animal feed (Sharma et al., 2023).

6.3 Public Health and Community Benefits

Remediation with hydrophytes results in cleaner water sources, which directly improve health outcomes. Hydrophytes aid in lowering exposure to harmful compounds that leads to the development of chronic illnesses such kidney damage, cardiovascular problems, and some types of cancer via the reduction of chemical and heavy metal pollution. In addition, lower aquatic pathogen loads prevent the spread of waterborne illnesses, including diarrhoea, which contribute to be a significant health burden in low- and middle-income nations (Karnwal et al., 2024).

Hydrophytes have immediate physiological health advantages as well as psychological ones. Both urban and rural residents benefit from the green and blue spaces produced by hydrophyte-rich ecosystems, which lower stress levels and promote social interaction (WHO, 2021). Because they promote both physical and mental health, these advantages highlight the multifaceted role that hydrophytes play in accomplishing SDG3.

6.4 The Role of Hydrophytes in SDG3 Approaches

Inclusion of hydrophyte phytoremediation approaches in environmental planning at national and regional levels has the ability to rapidly accelerate progress towards achieving SDG3 targets. Hydrophytes function as biotic filters to trap toxicants from food and drinking water (Priya et al., 2023; Sharma et al., 2023). This approach rejects cure after exposure to disease by preventing exposure by preventing toxicity exposure. Hydrophytes offer long-term resistance to toxin health problems via improved water quality, soil fertility, and ecological balance that promote biodiversity. Local initiatives using natural hydrophyte species can enhance resistance through development of environmental consciousness, local control, and adaptive management practices to help manage the health threat posed by pollution (Karnwal et al., 2024). Treated hydrophyte remediation allows broad themes of sustainability, including social equity, environmental protection, and health resilience.

Table 4. Health risks and benefits of hydrophytes with respect to SDG3

Health Risk/Benefit	Hydrophyte Role	References
Metal accumulation in food	Potential toxin pathway	(Pajević et al., 2008)
Ecosystem pathogen removal	Waterborne disease control	(Rai, 2008; Priya et al., 2023; Rai, 2009)
Exposure reduction	Biofiltration/Remediation	(Aziz, 2023; Uddin et al., 2021; Rai, 2008)

7. POLICY AND IMPLEMENTATION FRAMEWORK

The application of hydrophytes for sustainable soil reclamation is a nature-based solution that greatly aids in achieving global development goals, specifically Sustainable Development Goal (SDG 3): Good Health and Well-being (United Nations, 2015). A Policy and implementation framework is required to mainstream hydrophyte-based soil reclamation into agricultural and environmental, chemical-based, intensive methods to more sustainable approaches, such as phytoremediation, requires a strong policy framework and regulatory backing as well as an implementation strategy.

7.1 SDG 3 and the Health rationale for using hydrophytes

SDG 3 is "ensure healthy lives and promote well-being for all ages". A specific goal under this SDG is Target 3.9, which states, "By 2030, substantially reduce the number of deaths and illnesses from hazardous chemicals and air, water and soil pollution and contamination" (United Nations, 2015).

7.1(a) Soil Contamination and Public Health

Contaminated agricultural soils are a threat to human health due to residues of heavy metals, persistent organic pollutants, or excess chemical fertilizers (FAO, 2017). When food crops grow in contaminated soil, they unfortunately accumulate contaminants, which cause health consequences after consumption, such as neurological disorders and chronic conditions (Horrigan et al., 2002).

7.2(b) The Health Solution using Hydrophytes

Hydrophytes can be used for cleanup or stabilization of contaminated waterlogged soils (Ali et al., 2013) through diverse phytoremediation processes (i.e., phytoextraction, phytostabilization, etc.).

i) **Less toxicity in Food crops:** When cleaning up or stabilizing contaminated soil, hydrophytes reduce bioavailability of hazardous chemicals and subsequently reduce the hazardous chemicals from entering into staple food crops (Ali et al., 2013).
ii) **Ecosystem services:** A once-reclaimed area can improve the land for planting stakeholder crops, promote healthy ecosystem resiliency, improve overall water quality, and reduce exposure to environmental pollutants, thus relating to SDG 3 (United Nations, 2015).

7.2 Policy Framework for technology adoption

An effective policy framework is necessary to spur the uptake of hydrophyte technology in soil reclamation for sustainable agriculture. This framework should address the financial, regulatory, and knowledge gaps.

7.2(a) Legislative and Regulatory Integration

Existing environmental legislation and agricultural regulations (where applicable) must be updated to recognize or expand the support of using phytoremediation approaches, particularly for hydrophyte approaches.

i) **Official Recognition of phytoremediation:** National and regional environmental legislation should officially endorse and recognize hydrophyte-based remediation for contaminated agricultural land, or at least note it as a preferred or accepted remediation technology for contaminated land correction, this could be equivalent to declaring it a "Green Remediation" technology (U.S. Environmental Protection Agency, 2018).

ii) **Disposal Guidelines:** Policies must provide guidance on how to sustainably or safely dispose of the harvested, contaminated, hydrophyte biomass. This could include, incinerating biomass and obtaining energy from the contaminated material, volume reductions, or burying the contaminated biomass in a landfill designated for hazardous waste or at a contaminated site, along with guidelines that prevent secondary contamination in the process (Garbisu & Alkorta, 2001).

7.2(b) Economic Incentives and Financing

Implementing financial mechanisms is crucial for providing an economic rationale for farmers and land owners to use this environmentally sustainable but possibly slow technology.

i) **Green Investment Funds:** Green agricultural Investment Funds could be established to finance soil management for sustainable practices like phytoremediation specifically at a reduced interest rate or grant linked explicitly to the governing SDGs (FAO, 2020).

ii) **Payment for Ecosystem Services (PES):** Implementing PES in which landowners are rewarded for the public health and environmental benefits (Cleaner soil, better water quality, better biodiversity) of hydrophyte remediation, would provide a continuous means of revenue (Le et al., 2023)

8. CHALLENGES AND FUTURE DIRECTIONS

Hydrophytes provide significant opportunities for ecological restoration and phytoremediation, but their application and worldwide uptake will face a range of ecological, technological, and socio-economic challenges. Overcoming these challenges would require a transdisciplinary approach involving plant biology, biotechnology, ecological engineering, and environmental governance.

8.1 Challenges in Hydrophyte Use for Sustainable Agriculture

The use of hydrophytes to enhance soil quality for sustainable agriculture, especially to Sustainable Development Goal 3 (SDG 3), encounters some major technical and operational challenges.

8.1(a) Technical and Operational Constraints

i) **Biomass Harvest and Disposal:** Hydrophytes tend to generate huge volumes of biomass. Effective and economic ways of harvesting, drying, and using or disposing of the material safely, especially when it has stored heavy metals or impurities, are a challenge (Garbisu & Alkorta, 2001).
ii) **Climate and Site Specificity:** The effectiveness of hydrophyte-based soil reclamation is very site-specific, it is highly reliant on certain environmental variables (e.g., pH, water, quality, temperature, soil type), and so effective protocols can be non-transferable across different geographical sites without rigorous pre-testing (Pilon-smits, 2005).
iii) **Secondary Contamination Risk:** Inadequate handling of harvested contaminated vegetation can result in the re-release of contaminants into the environment, where they may contaminate water bodies or untreated soil (Salt et al., 1998).

8.1(b) Economic and Social Challenges

i) **Capital Costs of Initial Setup and Ongoing Maintenance:** Installation and maintenance of hydrophyte systems, particularly in engineered wetlands or large-scale remediation programmes, call for upfront capital expenditure or infrastructure, planting, and skilled personnel (Garbisu & Alkorta, 2001)
ii) **Shortage of Policy and Regulatory Framework:** The lack of standardized governmental policies and regulations directly dealing with the safe use of hydrophyte-derived products (e.g., bioenergy crops or feed stocks for animals) from the remediated soils will prevent market acceptance and commercial up-scaling (Mahar et al., 2016).
iii) **Public Acceptance and Perception:** There is often a lack of awareness or scepticism by the public and by farmers,concerning the safety and long-term viability of farming land thathasbeen phytoremediated, particularly in regards to the consumption of food crops produced on formerlycontaminatedland after remediation (Cunningham & Ow, 1996).

8.2 Future Directions for Research and Implementation

Inorderto address the challenges and torealize(fully) the potential of hydrophyte for soil remediation and for agricultural sustainability, future work should focus on specificdesignopportunities.

8.2(a) Biotechnological Advancements

i) **Genetic Engineering and Improvement:** Future research may develop genetically modified hydrophytes to be tolerant to a broader range of contaminants and improve biomass production to develop a more efficient and rapid remediation cycle (Pilon-Smits, 2005; Barathi et al., 2023)

ii) **Microbial Amendment:** Utilization of plant-microbe associations (e.g., endophytes or rhizobacteria) for increased contaminant degradation in the rhizosphere to improve overall plant uptake efficiency is a rapidly developing area of research (Khatoon et al., 2024).

8.2(b) Optimizing Process and Product Use

i) **Integrated Waste Management Systems:** It is vital to establish standard approaches for the use of contaminated hydrophyte biomass (e.g., through biochar pyrolysis/thermal conversion, or recovery of energy through incubation), that are safe and productive and can close the loop on the remediation (Li et al., 2023).

ii) **Development of Rapid Assessment Tools:** Develop and deploy low-cost real-time monitoring tools (e.g., biosensors) for contaminant quantification in soil and plants tissues will allow for quality control and management of the reclamation process (Williams et al., 2024)

iii) **Life Cycle Assessment (LCA):**Comprehensive assessments will require LCA studies to compare the overall environmental impact and economic viability of remediation using hydrophytes versus traditional remediation to build a scientific justification for scaling up (Lemming et al., 2010).

REFERENCES

Agunbiade, F. O., Olu-Owolabi, B. I., & Adebowale, K. O. (2009). Phytoremediation potential of Eichhorniacrassipes in metal-contaminated coastal water. *Bioresource Technology, 100*(19), 4521–4526. DOI: 10.1016/j.biortech.2009.04.011 PMID: 19414252

Ali, H., Khan, E., & Sajad, M. A. (2013). Phytoremediation of heavy metals—Concepts and applications. *Chemosphere, 91*(7), 869–881. DOI: 10.1016/j.chemosphere.2013.01.075 PMID: 23466085

Aziz, M. (2023). Heavy metal pollution in the aquatic environment: efficient and low-cost removal approaches to eliminate their toxicity: a review. *RSC Advances, 13*(26), 17595–17610. DOI: 10.1039/D3RA00723E PMID: 37312989

Barathi, S., Lee, J., Venkatesan, R., & Vetcher, A. A. (2023). Current status of biotechnological approaches to enhance the phytoremediation of heavy metals in India—A review. *Plants, 12*(22), 3816. DOI: 10.3390/plants12223816 PMID: 38005713

Beckett, D. C., Aartila, T. P., & Miller, A. C. (1992). Contrasts in Density of Benthic Invertebrates Between Macrophyte Beds and Open Littoral Patches in Eau-Galle Lake, Wisconsin. *American Midland Naturalist, 127*(1), 77–90. DOI: 10.2307/2426324

Brix, H. (1997). Do macrophytes play a role in constructed treatment wetlands? *Water Science and Technology, 35*(5), 11–17. DOI: 10.2166/wst.1997.0154

Buta, E., Torok, A., Zongo, B., Cantor, M., & Majdik, C. (2014). Comparative Studies of the Phytoextraction Capacity of Five Aquatic Plants in Heavy Metal Contaminated Water. *Notulae Botanicae Horti Agrobotanici Cluj-Napoca, 42*(1), 173–179. DOI: 10.15835/nbha4219341

Colmer, T. D., & Flowers, T. J. (2008). Flooding tolerance in halophytes. *The New Phytologist, 179*(4), 964–974. DOI: 10.1111/j.1469-8137.2008.02483.x PMID: 18482227

Cunningham, S. D., & Ow, D. W. (1996). Promises and challenges in the development of phytoremediation. *Plant Physiology, 110*(3), 715–719. DOI: 10.1104/pp.110.3.715 PMID: 12226213

De, M., Roy, C., Medda, S., Roy, S., & Dey, S. R. (2019). Diverse role of Macrophytes in aquatic ecosystems: A brief review. *International Journal of Experimental Research and Review, 19*, 40–48. DOI: 10.52756/ijerr.2019.v19.005

Ekoko, W. A., Yao, Y. L., Shan, Y. Q., Liu, B., & Shabani, I. E. (2021). Vascular plants from the Sanjiang Plain, Northeast China. *Applied Ecology and Environmental Research, 19*(4), 2689–2699. DOI: 10.15666/aeer/1904_26892699

FAO. (2020). *State of the World's Land and Water Resources for Food and Agriculture*. Food and Agriculture Organization of the United Nations.

Farid, N., Ullah, A., Khan, S., Butt, S., Khan, A. Z., Afsheen, Z., El-Serehy, H. A., Yasmin, H., Ayaz, T., & Ali, Q. (2023). Algae and Hydrophytes as Potential Plants for Bioremediation of Heavy Metals from Industrial Wastewater. *Water (Basel), 15*(12), 2142. DOI: 10.3390/w15122142

Food and Agriculture Organization of the United Nations (FAO). (2017). *The future of food and agriculture – Trends and challenges*. FAO.

Food and Agriculture Organization of the United Nations (FAO). (2020). *Financing sustainable agriculture*. FAO.

Garbisu, C., & Alkorta, I. (2001). Phytoextraction: A cost-effective plant-based technology for the removal of metals from the environment. *Bioresource Technology, 77*(3), 229–236. DOI: 10.1016/S0960-8524(00)00108-5 PMID: 11272009

Horrigan, L., Lawrence, R. S., & Walker, P. (2002). How sustainable agriculture can address the environmental and human health harms of industrial agriculture. *Environmental Health Perspectives, 110*(5), 445–456. DOI: 10.1289/ehp.02110445 PMID: 12003747

Houwing, E. J., Tánczos, I. C., Kroon, A., & de Vries, M. B. (2002). Interaction of Submerged Vegetation, Hydrodynamics and Turbidity: Analysis of Field and Laboratory Studies. In Winterwerp, J. C., & Kranenburg, C. (Eds.), *Proceedings in Marine Science* (Vol. 5, pp. 441–453). Elsevier.

Inelova, Z., Zayadan, B., Zaparina, Y., Aitzhan, M., & Boros, E. (2023). Perspectives for the application of aquatic and semi-aquatic plants in biomonitoring of freshwater, saline and soda aquatic ecosystems. *Pakistan Journal of Botany, 55*(3), 1099–1115. DOI: 10.30848/PJB2023-3(33)

Jachym, S., Uhlik, O., Viktorova, J., & Macek, T. (2018). Phytoextraction of Heavy Metals: A Promising Tool for Clean-Up of Polluted Environment? *Frontiers in Plant Science*, •••, 9. PMID: 30459775

Jaishankar, M., Tseten, T., Anbalagan, N., Mathew, B. B., & Beeregowda, K. N. (2014). Toxicity, mechanism and health effects of some heavy metals. *Interdisciplinary Toxicology, 7*(2), 60–72. DOI: 10.2478/intox-2014-0009 PMID: 26109881

Karnwal, A., Martolia, S., Dohroo, A., Al-Tawaha, A. R. M. S., & Malik, T. (2024). Exploring bioremediation strategies for heavy metals and POPs pollution: The role of microbes, plants, and nanotechnology. *Frontiers in Environmental Science, 12*, 1397850. DOI: 10.3389/fenvs.2024.1397850

Khatoon, Z., Orozco-Mosqueda, M. D. C., & Santoyo, G. (2024). Microbial contributions to heavy metal phytoremediation in agricultural soils: A review. *Microorganisms, 12*(10), 1945. DOI: 10.3390/microorganisms12101945 PMID: 39458255

Kim, Y., Park, K., Bak, J., & Choi, S. (2024). *Iris pseudacorus* and *Lythrumanceps* as Plants Supporting the Process of Removing Microplastics from Aquatic Environments—Preliminary Research. *Horticulturae, 10*(6), 631. DOI: 10.3390/horticulturae10060631

Kumar, A., Dadhwal, M., Mukherjee, G., Srivastava, A., Gupta, S., &Ahuja, V. (2024). Phytoremediation: Sustainable Approach for Heavy Metal Pollution. *Scientifica (Cairo)*.

Lal, R. (2008). Carbon sequestration. *Philosophical Transactions of the Royal Society of London. Series B, Biological Sciences, 363*(1492), 815–830. DOI: 10.1098/rstb.2007.2185 PMID: 17761468

Le, T. T., Vodden, K., Wu, J., Bullock, R., & Sabau, G. (2023). Payments for ecosystem services programs: A global review of contributions towards sustainability. *Heliyon, 10*(1), e22361. DOI: 10.1016/j.heliyon.2023.e22361 PMID: 38173537

Lemming, G., Hauschild, M. Z., & Bjerg, P. L. (2010). Life cycle assessment of soil and groundwater remediation technologies: Literature review. *The International Journal of Life Cycle Assessment, 15*(1), 115–127. DOI: 10.1007/s11367-009-0129-x

Li, Z., Huang, Y., Zhu, Z., Yu, M., Cheng, H., Shi, H., Zuo, W., Zhou, H., & Wang, S. (2023). Co-pyrolysis of industrial sludge with phytoremediation residue: Improving immobilization of heavy metals at high temperature. *Fuel, 351*, 128942. DOI: 10.1016/j.fuel.2023.128942

Mahar, A., Wang, P., Ali, A., Awasthi, M. K., Lahori, A. H., Wang, Q., Li, R., & Zhang, Z. (2016). Challenges and opportunities in the phytoremediation of heavy metals contaminated soils: A review. *Ecotoxicology and Environmental Safety, 131*, 218–228. DOI: 10.1016/j.ecoenv.2015.12.023 PMID: 26741880

Muramoto, S., & Oki, Y. (1983). Removal of some heavy metals from polluted water by water hyacinth (*Eichhorniacrassipes*). *Bulletin of Environmental Contamination and Toxicology, 30*(1), 170–177. DOI: 10.1007/BF01610117 PMID: 6839042

New Hampshire Department of Environmental Services. (2019). *Aquatic Plants and Their Role in Lake Ecology.*

Nzengung, V. A., & Jeffers, P. (2001). Sequestration, Phytoreduction, and Phytooxidation of Halogenated Organic Chemicals by Aquatic and Terrestrial Plants. *International Journal of Phytoremediation, 3*(1), 13–40. DOI: 10.1080/15226510108500048

Pajević, S., Borišev, M., Rončević, S., Vukov, D., & Igić, R. (2008). Heavy metal accumulation of Danube River aquatic plants — Indication of chemical contamination. *Open Life Sciences, 3*(3), 285–294. DOI: 10.2478/s11535-008-0017-6

Pilon-Smits, E. (2005). Phytoremediation. *Annual Review of Plant Biology, 56*(1), 15–39. DOI: 10.1146/annurev.arplant.56.032604.144214 PMID: 15862088

Priya, A. K., Muruganandam, M., Ali, S. S., & Kornaros, M. (2023). Clean-Up of Heavy Metals from Contaminated Soil by Phytoremediation: A Multidisciplinary and Eco-Friendly Approach. *Toxics, 11*(5), 422. DOI: 10.3390/toxics11050422 PMID: 37235237

Rai, P. K. (2008). Heavy Metal Pollution in Aquatic Ecosystems and its Phytoremediation using Wetland Plants: An ecosustainable approach. *International Journal of Phytoremediation, 10*(2), 133–160. DOI: 10.1080/15226510801913918 PMID: 18709926

Rai, P. K. (2009). Heavy Metal Phytoremediation from Aquatic Ecosystems with Special Reference to Macrophytes. *Critical Reviews in Environmental Science and Technology, 39*(9), 697–753. DOI: 10.1080/10643380801910058

Raj, M. (2021). A Review on the Effect of Heavy Metal Contamination and its Impact on the Environment. *International Journal of Zoological Investigations., 7*(2), 762–771. DOI: 10.33745/ijzi.2021.v07i02.061

Rashid, A., Schutte, B. J., Ulery, A., Deyholos, M. K., Sanogo, S., Lehnhoff, E. A., & Beck, L. (2023). Heavy Metal Contamination in Agricultural Soil: Environmental Pollutants Affecting Crop Health. *Agronomy (Basel), 13*(6), 1521. DOI: 10.3390/agronomy13061521

Saha, A., Mukherjee, P., Roy, K., Sen, K., & Sanyal, T. (2022). A review on phytoremediation by aquatic macrophytes: A natural promising tool for sustainable management of ecosystem. *International Journal of Experimental Research and Review, 27*, 9–31. DOI: 10.52756/ijerr.2022.v27.002

Salt, D. E., Smith, R. D., & Raskin, I. (1998). Phytoremediation. *Annual Review of Plant Physiology and Plant Molecular Biology, 49*(1), 643–668. DOI: 10.1146/annurev.arplant.49.1.643 PMID: 15012249

Shahid, S. A., Zaman, M., & Heng, L. (2018). *Soil Salinity: Assessment, Management, and Environmental Impact*. Springer.

Sharma, J. K., Kumar, N., Singh, N. P., & Santal, A. R. (2023). Phytoremediation technologies and their mechanism for removal of heavy metal from contaminated soil: An approach for a sustainable environment. *Frontiers in Plant Science*, *14*, 14. DOI: 10.3389/fpls.2023.1076876 PMID: 36778693

Sondergaard, M., & Moss, B. (1998). *Impact of Submerged Macrophytes on Phytoplankton in Shallow Freshwater Lakes* (Vol. 131). Ecological Studies. DOI: 10.1007/978-1-4612-0695-8_6

Tchounwou, P. B., Yedjou, C. G., Patlolla, A. K., & Sutton, D. J. (2012). Heavy metal toxicity and the environment. *EXS*, *101*, 133–164. DOI: 10.1007/978-3-7643-8340-4_6 PMID: 22945569

Tront, J. M., Reinhold, D. M., Bragg, A. W., & Saunders, F. M. (2007). Uptake of Halogenated Phenols by Aquatic Plants. *Journal of Environmental Engineering*, *133*(10), 955–961. DOI: 10.1061/(ASCE)0733-9372(2007)133:10(955)

Uddin, M. M., Zakeel, M. C. M., Zavahir, J. S., Marikar, F. M. M. T., & Jahan, I. (2021). Heavy Metal Accumulation in Rice and Aquatic Plants Used as Human Food: A General Review. *Toxics*, *9*(12), 360. DOI: 10.3390/toxics9120360 PMID: 34941794

United Nations Development Programme (UNDP). (2023). *Sustainable Development Goals*.

United Nations (UN). (2015). *Transforming our world: The 2030 Agenda for Sustainable Development*. UN.

U.S. Environmental Protection Agency. (2018). *Green remediation: Best management practices for phytoremediation*. US EPA.

Vymazal, J. (2011). Plants used in constructed wetlands with horizontal subsurface flow: A review. *Hydrobiologia*, *674*(1), 133–156. DOI: 10.1007/s10750-011-0738-9

Wetzel, R. G. (2001). *Limnology: Lake and River Ecosystems* (3rd ed.). Academic Press.

Williams, A., Aguilar, M. R., Pattiya Arachchillage, K. G. G., Chandra, S., Rangan, S., Ghosal Gupta, S., & Artes Vivancos, J. M. (2024). Biosensors for public health and environmental monitoring: The case for sustainable biosensing. *ACS Sustainable Chemistry & Engineering*, *12*(28), 10296–10312. DOI: 10.1021/acssuschemeng.3c06112 PMID: 39027730

World Health Organization (WHO). (2021). *Green and blue spaces and mental health: new evidence and perspectives*. WHO Regional Office for Europe.

World Health Organization (WHO). (2021). *Water, Sanitation and Hygiene: Guidelines on Heavy Metal Limits*. WHO Press.

Zhakypbek, Y., Kossalbayev, B. D., Belkozhayev, A. M., Murat, T., Tursbekov, S., Abdalimov, E., Pashkovskiy, P., Kreslavski, V., Kuznetsov, V., & Allakhverdiev, S. I. (2024). Reducing Heavy Metal Contamination in Soil and Water Using Phytoremediation. *Plants*, 13.

Karwal, T., Patel, Y., Bartlett, A., Poulin, V., Smith, T. L., & Pfeffer, D. N. (2024). Procoli: Profiles of cosmological likelihoods. arXiv preprint arXiv:2401.14225.

Chapter 7
Hydrophytes and Climate Change:
Bioremediation of Heavy Metals and SDG13

Padma Sharma
https://orcid.org/0000-0001-5098-2725
Paryavaran Bhawan Transport, India

Vijay Kumar
Patanjali Research Institute, Haridwar, India

Majid Farooq
Paryavaran Bhawan Transport, India

Mansi Sharma
https://orcid.org/0000-0001-6536-1075
Sharda University, Greater Noida, India

Shagufta Rashid
https://orcid.org/0000-0002-2063-1323
Annamalai University, India

Raashi Gupta
Panjab University, Chandigarh, India

Kuljit Kour
Paryavaran Bhawan Transport, India

Daizy Rani Batish
Panjab University, Chandigarh, India

Harminder Pal Singh
Panjab University, Chandigarh, India

ABSTRACT

Climate change brings more challenges to agriculture, ecosystems, and human health, with increasing incidences of soil degradation, water scarcity, and contamination from heavy metals. Hydrophytes are water-loving plants which offer a sustainable, environment friendly solution through their natural capacity for phytoremediation. The study evaluates the role of hydrophytes in climate change mitigation, and Sustainable Development Goal 13 (SDG13), mainly focusing on the role of hydrophytes in the bioremediation of heavy metals. By utilizing hydro-

DOI: 10.4018/979-8-3373-4037-1.ch007

phytes in bioremediation practices, the study explores innovative pathways for developing resilient food systems and restoring aquatic ecosystems amid changing climatic conditions. Integration of hydrophyte-based bioremediation systems into climate change adaptation and mitigation strategies offers a low-cost, nature-based solution, particularly for developing regions. The chapter emphasizes the need for interdisciplinary approaches, supportive policies, and public awareness to harness the full potential of hydrophytes in achieving SDG13.

1. INTRODUCTION

Climate change is increasing the pressure on agricultural productivity, ecosystem stability, and human health (Huang et al., 2025). In the agriculture system, warmer weather, changing precipitation patterns, and enhanced incidences of extreme weather patterns led to land degradation, water shortage, and decreased crop productivity, as well as greater exposure to pests and diseases (IPCC, 2022; Lobell et al., 2011). Ecosystems are losing habitat, becoming fragmented, and experiencing shift in species distributions and loss of biodiversity, as well as being more vulnerable to invasive species by alien flora, all of which reduces ecosystem services, including carbon sequestration, water purification, and pollination. Climate change elevates the risk of heat-related illness, respiratory diseases, malnutrition, and transmission of vector-borne diseases to the health of humans (WHO, 2018). The fact that these three areas are interconnected implies that a failure in one area can be catalyzed by effects in the other, such as how ecosystem degradation could later lead to agricultural catastrophe and health care crises, which is why a united front on climate action is necessary. In this regard, Sustainable Development Goal 13 (SDG13) identifies the particular measures needed to address climate change and its impacts by increasing resilience, the integration of climate actions into legislation, and climate education and awareness-raising (United Nations, 2015; Hussain et al., 2024). The direct benefits of realizing SDG13 are in the agriculture sector in securing crop production in changing climatic conditions, in ecosystems through the promotion of biodiversity conservation, and human health by reducing climate-related disease burdens. Notably, the cross-sectoral nature of SDG13 has ensured that climate change adaptation and mitigation solutions bring co-benefits cutting across the mentioned diverse sectors, in line with other SDGs like Zero Hunger (SDG2), Life on Land (SDG15), and Good Health and Well-Being (SDG3). The direct benefits of realizing SDG13 are in the agriculture sector in securing crop production in changing climatic conditions, in ecosystems through the promotion of biodiversity conservation, and human health by reducing climate-related disease burdens. Moreover, nature-based solutions (NbS) utilize natural processes to solve environmental, social, and economic problems and provide benefits such as biodiversity and climate (Cohen-Shacham et al., 2016). In agriculture, NbS like agroforestry and wetland rehabilitation are used as measures to sustain the soil fertility, control water runoffs, and buffer crops against climate extremes. In the case of ecosystems, such actions as reforestation and habitat restoration increase biodiversity, stabilize the landscape, and increase ecological resilience. NbS helps generate cleaner air and water, cool urban areas, and cut down transmission of diseases in human health with healthier environments. Particular treatment methods like phytoremediation with hydrophytes can offer alternative, inexpensive, and low-carbon-based solutions to engineered systems and enhance the quality of water and community health (Wentzell, 2025). The inclusion of NbS in climate strategies therefore reinforces both adaptation and mitigation actions at once, directly aligning with SDG13 goals and offering long-term and sustainable resilience (Figure 1).

Figure 1. Role of hydrophytes in environment sustainability and climate resilience

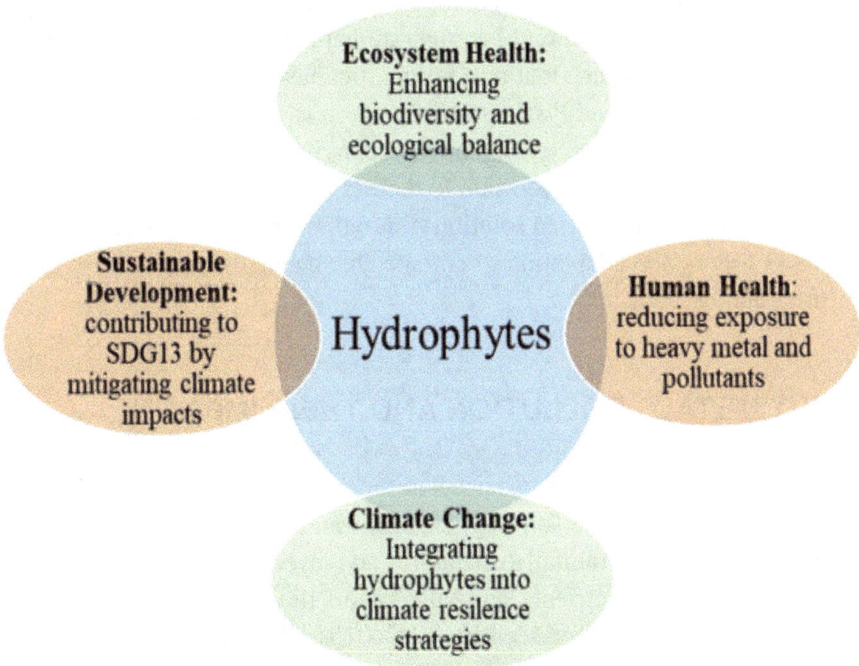

2. HYDROPHYTES: ECOLOGICAL AND FUNCTIONAL OVERVIEW

Hydrophytes are plants with structural and physiological systems that help them to grow in aquatic conditions, either totally or partially submerged, or floating upon the surface water (Huang et al., 2025). They have developed specialized traits including aerenchyma tissue to provide buoyancy and gaseous exchange, little or no cuticle, thin leaves to allow nutrients and gases can be diffused, and flexible stems to resist being pushed along by water currents (Cronk and Fennessy, 2016). Physiologically, hydrophytes generally exhibit a high capacity for adaptation, including low content of lignin content, efficient photosynthesis under low light, and anoxic tolerance of root zones. These adaptations enable them to survive across different freshwater environments, such as in lakes, rivers, wetlands and constructed ponds. Hydrophytes are distributed on all continents except Antarctica and are restricted according to the climatic zones, hydrological regimes, and nutrient availability (Chambers et al., 2008). They are important ecological components in aquatic systems, stabilizing the bottom, nitrogen cycling, fish and invertebrate habitat, and pollutant sorption. In tropical and subtropical areas, floating communities have been dominated by species like *Eichhornia crassipes* and *Pistia stratiotes*, whereas in temperate water, there are submerged species like *Potamogeton* and *Myriophyllum* (Villamagna and Murphy, 2010; Coetzee et al., 2009). In addition to their ecological roles, hydrophytes play an important role in nature-based solutions through the purification and removal of pollutants in waters and adaptation to climate change challenges because they act as biofilters and carbon sinks.

3. HEAVY METALS POLLUTION AND THEIR IMPACTS

Climate change is a cause and an accelerator of environmental degradation and increases other problems like soil degradation, water scarcity, and pollution. The warming climate, sporadic rainfall distribution, and severe weather conditions accelerate soil erosion, loss of nutrients, and salinization, thus lowering the productivity of the land and endangering food security (Lal, 2015). Climate-induced droughts are increasing water scarcity in the world by disrupting hydrological cycles, overusing groundwater, and diminishing snow and glacier melting, among others (WWAP, 2020). Contamination of soil and water bodies with heavy metals (cadmium (Cd), lead (Pb), arsenic (As), and chromium (Cr)) is one of the most worrying aspects of environmental degradation. The most common sources of these pollutants are industrial effluents, mining, agricultural runoff, and poor waste disposal (Jaishankar et al., 2014). When non-biodegraded heavy metals can bioaccumulate into plants and animals, the heavy metals are incorporated into the food chain and create serious risks to human health, such as carcinogenic effects, as well as neurotoxic and teratogenic effects. When present in high amounts, heavy metals can interfere with trophic interactions, lower biodiversity, and impair ecosystem functioning in aquatic ecosystems, and impact microbial communities and lower fertility in soils. The complexity of these related challenges needs combined management solutions involving pollution prevention, remediation technologies, and climate-resilient land and water management practices.

Some naturally occurring earth elements include heavy metals, which are cadmium (Cd), lead (Pb), arsenic (As), and chromium (Cr), have been raised to unsafe levels in the environment due to human activities. The metal smelting and electroplating, and cement production are industrial processes that emit huge amounts of heavy metals into the air, water, and soil (Tchounwou et al., 2012). Mining activities are known to cause heavy metals to enter groundwater and river systems; agricultural activities inject Cd, Pb, and As into soil, due to overuse of phosphate fertilizers, pesticides, and sewage sludge. The dumping of waste, especially e-waste recycling and battery manufacturing, contributes to the size of the contamination load, especially in cities and peri-urban zones (Ali et al., 2019). The effects of the heavy metals on the environment are very severe and long-lasting since they are non-biodegradable. Heavy metals are deposited and biomagnified in water bodies and are transported up the food chain, disrupting fish reproduction and immune functioning and changing species composition (Ali et al., 2019). Heavy metals cause soil microbial diversity to decline and enzyme activity to be hampered in terrestrial systems, as well as limit nutrient cycling, reducing soil fertility (Luo et al., 2012). Chronic exposure to Cd may lead to kidney dysfunction and bone demineralization; Pb affects the nervous system and cognitive growth among children; As to cancer (skin, lungs, and bladder); and Cr (and especially hexavalent Cr^{6-}) to respiratory disease and cancer (Jaishankar et al., 2014).

4. PHYTOREMEDIATION POTENTIAL OF HYDROPHYTES

Hydrophytes show significant potential for phytoremediation, particularly in the removal of heavy metals from contaminated water bodies through various physiological mechanisms (Kristanti and Hadibarata, 2023). These plants can accumulate pollutant concentrations up to nearly 1450 times higher than terrestrial species, owing to their rapid growth rates, extensive root systems, and direct contact with polluted water (Anna et al., 2023). Their role in ecosystem restoration is multi-dimensional, offering advantages like oxygen regulation, food chain maintenance, and significant heavy metal uptake capabilities.

4.1 Mechanisms of Heavy Metal Uptake and Its Accumulation

Aquatic macrophytes or hydrophytes have special physiological and morphological features, which allow them to grow in wet and saturated environments with water, which is why they are considered good agents of phytoremediation of heavy metals (Figure 2). Hydrophytes can remove heavy metals mainly through the process of rhizofiltration, wherein dissolved contaminants are either absorbed or adsorbed by their roots. Metal ions that exist in contaminated water are attracted to the negatively charged sites on the plant cell wall and are either bound there by strong, non-exchangeable bonds or by weaker, exchangeable bonds. The resultant electrochemical gradient facilitates the passage of metal ions into the cells, and secondary uptake could be mediated by specific carrier proteins, as well as chelating agents such as organic acids and phytochelatins. Once present in the root, heavy metals can be sequestered locally or translocated to the shoot via the vascular tissues. Uptake kinetics are frequently described with the Michaelis-Menten model (concentration-dependent uptake) with higher uptake rates and higher transporter density on root membrane in hyperaccumulator species (Prasad & De Oliveira Freitas, 2003). The main processes are phytoextraction, presuming that plants absorb heavy metals in contaminated water and sediments in the tissues; rhizofiltration, which presupposes the roots to filter out and adsorb the dissolved metals; and phytostabilization, the contaminants are immobilized in the sediments by way of root exudates and complexation (Ali et al., 2013; Rai et al., 2019). Species-specific characteristics of these processes include root surface area, aerenchymatous tissue, and the capacity to make metal-chelating substances, among other factors, which determine their efficiency (Rai et al., 2019). When absorbed, metals can be sequestered in root tissues to eliminate toxicity to aerial parts or translocated to shoots to be further stored in vacuoles. Translocation efficiency varies by species and metal type. An example is Cd and Pb tend to accumulate in roots to reduce phytotoxicity, but other elements, such as Zn and Ni, can be more easily translocated to stems and leaves (Vymazal, 2011). The aerenchyma tissue of roots also improves the transport of oxygen, and therefore behaves as an oxidation of the rhizosphere, which may precipitate metals and restrict their mobility in the sediments. Metals are compartmentalized in shoots in vacuoles or are attached to phytochelatins and metallothioneins, which make them less reactive and less toxic (Salt et al., 1998).

Figure 2. Mechanisms of heavy metal contamination, and soil reclamation by hydrophytes

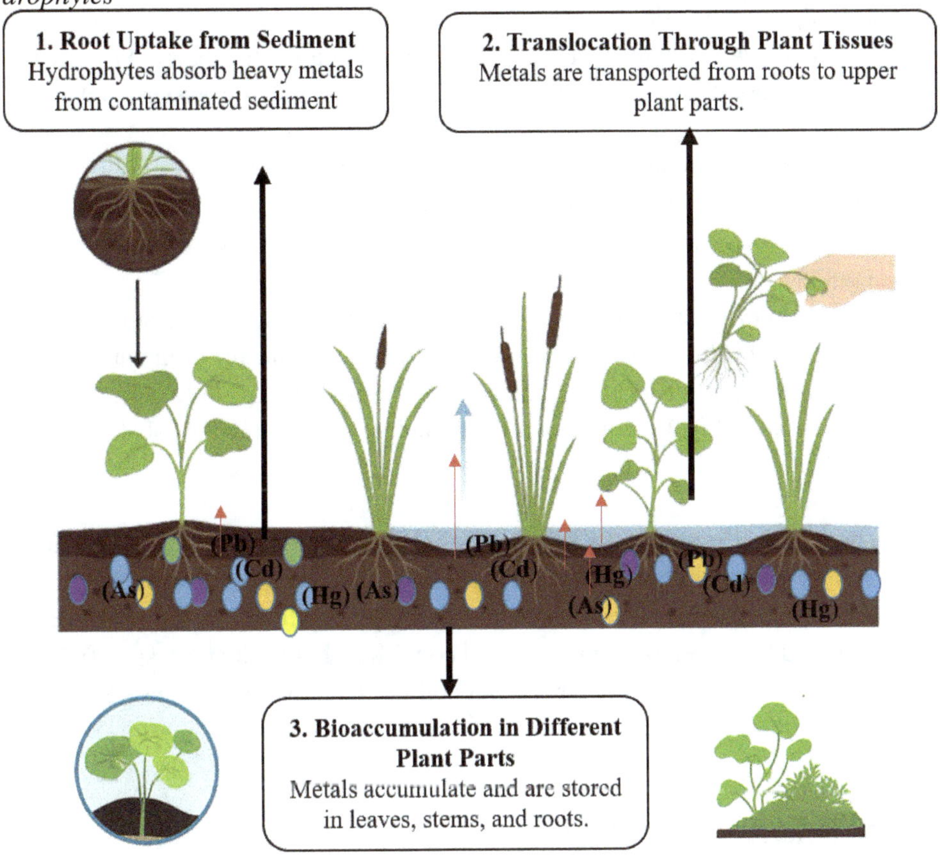

4.2 Case Examples of Effective Hydrophyte Species

Several hydrophyte species have demonstrated exceptional capacity for heavy metal remediation.

- *Eichhornia crassipes* (water hyacinth) is mostly known for its greater biomass productivity and ability to accumulate Cd, Pb, and Cr from wastewater (Zhu et al., 2014).
- *Lemna minor* (duckweed) can speedily uptake As and Hg from contaminated waters due to its high surface-to-volume ratio and fast growth rate (Mkandawire & Dudel, 2005).

- *Phragmites australis* (common reed) is efficient in constructed wetlands for removing Zn, Cu, and Ni owing both phytostabilization and rhizofiltration techniques (Brix, 1997).
- *Typha latifolia* (cattail) has exhibited the potential to absorb Pb and Cd through root tissues, mitigating proliferation in aquatic systems (Weis & Weis, 2004).
- *Typha latifolia* (cattail) and *Eichhornia crassipes* (water hyacinth): A recent study examined the phytoremediation ability of *Typha latifolia* and *Eichhornia crassipes* for the removal of cadmium (Cd) and arsenic (As) from a polluted water system (Khan et al., 2024).
- *Typha latifolia* (cattail) and *Carex rostrata* (bottle sedge): This study proved the applicability of *Typha latifolia* and *Carex rostrata* for eliminating engineered nanoparticles (NPs) such as silver (Ag), zinc oxide (ZnO), and titanium dioxide (TiO_2) from water of the industrial area (Ebrahimbabaie et al., 2023).

Such examples exemplify that hydrophytes are mostly cost-effective, environment-friendly, and sustainable solution for the removal of heavy metals, especially from wastewater plants and ecosystem restoration projects. However, the selection of species should be driven by environmental conditions in the site (site-specific), the type of target pollutants, and the potential end-use of harvested biomass to reduce the risk of secondary contamination.

5. INTEGRATION OF HYDROPHYTE-BASED SYSTEMS INTO CLIMATE CHANGE ADAPTATION

Hydrophyte-based systems are a viable nature-based solution to climate change adaptations as they stabilize, preserve the quality of water, biodiversity, and community resilience simultaneously. Such systems, be it in the form of constructed wetlands or floating treatment wetlands or a natural water body, can be deliberately installed in the city, suburban, as well as rural environments to offset the effects of the various stressing factors caused by climate like extended droughts, floods, and increased temperature (Masi et al., 2017a). Hydrophytes are also very important in the adjustment to climate as they enhance the retention of water, lead to the stabilization of high and low temperatures in water bodies, and offer the shelter of aquatic animals. They contribute to the minimization of climate susceptibility of agricultural systems because of the opportunity to provide year-round irrigation with treated wastewater and reduce dependence on rainfall that cannot be predicted (Saeed and Sun, 2012). Also, they have a carbon sequestration capability, both within the biomass and beneath the sediments, which in itself leads to the mitigation of climate change and supports the Sustainable Development Goal 13 (Climate Action) (Jones et al., 2024; Mohanty et al., 2024). There are also socio-economic co-benefits in the use of hydrophytes. Wetland systems that are controlled by communities can be used to yield livelihoods by harvesting biomass to produce bioenergy, compost, or handicrafts, especially in developing countries where the financial resources to fund sophisticated wastewater treatment are scarce (Konnerup et al., 2009). In addition, incorporation of hydrophyte-based treatment into the current infrastructure minimizes the operational energy needs relative to traditional treatment plants, thus decreasing greenhouse gases. Hydrophyte-based systems are likely to be most effective when included in national and regional climatic policies and supported by supportive frameworks, including community training, frequent monitoring, and species selection depending on local climatic and hydrological conditions. The expansion of these systems can make them not an ecological intervention of small scale, but a mainstream climate change adaptation intervention, especially in areas with water stress or in pollution-impacted areas.

5.1 Hydrophytes in Climate Change Mitigation and Adaptation

5.1.1 Contribution to Sustainable Development Goal 13 (SDG13)

Hydrophytes play a key role in the attainment of Sustainable Development Goal 13 (Climate Action) through the delivery of nature-based solutions that respond to climate mitigation and adaptation (Mohanty et al., 2024). Hydrophytes reduce greenhouse gas emissions in the environment and stabilize aquatic environments by sequestrating greenhouse gases indirectly through the enhancement of water quality, lessening eutrophication, and encouraging carbon sequestration in sediments (Figure 3). Their contribution to the recovery of deteriorated water bodies will also boost the resilience of the ecosystem to extreme weather patterns, including floods and droughts, which are bound to escalate in conditions of climate change. Policymakers can support climate-related goals and enhance ecological sustainability by including the hydrophyte-based systems in the wastewater treatment, stormwater management, and restoration programs of wetlands (Pan et al., 2011a; Masi et al., 2017b).

Figure 3: Hydrophyte Phytoremediation: Supporting SDG13 Climate Action

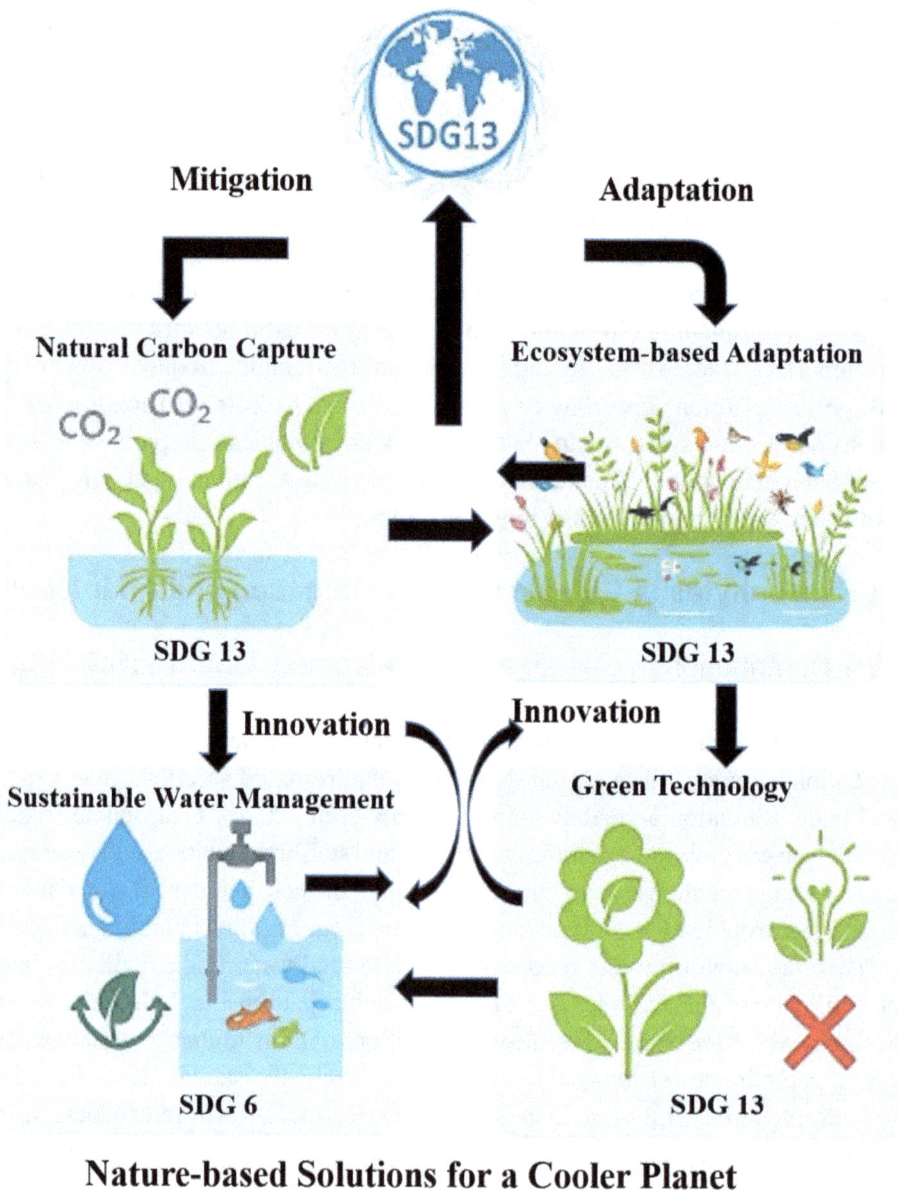

5.1.2 Enhancing Food Security and Resource Efficiency

Hydrophyte-based treatment systems have the potential to indirectly augment food security by using treated wastewater to irrigate crops, especially in water-restricted regions. This reduces dependence on unpredictable rain patterns, and it also eliminates the risks of water shortage. The hydrophyte systems are also capable of utilizing the effluent that is rich in nutrients and contains bioavailable nitrogen and phosphorus, and this can be applied in agriculture, thus saving the production costs and environmental impact due to the use of synthetic fertilizers (Konnerup et al., 2009). The hydrophytes will improve resilient food production systems in that they will improve the efficiency of water and nutrient cycles that enable communities to remain productive in agricultural activities despite varying climatic changes.

5.1.3 Reducing Environmental Footprint in Water Systems

Hydrophytes also minimize the water system environmental footprint as low-energy and low-maintenance alternatives to traditional wastewater treatment technologies. The constructed wetlands and floating systems are based on hydrophytes as opposed to mechanical treatment plants that consume a lot of energy to treat their contaminants through mechanical methods, including phytoremediation, sedimentation, and microbial degradation. This significantly lowers the emissions of greenhouse gases and the cost of operation, and enhances biodiversity and surrounding water bodies. In addition, hydrophytes can also be used to replenish water bodies to a healthy ecological condition by eliminating heavy metals, surplus nutrients, and pathogens, thereby making it a sustainable and climate-friendly constituent of integrated water management measures (Pan et al., 2011b; Saeed and Sun, 2012).

6. INTEGRATION INTO WASTEWATER AND ECOSYSTEM MANAGEMENT

6.1 Hydrophytes in Constructed Wetlands

Constructed wetland (CW) systems are considered to be engineered to utilize natural processes to purify water by use of hydrophytes. These are plants that are designed to be used either in aquatic or semi-aquatic conditions, where they increase pollutant removal through sedimentation, filtration, microbial degradation, and phytoremediation. Plant roots and rhizomes are surfaces on which microbial biofilms develop, which facilitates the degradation of organic matter and nutrient transformation. Hydrophytes are known to actively absorb and capture contaminants, such as heavy metals, and thus enhance the quality of effluent. *Phragmites australis, Typha latifolia,* and *Canna indica* are mostly in demand because they exhibit high biomass production, pollution tolerance, and adaptation to the changing climatic conditions (Brix, 1997; Vymazal, 2011).

6.2 Applications in Rural and Urban Wastewater Treatment

Constructed wetlands based on hydrophytes have been effective in rural and urban wastewater control because of their flexibility and low operation costs. They are applied in the rural settings to provide decentralized treatment systems of waste water where the conventional sewage systems are absent. They are also increasingly being added to the urban green infrastructure designs such as treatment wetlands in parks or river banks or in peri-urban settings. The systems can be used to treat domestic sewage, agricultural runoff, and industrial effluents and enhance the biodiversity and provide aesthetic values. They are especially applicable in developing regions that lack resources since they can be used using the least amount of energy sources (Kivaisi, 2001; Dotro et al., 2017).

6.3 Nature-Based, Low-Cost Implementation Models

The nature-based solutions (NbS) such as use of hydrophytes offer low-cost model of applying wastewater treatment and ecosystem restoration at a small scale. These models focus on the use of the native plants, natural hydrological streams, and passive treatment processes to minimize the construction and maintenance cost. Such systems have low costs and are socially acceptable to the community-led wetland projects in the rest of the world such as in India, Africa, and Southeast Asia and can exploit the local knowledge and use the locally available resources. Additionally, NbS approaches coincide with those of climate adaptation by enhancing the ability of ecosystems, water, and co-benefits of carbon capture and habitat development (Masi et al., 2017b; Bauer, 2020).

7. POLICY, AWARENESS, AND INTERDISCIPLINARY APPROACHES

7.1 Supportive Governance for Hydrophyte-Based Solutions

Hydrophyte-based systems are successful only with the help of enabling policies and governance structures that help to treat wastewater and restore ecosystems. Government can achieved by NbS into climate adaptation strategies, wastewater management legislations, and environmental restoration plans. Financial incentives of good governance: subsidies, grants and carbon credits of community based or industrial scale construction of constructed wetlands. When hydrophyte-based approaches are institutionalized as it has been observed more recently with policy frameworks in such places as China, Denmark, and India, they are not only investor and activator of ecosystem services that can be quantified but also investment by both the state and the investor (Liu et al., 2016; UN-Habitat, 2020).

7.2 Public Engagement and Education

The participation of the population and reporting of the process is important in order to render the hydrophyte-based wastewater management projects sustainable in the long run. Through the involvement of the community, it is possible to create ownership and manage the wetlands should thematically and take a regular maintenance of the wetlands and also encourage protection efforts. Visibility of the services provided by hydrophytes can be enhanced with educational efforts like school-based environmental education, citizen science monitoring, and demonstration sites. The examples of successful cases are participatory wetland management projects in East Kolkata Wetlands in India, where awareness campaigns resulted in better stewardship and decreased the flow of pollution (Finlayson et al., 2017; Ramsar Convention Secretariat, 2021).

7.3 Cross-Sectoral Collaboration for Scaling Up

To scale up to hydrophyte-based solutions, various sectors will have to work together to achieve this objective, with the involvement of environmental agencies, water utilities, agriculture, urban planning, and academia. The cross-sectoral partnerships would help in closing knowledge gaps, resource pooling, and innovation in construction wetlands design, implementation, and monitoring. Integration of hydrophyte-based solutions in larger sustainability policies, including circular economies and integrated water resource management (IWRM), can also be achieved through collaborative frameworks. Multi-stakeholder engagement has proven to be able to accelerate the mainstreaming of hydrophyte-based systems by regional networks such as the European Constructed Wetland Association (ECWA) and programs in the Global Wetland Outlook (Kadlec & Wallace, 2009; McInnes, 2020).

8. FUTURE RESEARCH DIRECTIONS

8.1 Genetic Improvement for Enhanced Phytoremediation

The next generation of research in the area of hydrophyte-based phytoremediation needs to be directed at the genetic enhancement of the efficiency of the heavy metal uptake, tolerance, and biomass generation. Innovation in plant genomics, transcriptomics, and CRISPR-Cas9 gene editing has the potential to create hydrophyte cultivars that have greater pollutant binding capacity, accelerated growth, and enhanced resistance to environmental stresses (Wentzell, (2025). Iron transporter metal transporters (e.g., ZIP, NRAMP families) and antioxidant defense technologies may be controlled to maximize the efficacy of the phytoremediation process. Besides, the combination of bioengineering and ecological attention will be relevant in the context of making genetically enhanced hydrophytes unable to impact the native biodiversity or disturb the ecosystem balance (Ali et al., 2013; Salt et al., 2018).

8.2 Scalability and Large-Scale Deployment

Although hydrophyte-based systems have been successful at small and pilot-scale, studies are required to overcome challenges in scaling up to regional and national wastewater treatment networks. The main ones are the optimization of hydrophyte density, planning of the system in various climatic areas, compatibility with the current water treatment units, and cost-benefit considerations in large-scale applications. The necessary conditions to ensure efficiency and ecological balance will involve long-term monitoring and adaptive management. Co-operation with ecologists, policy makers, and engineers is extremely important to make sure that the logistical, economic and regulatory constraints of this technology are broken to permit the use of hydrophyte systems both in rural and urban relation (Brix, 2017; Vymazal, 2011).

8.3 Climate-Resilient Aquatic Ecosystem Restoration

The research that must be done in the future should consider how to employ hydrophytes in future, not only to control pollution, but also to make water bodies more climate resilient. The areas of research are given priority on identification of hydrophytes species that can withstand extreme weather conditions, temperatures, and fluctuation of salt that may take place in the changing climate conditions. In addition, hydrophyte-based restoration with blue carbon methods could also be included to assist in reducing the emission of greenhouse gases as more carbon would be captured in the wetland sediments. They can be used to improve ecological services like flood control, habitat support, and nutrient cycling that are required to render both natural and constructed wetlands in the altered climate environment long-term viable (Mitsch and Gosselink, 2015; Davidson, 2014).

9. CONCLUSION

Hydrophytes, and their physiological adaptations to the water environment, are also a powerful and nature-oriented solution to some of the most acute problems of the current age of the environment. All of this is the testimony to their ability to clean up heavy metals, restore life in the aquatic environment, and manage water sustainably, and it is directly proportional to the goals of Sustainable Development Goal 13 (Climate Action). The control of pollution, adaptation to changes of climate, and protection of biodiversity may be promoted by the innovation of hydrophytes in the wastewater treatment system, as well as the implementation of wetlands and the reconstruction of ecosystems. In addition, the hydrophyte-based systems have co-benefits, including food security since the systems produce water efficiently, carbon prints since natural filters are used, climate resilience in rural and urban regions. Their potential is however achievable through collaborating in research, policy and citizen engagements. The studies of the genetic progress, high-volume application, and selection of the species that can resist the climate will make it possible to optimize their role in the ecological regulation. Hydrophytes due to their low cost, sustainability, and scalability is an effective climate action solution, as the impacts of climate change are getting worse. With the aid of science and governance, and even community participation, the full potential of these aquatic plants can be ensured so as to restore the degraded ecosystems, guarantee water resources and establish a healthy future that could sustain the future generations. Finally, the exploitation of the hydrophytes can be seen as synergistic entry to the problem of pollution control, biodiversity preservation, and climate resilience, highlighting the hydrophytes as a key driver of the creation of sustainable and climate-resilient world.

REFERENCES

Ali, H., Khan, E., & Ilahi, I. (2019). Environmental chemistry and ecotoxicology of hazardous heavy metals: Environmental persistence, toxicity, and bioaccumulation. *Journal of Chemistry*, *2019*, 1–14. DOI: 10.1155/2019/6730305

Ali, H., Khan, E., & Sajad, M. A. (2013). Phytoremediation of heavy metals—Concepts and applications. *Chemosphere*, *91*(7), 869–881. DOI: 10.1016/j.chemosphere.2013.01.075 PMID: 23466085

Anna, MK., Vimala, KS., Raiby, PP., Priyalatha, B., & Priya, S. (2023). Phytoremediation potential of a few hydrophytic medicinal plants: A review. International Journal of Scholarly Research in Biology and Pharmacy, 2, 005-009.

Bauer, LH. (2020). Nature-based solutions: the use of a floating wetland as an alternative to wastewater treatment.

Brix, H. (1997). Do macrophytes play a role in constructed treatment wetlands? *Water Science and Technology*, *35*(5), 11–17. DOI: 10.2166/wst.1997.0154

Brix, H. (2017). Do macrophytes play a role in constructed treatment wetlands? *Water Science and Technology*, *35*(5), 11–17. DOI: 10.2166/wst.1997.0154

Chambers, P. A., Lacoul, P., Murphy, K. J., & Thomaz, S. M. (2008). Global diversity of aquatic macrophytes in freshwater. *Hydrobiologia*, *595*(1), 9–26. DOI: 10.1007/s10750-007-9154-6

Coetzee, J. A., Hill, M. P., Julien, M. H., Center, T. D., & Cordo, H. (2009). *Eichhornia crassipes* (Mart.) Solms–Laub. (Pontederiaceae). In Muniappan, R., Reddy, G. V. P., & Raman, A. (Eds.), *Biological control of tropical weeds using arthropods* (pp. 183–210). Cambridge University Press., DOI: 10.1017/CBO9780511576348.011

Cohen-Shacham, E., Walters, G., Janzen, C., & Maginnis, S. (2016). Nature-based Solutions to address global societal challenges. Gland, Switzerland: International Union for Conservation of Nature (IUCN). DOI: 10.2305/IUCN.CH.2016.13.en

Cronk, J. K., & Fennessy, M. S. (2016). *Wetland plants: Biology and ecology* (2nd ed.). CRC Press., DOI: 10.1201/9781420032925

Davidson, N. C. (2014). How much wetland has the world lost? *Marine and Freshwater Research*, *65*(10), 934–941. DOI: 10.1071/MF14173

Dotro, G., Langergraber, G., Molle, P., Nivala, J., Puigagut, J., Stein, O., & Von Sperling, M. (2017). *Treatment Wetlands. Biological Wastewater Treatment Series*. IWA Publishing., DOI: 10.2166/9781780408774

Ebrahimbabaie, P., Smith, A., Zahran, E. M., & Pichtel, J. (2023). Phytoremediation of Engineered Nanoparticles Using *Typha latifolia* and *Carex rostrata.* Journal of Nanotechnology. *Applied and Environmental Soil Science, 3417525,* 1–16. Advance online publication. DOI: 10.1155/2023/3417525

Finlayson, C. M., Arthington, A. H., & Pittock, J. (2017). Freshwater ecosystems in protected areas: From 'ripples' to 'rivers' of conservation responsibility. *Aquatic Conservation, 27,* 96–113. DOI: 10.1002/aqc.2798

Huang, Q., Zheng, X., Pan, Z., He, J., Ma, J., Wang, J., Chen, B., & Chu, C. (2025). A synthetic hydrophyte for sunlight-driven water purification. *Environmental Science & Technology, 59*(28), 14597–14604. Advance online publication. DOI: 10.1021/acs.est.5c00765 PMID: 40631453

Hussain, S., Hussain, E., Saxena, P., Sharma, A., Thathola, P., & Sonwani, S. (2024). Navigating the impact of climate change in India: A perspective on climate action (SDG13) and sustainable cities and communities (SDG11). *Frontiers in Sustainable Cities, 5,* 1308684. DOI: 10.3389/frsc.2023.1308684

IPCC. (2022). *Climate Change: impacts, adaptation and vulnerability, contribution of working group ii to the sixth assessment report of the intergovernmental panel on climate change.* Cambridge University Press., DOI: 10.1017/9781009325844

Jaishankar, M., Tseten, T., Anbalagan, N., Mathew, B. B., & Beeregowda, K. N. (2014). Toxicity, mechanism and health effects of some heavy metals. *Interdisciplinary Toxicology, 7*(2), 60–72. DOI: 10.2478/intox-2014-0009 PMID: 26109881

Jones, S. F., Arias-Ortiz, A., Baldocchi, D., Eagle, M., Friess, D. A., Gore, C., Noe, G., Nolte, S., Oikawa, P., Paytan, A., Raw, J. L., Roberts, B. J., Rogers, K., Schutte, C., Stagg, C. L., Thorne, K. M., Ward, E. J., Windham-Myers, L., & Yando, E. S. (2024). When and where can coastal wetland restoration increase carbon sequestration as a natural climate solution? *Cambridge Prisms: Coastal Futures, 2,* e13. Advance online publication. DOI: 10.1017/cft.2024.14 PMID: 40881330

Kadlec, R. H., & Wallace, S. D. (2009). *Treatment Wetlands* (2nd ed.). CRC Press.

Khan, S., Kamal, M., Noor, S., & Afzal, S. M. (2024). Assessment of heavy metals and its treatment through phytoremediation in groundwater along River Kabul in district Charsadda. *Frontiers in Environmental Science, 12,* 1392892. Advance online publication. DOI: 10.3389/fenvs.2024.1392892

Kivaisi, A. K. (2001). The potential for constructed wetlands for wastewater treatment and reuse in developing countries: A review. *Ecological Engineering, 16*(4), 545–560. DOI: 10.1016/S0925-8574(00)00113-0

Konnerup, D., Koottatep, T., & Brix, H. (2009). Treatment of domestic wastewater in tropical, subsurface flow constructed wetlands planted with Canna and Heliconia. *Ecological Engineering*, *35*(2), 248–257. DOI: 10.1016/j.ecoleng.2008.04.018

Kristanti, R. A., & Hadibarata, T. (2023). Phytoremediation of contaminated water using aquatic plants, its mechanism and enhancement. *Current Opinion in Environmental Science & Health*, *32*, 100451. DOI: 10.1016/j.coesh.2023.100451

Lal, R. (2015). Restoring soil quality to mitigate soil degradation. *Sustainability (Basel)*, *7*(5), 5875–5895. DOI: 10.3390/su7055875

Liu, H., Zhang, J., Zhang, J., & Chen, J. (2016). Constructed wetlands in China: Recent developments and future challenges. *Frontiers of Environmental Science & Engineering*, *10*, 478–490. DOI: 10.1007/s11783-016-0831-0

Lobell, D. B., Schlenker, W., & Costa-Roberts, J. (2011). Climate trends and global crop production since 1980. *Science*, *333*(6042), 616–620. DOI: 10.1126/science.1204531 PMID: 21551030

Luo, L., Ma, Y., Zhang, S., Wei, D., & Zhu, Y. G. (2012). An inventory of trace element inputs to agricultural soils in China. *Journal of Environmental Management*, *100*, 124–129. DOI: 10.1016/j.jenvman.2011.11.013 PMID: 19246150

Masi, F., Rizzo, A., & Regelsberger, M. (2017a). The role of constructed wetlands in a new circular economy, resource oriented, and ecosystem services paradigm. *Journal of Environmental Management*, *216*, 275–284. DOI: 10.1016/j.jenvman.2017.11.086 PMID: 29224716

Masi, F., Rizzo, A., Regelsberger, M., & Bresciani, R. (2017b). Constructed wetlands for combined sewer overflow treatment: Ecosystem services at Gorla Maggiore, Italy. *Ecological Engineering*, *98*, 427–438. DOI: 10.1016/j.ecoleng.2016.03.043

McInnes, R. (2020). *Global Wetland Outlook* (Special Edition 2021). Ramsar Convention Secretariat.

Mitsch, W. J., & Gosselink, J. G. (2015). *Wetlands* (5th ed.). Wiley.

Mkandawire, M., & Dudel, E. G. (2005). Accumulation of arsenic in Lemna gibba L. (duckweed) in tailing waters of two abandoned uranium mining sites in Saxony, Germany. *The Science of the Total Environment*, *336*(1-3), 81–89. DOI: 10.1016/j.scitotenv.2004.06.002 PMID: 15589251

Mohanty, S., Pandey, P. C., Pandey, M., Srivastava, P. K., & Dwivedi, C. S. (2024). Wetlands contribution and linkage to support SDGs, its indicators and targets-A critical review. *Sustainable Development (Bradford), 32*(5), 5348–5392. DOI: 10.1002/sd.2948

Pan, J., Zou, X., Wang, Q., Song, Y., & Liu, Y. (2011b). Carbon sequestration through constructed wetlands: A meta-analysis. *Ecological Engineering, 37*, 1083–1091. DOI: 10.1016/j.ecoleng.2011.02.010

Pan, Y., Birdsey, R. A., Fang, J., Houghton, R., Kauppi, P. E., Kurz, W. A., Phillips, O. L., Shvidenko, A., Lewis, S. L., Canadell, J. G., Ciais, P., Jackson, R. B., Pacala, S. W., McGuire, A. D., Piao, S., Rautiainen, A., Sitch, S., & Hayes, D. (2011a). A Large and Persistent Carbon Sink in the World's Forests. *Science, 333*(6045), 988–993. DOI: 10.1126/science.1201609 PMID: 21764754

Prasad, M. N. V., & De Oliveira Freitas, H. M. (2003). Metal hyperaccumulation in plants—Biodiversity prospecting for phytoremediation technology. *Electronic Journal of Biotechnology, 6*(3), 110–146. DOI: 10.2225/vol6-issue3-fulltext-6

Rai, P. K., Lee, S. S., Zhang, M., Tsang, Y. F., & Kim, K. H. (2019). Heavy metals in food crops: Health risks, fate, mechanisms, and management. *Environment International, 125*, 365–385. DOI: 10.1016/j.envint.2019.01.067 PMID: 30743144

Ramsar Convention Secretariat. (2021). *Ramsar Wetland Conservation Awards*. Ramsar Convention.

Saeed, T., & Sun, G. (2012). A review on nitrogen and organics removal mechanisms in subsurface flow constructed wetlands: Dependency on environmental parameters, operating conditions and supporting media. *Journal of Environmental Management, 112*, 429–448. DOI: 10.1016/j.jenvman.2012.08.011 PMID: 23032989

Salt, D. E., Baxter, I., & Lahner, B. (2018). Ionomics and the study of the plant ionome. *Annual Review of Plant Biology, 69*, 481–504. DOI: 10.1146/annurev-arplant-042817-040212 PMID: 18251712

Salt, D. E., Blaylock, M., Kumar, N. P., Dushenkov, V., Ensley, B. D., Chet, I., & Raskin, I. (1998). Phytoremediation: A novel strategy for the removal of toxic metals from the environment using plants. *Biotechnology : An Indian Journal, 13*(5), 468–474. DOI: 10.1038/nbt0595-468 PMID: 9634787

Tchounwou, P. B., Yedjou, C. G., Patlolla, A. K., & Sutton, D. J. (2012). Heavy metal toxicity and the environment. Molecular. *EXS, 101*, 133–164. DOI: 10.1007/978-3-7643-8340-4_6 PMID: 22945569

UN-Habitat. (2020). *Nature-based solutions for water*. UN-Habitat.

United Nations. (2015). Transforming our world: The 2030 agenda for sustainable development. United Nations General Assembly. https://sdgs.un.org/2030agenda

Villamagna, A. M., & Murphy, B. R. (2010). Ecological and socio-economic impacts of invasive water hyacinth (*Eichhornia crassipes*): A review. *Freshwater Biology, 55*(2), 282–298. DOI: 10.1111/j.1365-2427.2009.02294.x

Vymazal, J. (2011). Constructed wetlands for wastewater treatment: Five decades of experience. *Environmental Science & Technology, 45*(1), 61–69. DOI: 10.1021/es101403q PMID: 20795704

Weis, J. S., & Weis, P. (2004). Metal uptake, transport and release by wetland plants: Implications for phytoremediation and restoration. *Environment International, 30*(5), 685–700. DOI: 10.1016/j.envint.2003.11.002 PMID: 15051245

Wentzell, B. M. (2025). Phytoremediation for water quality improvement: Current advances and future prospects. *Biotechnology for the Environment, 2*(1), 7. DOI: 10.1186/s44314-025-00022-9

WHO. (2018). Climate change and health. World Health Organization. https://www.who.int/news-room/fact-sheets/detail/climate-change-and-health

WWAP (United Nations World Water Assessment Programme). (2020). The United Nations World Water Development Report 2020: Water and Climate Change. UNESCO. https://unesdoc.unesco.org/ark:/48223/pf0000372985

Zhu, Y. L., Zayed, A. M., Qian, J. H., De Souza, M., & Terry, N. (2014). Phytoaccumulation of trace elements by wetland plants: III. Uptake and accumulation of ten trace elements by twelve plant species. *Journal of Environmental Quality, 28,* 1448–1455. DOI: 10.2134/jeq1999.00472425002800050029x

Chapter 8
Hydrophytes and Circular Economy for Heavy Metal Remediation

Mustapha Saidu
https://orcid.org/0009-0005-5951-9258
University of St. Andrews, UK

Abdulhakeem Idris Abdulhakeem
Al-Farabi Kazakh National University, Kazakhstan

Muhammad Usman Kabir
https://orcid.org/0000-0003-2129-5242
University of Maiduguri, Nigeria

Mariam Oyinde Bakare
Olabisi Onabanjo University, Ogun, Nigeria

ABSTRACT

Heavy metal pollution poses a significant global environmental challenge, affecting the ecosystem and the health of humans due to different industrial and anthropogenic activities. Traditional remediation methods are ineffective, expensive, and causes secondary pollution. This chapter discusses how hydrophytes, and aquatic plants can be used to provide a sustainable and eco-friendly solution for heavy metal remediation. We explore the variances on how hydrophytes absorb, accumulate and detoxify heavy metals released into water, including phytoextraction, phytostabilization, and rhizofiltration. In addition to what they do in terms of remediation, the chapter points out the importance of integrating the use of hydrophyte-based systems with the philosophy of circular economy.

DOI: 10.4018/979-8-3373-4037-1.ch008

1. INTRODUCTION

Heavy metals pollutants are one of the most difficult to treat environmental issues and they are widespread (Xie *et al.*, 2022). Unlike organic pollutants, heavy metals are not biodegradable and thus have long-term detrimental effects to ecosystems and human health due to their potentially high levels of accumulation into the environment (Farid *et al.*, 2023). Current remediation approaches to address heavy metal contamination, including chemical precipitation, ion exchange, and membrane filtration, have the disadvantage of being quite expensive, energy-intensive, and producing secondary wastes, which underlines the fact that they become less sustainable over the long term (Iyyappan *et al.*, 2023). This has encouraged the quest of more environmental friendly and cost-efficient alternatives.

The ability to use plants as a medium of cleaning up polluted environments, known as phytoremediation, has become one of the emerging green technologies. Among the plants that have been being used to do this, the use of the hydrophytes has emerged as a very great potential in removing heavy metals in water (Khan *et al.*, 2024). They are good candidates in purifying water because they can absorb and concentrate the metals that they find around their surroundings. Nonetheless, one of the bottlenecks of phytoremediation has been the issue of management of the metal-laden biomass left after the cleanup process is completed. Otherwise, this biomass may end up providing secondary pollution and ruin all the effectiveness of the entire approach of sustainability.

Here is where the rules of the circular economy are applied. A circular economy does not leave any waste and focuses on the fact that resources can be kept in the merry-go-round, where they can be used all the time, and does not presuppose the traditional straight line of take-make-dispose (Sarah Lee, 2025). By connecting such hydrophyte-based remediation to a sustainable circular economy and assessing its overall health impacts, the possibility of turning the contaminated plant biomass liability into a resource can in fact become a reality. It can be viewed as developing new techniques to produce bioenergy, extract metals, or add value to the harvested biomass (Iyyappan *et al.*, 2023). In addition to addressing the issue of heavy metal pollution, this combination of tactics can support sustainable development and resource efficiency.

This chapter will provide a comprehensive overview on the utilisation of hydrophytes in heavy metals remediation in the aspect of circular economy. We will explore the biochemistry of extreme metal uptake in hydrophytes, how the ensuing biomass can be valorized and potential applications and problems of this coupled strategy. Envisioning the transformation of the environment remediation process to resource recovery, we can offer a comprehensive and sustainable solution to the burning issue of heavy metal pollution.

2. HEAVY METAL POLLUTION: SOURCES, IMPACTS, AND CHALLENGES

The heavy metals refer to a set of metallic elements which occur naturally in the Earth's crust. Nevertheless, human activities have significantly increased their concentrations in various environmental compartments resulting in pollution (Madeła and Skuza, 2021). These elements are particularly concerning because they are non-biodegradable (i.e. unable to be transformed into less harmful materials by natural processes) and they can remain in the environment over long period of time (Khan *et al.*, 2023).

2.1. Major Sources of Heavy Metal Contamination

Numerous anthropogenic activities contribute to environmental heavy metal contamination. It is essential to comprehend these sources in order to create remediation and prevention plans that work.

2.1.1. Industrial Activities

The heavy metal pollution could be caused by industrial processes most likely in their largest proportions. The processes of mining and smelting of these metals move enormous amounts of Fs into soil, water and air, i.e., as lead (Pb), cadmium (Cd), arsenic (As), copper (Cu), and zinc (Zn) (Zulkernain *et al.*, 2023). Extraction and processing of ores can produce a lot of wastewater and solid waste in high concentrations of these poisonous elements. Other examples of heavy materials that are discharged into the environment by industries include electroplating employment, battery production, textile dying and chemical industries (Ahmed *et al.*, 2022). Industrial wastes disposed in such a way that they have not been properly treated may pollute surface and even ground water eventually causing the contamination of soil through irrigation or deposit.

2.1.2. Agricultural Practices

Agriculture is essential for food production, but can heavily contribute to metal pollution. The extensive application of phosphate-based fertilizers, for instance will introduce cadmium to the agricultural soil as phosphate rocks exists in different amounts of this heavy metal (Wan *et al.*, 2024). Older formulations of pesticides and herbicides in some cases include heavy metals such as arsenic and lead. Manure deposited by animals fed with contaminated food items or drugs administered can also be a contributor to heavy metals in soils once the manure is used as fertilizers.

Another route is irrigation using wastewater that is contaminated, such as in some areas where people have no other source of water to continue cultivation of crops (Xie *et al.*, 2022).

2.1.3. Urban and Domestic Waste

Urbanization and population growth contribute to increase in domestic and municipal solid solid waste generation. Improperly managed landfills have the potential to emit heavy metals into the ground and soil (Tsai *et al.*, 2021). A fast expanding waste stream is electronic waste (e-waste), which is of particular concern. E-waste has a diverse range of highly valuable but toxic heavy metals, whose abundance includes lead, mercury, cadmium and chromium. Failure to dispose of e-waste or formal recycling exposes these metals to the environment causing serious health hazards to the environment and the humans (Srivastav *et al.*, 2023). Metals that enter a waste stream may be in paints, batteries and in some cleaners that are household products and would be used in the house hold.

2.1.4. Atmospheric Deposition

Industrial emissions, automobile exhaust and burning of fossil energy are known to emit heavy metals into the atmosphere in the form of particulate matter. These air particles are transferable through long distances where they end up on the land and water surface by the process of wet or dry deposition (rain and snow). Such a nonpoint source of pollution will contaminate soils and waterways, and in areas far removed and in smaller amounts than immediate industrial activity (Heisi *et al.*, 2023).

Figure 1. Sources of heavy metal contamination

2.2. Environmental and Health Impacts of Heavy Metals

Both human health and ecological balance are seriously impacted by the presence of heavy metals in the environment. They are especially harmful because of their tenacity and propensity to bioaccumulate and biomagnify throughout the food chain.

2.2.1. Ecological Impacts

Heavy metals are harmful to most living organisms, including, plants, micro-organisms, and animals. High concentrations of heavy metals in the soil can hinder crop growth, slow down plant development, and even kill plants by interfering with vital functions like photosynthesis, water uptake, and nutrient assimilation (Wan *et al.*, 2024). Additionally, they can alter the soil's microbial communities, reduce biodiversity, and jeopardize vital ecosystem services like the breakdown of organic matter and nutrient cycling (Farid *et al.*, 2023).

Aquatic systems are also at high risk; heavy metals may be deposited in the sediments where they may be adsorbed to the benthic life-forms, or they might be ingested by aquatic plant and animal life resulting in their accumulation throughout the food chain. This may lead to impaired reproduction, high embryonic abnormality and mortality rates of aquatic life (Rezania *et al.*, 2016). The overall consequence is that there is diminishing ecosystem health, resilience and biodiversity.

2.2.2. Human Health Impacts

The exposure of humans to heavy metals can occur through different pathways, such as the intakes of contaminated food and water, breathing in the contaminated dust or air, and contact with the skin (El Awady *et al.*, 2021). Even low doses of heavy metals may cause numerous health effects after chronic exposure since heavy metals can be accumulated in different organs and disrupt their biological processes. As an example, lead can result in neurological damage, developmental abnormalities in children and kidney dysfunction. Cadmium is a well-established carcinogen and may cause renal damage, bone demineralization and respiratory difficulties. Neurological and developmental disorders may be caused by exposure to mercury. Arsenic is also cancerous and may lead to lesions on the skin, cardiovascular disease and neuro-problems (Jaishankar *et al.*, 2014). Different health effects of exposure to heavy metals are related to the type of metal, amount and length of exposure, as well as the vulnerability of individuals (Jaishankar *et al.*, 2014). Vulnerable populations, such as children and pregnant women, are particularly at risk due to their physiological sensitivity and potential for developmental harm (Ali *et al.*, 2019).

2.3. Conventional Heavy Metal Remediation Methods and Their Limitations

Over the years, various conventional methods have been employed for heavy metal remediation from contaminated sites. While effective to some extent, many of these techniques come with significant limitations.

2.3.1. Physical and Chemical Methods

Physical method often involves excavation and disposal of contaminated soil in secure landfills, or soil washing, where the contaminants are washed or leached out using chemical solutions (Rezania *et al.*, 2016). Chemical methods include precipitation, coagulation-flocculation, ion exchange, and adsorption. Although these techniques can be useful in the removal of heavy metals, such procedures are usually costly, energy-consuming, and potentially create a large amount of secondary waste products (e.g., contaminated sludge or wastewater) that must be further treated and disposed of (Ahmed *et al.*, 2022). An example of Chemical precipitation is that it may necessitate the large volume of reagents and create dangerous sludge that can be difficult to dispose of and present environmental threats (Ahmed *et al.*, 2022). In the same manner, ion exchange resins must be regenerated periodically producing concentrated waste outputs which must be subject to additional treatment (Bolan *et al.*, 2014). These approaches can be effective in use when there is severe contamination of a local area but are logistically infeasible in large scale or diffuse contamination mainly because of high operational costs and their ecological impact (Liu *et al.*, 2018).

2.3.2. Limitations of Conventional Methods

The major disadvantages of the traditional approaches to heavy metal removal involve their high cost, which often makes large-scale application economically unfeasible (Ahmed *et al.*, 2022). Most processes including chemical precipitation and ion exchange are energy-intensive thereby making it to their carbon-intensive footprint (Bolan *et al.*, 2014). Additional disposal fees and costs are associated with the production of secondary wastes, e.g. hazardous sludge during precipitation or concentrated waste streams in resin regeneration (Liu *et al.*, 2018). Moreover, other techniques such as soil washing, or thermal decontamination may cause the loss of natural composition and fertility of the land, which makes the restoration process more difficult (Wan *et al.*, 2024). In the water treatment industry, it is possible that the traditional methods of analysis are not up to the task of attaining the ultra-low detection limits necessary to guarantee safe reuse or discharge of treated water. Furthermore, the effectiveness of these traditional approaches tends to be negatively dependent on the concrete physicochemical properties of the water, and thus practically nothing can be done to them in theory (Rezania *et al.*, 2016). These restrictions highlight the sense of urgency behind the increased consideration of cost-effective, ecologically clean, and environmentally friendly approaches to remediation, which biological methods can offer, such as phytoremediation (Farid *et al.*, 2023).

3. HYDROPHYTES: BIOLOGY, MECHANISMS, AND APPLICATIONS IN HEAVY METAL REMEDIATION

Hydrophytes or aquatic plants are one of the most diverse assemblages that have developed mechanisms to survive all through an aquatic lifecycle (Kamal *et al.*, 2004). These morphological and physiological features make them extremely efficient in absorbing and changing heavy metals in the background and in the aquatic environment (Kamal *et al.*, 2004; Farid *et al.*, 2023). The following sections outline their taxonomic group, complex remediation methods, and their applications (Rezania *et al.*, 2016; Fletcher *et al.*, 2022).

3.1. Classification and Characteristics of Hydrophytes

Hydrophytes can be broadly classified into several categories based on their growth habit and relationship with the water body:

- **Emergent Hydrophytes:** These are plants that have their roots in the sediment, but stems and leaves are above the water (Rezania *et al.*, 2016). They have strong roots and thus absorb the contaminants in the overlaying water column as well as absorb the contaminants in the sediment that is contaminated (Kamal *et al.*, 2004; Fletcher *et al.*, 2022).
    - **Submerged Hydrophytes:** Submerged hydrophytes are those plants that develop and grow solely beneath the water surface and are usually rooted at the substrate such as *Hydrilla verticillata* and *Myriophyllum spicatum*. They obtain the nutrients and pollutants via their leaves and stems (Tanko *et al.*, 2025).
    - **Floating-leaved Hydrophytes:** These are plants with floating leaves, they are rooted in the bottom of submerged environments (e.g., *Nymphaea spp. (water lilies)*, *Nelumbo nucifera* (lotus)). They are able to uptake the contaminants that are present in the sediment as well as water column (Farid *et al.*, 2023).
    - **Free-floating Hydrophytes:** These plants do not grow into the sedment and are free floating on the water surface (e.g., *Eichhornia crassipes* (water hyacinth), *Pistia stratiotes* (water lettuce), *Lemna spp.* (duckweed)). They are in direct contact with water and absorb contaminants in it well (Ali *et al.*, 2020).

Figure 2. Classes of hydrophytes

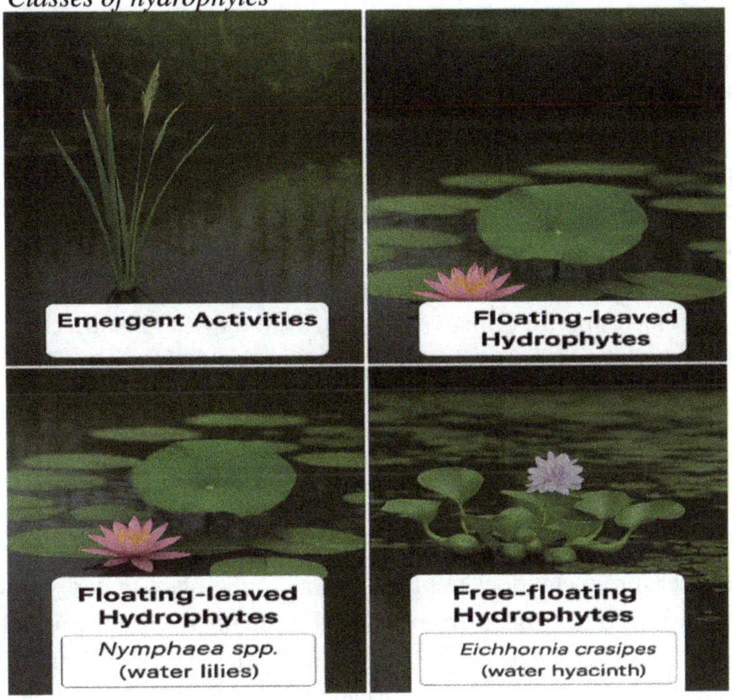

The most important features of the hydrophytes that make them effective towards heavy metal remediation are their ability to grow rapidly and produce high biomass rates, elaborate root system (powerful in emergent and rooted floating species), and more so their ability to afford heavy metal concentrations without significant toxicity (Rezania *et al.*, 2016). Most of the hydrophytes are also able to survive through mechanisms suited to handle the presence of anoxic conditions in saturated soils and this can be useful in cases of contaminated soils.

3.2. Mechanisms of Heavy Metal Uptake and Detoxification in Hydrophytes

Hydrophytes employ a variety of mechanisms to take up, transport, and detoxify heavy metals. These processes can be broadly categorized as follows:

3.2.1. Phytoextraction

Phytoextraction refers to the process of absorbing the heavy metals by plant roots in the growth medium (water or sediment) and then translocating it to above ground parts, which can be harvested (shoots and leaves) (Vikraman et al., 2022).

This process works particularly well with hyperaccumulator species, which are capable of concentrating the metal up to one hundred times higher than non-accumulator species without showing any sign of toxicity. After growth, the biomass that has been harvested is then processed to either recover or dispose the metals accumulated. According to Suman *et al.* (2018), the method is effective in eliminating metals in water bodies and lowering the levels of sediments.

3.2.2. Phytostabilization

Phytostabilization involves the use of plants that minimize the mobility and bioavailability of the heavy metal in the polluted environment (Abdelaal *et al.*, 2021). Plants ensure metal does not find its way to the plant-underlying waters or in the food chain of other organisms because they do not release them back into the environment, but instead, they store them in their root system or internal systems (Sharma *et al.*, 2023). Such sequestration can be by precipitation in the rhizosphere, adsorption onto root surfaces, or by inclusion in plant tissues (Bolan et al., 2014). Phytostabilization is also applicable to large sites with low to moderate contamination in which the main goals are to stop the spread of pollutants and to minimize the risk of exposure (Liu *et al.*, 2018; Wan *et al.*, 2024).

3.2.3. Rhizofiltration

Rhizofiltration is a type of phytoremediation specifically designed for treating non-point-source water pollution in which the roots of water plants absorb or adsorb the pollutants in the water column (Rezania *et al.*, 2016). Dilute concentrations of heavy metals in a large volume of water, are perfect examples of contaminated effluents, or potentially polluted surface water sources, and can be addressed extremely well through this mechanism (Rezania *et al.*, 2016). Fibrous and extensive root systems in plants (such as sunflowers or Indian mustard) are frequently implemented, although the rhizofiltration potential of plenty of hydrophytes is similarly great (Farid *et al.*, 2023). The plants would be normally planted in hydroponic systems or floating rafts so that their roots are directly exposed to the contaminated water (Ali *et al.*, 2020).

3.2.4. Phytovolatilization

Phytovolatilization is the process in which plants take in heavy metals and then releases the volatile form into the atmosphere (Sharma *et al.*, 2023). This type of mechanism is less prevalent amongst most of the heavy metals, but is significant with mercury (Hg) and selenium (Se) (Lone *et al.*, 2008). Metals can be taken up by plants and then be converted to gaseous releases (e.g. elemental mercury or dimethyl selenide) that can be transpired to leave plants (Yadav *et al.*, 2022). While it offers a way to remove metals from the contaminated site, there are concerns that it has the potential of causing atmospheric pollution and, therefore, it is mostly performed under controlled conditions (Bolan *et al.*, 2014).

3.3. Key Hydrophyte Species for Heavy Metal Remediation

A great number of hydrophyte species have been acquired and examined on the basis of their future in the remediation of heavy metals. Some of the most commonly researched and effective species include:

- ***Eichhornia crassipes* (Water Hyacinth):** This free-floating vegetation is well-known due to its rapid growth rate and biomass that makes it a perfect candidate in phytoextraction and rhizofiltration of the various heavy metals such as Cd, Cr, Cu, Pb, and Zn (Ali *et al.*, 2020; Kumar *et al.*, 2024). Its extensive fibrous root system provides a large surface area for metal adsorption.
  - ***Pistia stratiotes* (Water Lettuce):** The empirical evidence shows that the free-floating species has a great ability to bioaccumulate heavy metals like lead (Pb), cadmium (Cd), and copper (Cu) in wastewater (Xie *et al.*, 2022). It has a high growth rate and ease of growth, which makes it equally efficient when it comes to phytoremediation especially when it comes to the use of the water hyacinth, which is widely used.
  - ***Lemna minor* (Common Duckweed):** Duckweed is a small, floating plant which grows in masses all over the surface of the water. The characteristics of these plants include high rates of metal absorption, in particular, Cd, Cu, and zinc (Zn), and a short breeding cycle (Rezania *et al.*, 2016). Their low size makes them easy to manipulate and harvest, which adds to their feasibility of solution in remediation.
  - ***Typha latifolia* (Cattail):** It is an emergent hydrophyte that is capable of stabilizing the sediments and at the same time trapping the heavy metals in the soil and water (Khan *et al.*, 2024). Because of its high biomass and resistance to various pollutants, it is frequently utilized as a waste treatment mechanism in artificial wetlands.

- ***Phragmites australis (Common Reed):*** The common reed is a ubiquitous emergent species with a high rhizome density and hence looks like cattails. It is commonly incorporated into constructed large-scale wetland infrastructures and has proven to be very effective in phytostabilization as well as phytoextraction of heavy metals in the wetland (Heisi *et al.*, 2023).
- ***Giant Salvinia, or Salvinia molesta:*** This free-floating fern has demonstrated the ability to absorb heavy metals, namely Cu, Cd, and Pb, but the process can be explained by the fact that it grows a large mass of biomass and its fronds have the capacity to absorb heavy metals (Farid *et al.*, 2023).

3.4. Factors Influencing Hydrophyte-Mediated Remediation Efficiency

The efficiency of hydrophyte-based heavy metal remediation is influenced by a complex interplay of environmental factors and plant characteristics:

- **Heavy Metal Concentration and Speciation:** The high concentration of heavy metal in water or sediment has a substantial influence on absorption rates. Higher concentrations are sometimes toxic to the plants, but very low concentration could make the uptake to be slow. Speciation (chemical form) of the metal is also highly critical as bioavailability differs between free ionic, complexes, and the adsorption of metals to particles (El Awady *et al.*, 2021).
- **pH and Redox Potential:** Both the water and sediment pH have a strong effect on solubility and the bioavailability of the metals. It is generally thought that many of the heavy metals become more available at lower pH in an easy state to be absorbed by the plants and at the same time more toxic. Redox potential influences the oxidation status of metals, and thus the mobility of metals and transfer into plants (Lone *et al.*, 2008).
- **Nutrient Availability:** The ability of any plant to take in contaminants depends on the presence of the required macronutrients, especially potassium, phosphorus, and nitrogen, the absence of which would lead to unhealthy vegetative growth (Rezania *et al.*, 2016). However, some instances of these nutrients can occasionally lead to inhibition of metal uptake or rapid algal growth hence preventing the remediation process (Farid *et al.*, 2023).
- **Temperature and Light Intensity:** Temperature and intensity of light moderate plant growth, metabolic activities and photosynthetic efficiency (Kamal *et al.*, 2004). The most ideal conditions of temperature and light stimulate

the active development of plants, raise the sale of biomass, and augment the ability of plants to absorb metals (Fletcher *et al.*, 2022).

- **Presence of Other Contaminants:** Additional contaminants, organic compounds or other heavy metals, may also adjust the degree of accumulation of metals by hydrophytes either increasing or reducing it (Liu *et al.*, 2018). Remediation of mixed contamination is quite a challenge because the interaction between contaminants may be antagonistic or synergistic (Wan *et al.*, 2024).
- **Plant Biomass and Age**: Plants that are actively growing, especially young are usually more susceptible to metal uptake (Abdelaal *et al.*, 2021). The overall amount of metal that can be obtained out of a system is directly proportional to the biomass forming the hydrophyte population (Khan *et al.*, 2023). In order to maintain high removal rates and prevent the re-release of metals, it is urgent to perform regular harvests once the biomass grew up and reached a high level of metal concentration (Fletcher *et al.*, 2022).

4. CIRCULAR ECONOMY

4.1. Understanding the Circular Economy Framework

The circular economy is built upon several core principles that aim to decouple economic growth from resource depletion and environmental impact. These principles include:

4.1.1 Design out waste and pollution: This involves designing goods and procedures to stop the production of waste and pollution emissions (Sarah Lee, 2025). This refers to the creation of remediation methods unrelated to secondary hazardous wastes in order to remove heavy metals like lead (Ahmed *et al.*, 2022)

4.1.2 Keep products and materials in use: This principle emphasizes reusing products and materials, by repairing and remanufacturing, reusing, and recycling them (Srivastav *et al.*, 2023). In the case of hydrophyte-based remediation, this can be done through finding means to put the remediated materials or recovered metals back into the economy (Khan *et al.*, 2023; Amlon Group, 2024).

4.1.3 Regenerate natural systems: The CE aims to enhance natural systems to be able to benefit natural capital by giving back precious resources to the biosphere and improving the health of ecosystems (Sarah Lee, 2025). Phytoremediation, through its ability to clean up of the water and soil, has a direct positive influence on the restoration of natural systems (Fletcher *et al.*, 2022).

4.2. Resource Recovery from Contaminated Biomass

In a hydrophyte-based treatment, biomass harvest with concentrated heavy metal levels is not just waste biomass (Khan *et al.*, 2023). According to the cycle economy idea, it is a rich base that may be processed to produce metals or energy (Amlon Group, 2024; Zulkernain *et al.*, 2023). This conversion results in a reduced overall environmental effect and financial benefits for the rehabilitation procedure (Sarah Lee, 2025).

4.2.1. Bioenergy Production

Converting biomass into bioenergy is one of the potentially lucrative routes of valorization of metal-laden biomass (Khan *et al.*, 2023). Subsequent techniques can convert the organic matter present in the biomass to biogas using anaerobic digestion, bio-oil using pyrolysis and syngas using gasification (Kumar *et al.*, 2024). In an anaerobic process, biogas is generated through the use of methane-enriched fuels that can either be used in the generation of electricity or heating (Madela & Skuza, 2021). A thermochemical process of pyrolysis transforms biomass into bio-oil, biochar and non-condensable gases (Zulkernain *et al.*, 2023). Gasification yields syngas, a combustion of hydrogen and carbon monoxide that can be both a fuel and a chemical raw material (Soo *et al.*, 2024). The crucial issue of this technology is the handling of highly concentrated heavy metals found in the ash or biochar left after the energy conversion (Khan *et al.*, 2023). Nevertheless, this highly concentrated form makes the next step of metal recovery more efficient or it can be discarded in a safe way (Amlon Group, 2024). The energy produced can reduce the cost of operations of the remediation system making the whole system economical and sustainable (Sarah Lee, 2025).

4.2.2. Phytomining and Metal Recovery

Agromining, or phytomining, is a variation of phytoextraction, in which hyperaccumulator plants are cultivated on sites with a metal-rich soil or tailings to recover valuable metals (Suman *et al.*, 2018). Following harvest, the biomass of this type of production is burnt or undergoes other forms of thermochemical processing, and the resulting metals are retrieved out of the ash (Kumar *et al.*, 2024). Valuable metals such as nickel, cobalt, gold can be recovered through this process and this can make what had been a contaminated site economically productive. In heavy metal clean up by phytomining method, metals may be extracted out of the collected biomass using phytomining concepts. As another example, in case the hydrophytes store valuable metals such as copper or zinc in large quantities, these can be recovered in the ash following combustion (Zulkernain *et al.*, 2023). Not only does this eliminate the dangerous metals into the environment, but also introduces a new supply of raw materials and therefore less conventional mining is necessary and its associated environmental effects (National Material, 2025).

4.2.3. Biochar Production and Application

Biochar is a form of charcoal manufactured through the process of biomass pyrolysis and has gained significant attention as an industry of environmental remediation and carbon sequestration. Hydrophyte biomass containing metal input is largely retained in the biochar matrix when this is pyrolyzed (Kumar *et al.*, 2024). These biochar can be made into metal-enriched form and thus applied in a variety of ways. An example would be the use of biochar in polluted soils due to phytostabilization where due to the high adsorption, residual heavy metals are immobilised, thus lowering the mobility and bioaccessibility of the metals. It may also enhance soil quality, moisture and microbe (Liu *et al.*, 2018). Otherwise, the metal concentration, when it is sufficient, will allow re-extraction of the metals contained into the biochar. Repurposing the resulting biochar in effect entrusts the biomass with both a lasting resting place and a means of carbon sequestration, which is of benefit to the processes and endeavors of climate change mitigation. Such a twofold advantage predisposes biochar to be chosen as an attractive element of a circular economy strategy, aiming at heavy metal cleaning (Srivastav *et al.*, 2023).

4.3. Waste Valorization and By-product Utilization

Besides direct metal recovery and the production of bioenergy, the concept of the circular economy will prompt the reuse and recycling of all by-products and waste streams that are produced during the process of the hydrophyte-based remediation of the environment (Sarah Lee, 2025).

A good example is that the digestate left after the anaerobic digestion of biomass can serve as a biofertilizer, returning nutrients vital to boosting crop growth to arable land, cutting down on the use of synthetic fertilizers (Fletcher *et al.*, 2022). This makes the nutrient loop circular thus allowing sustainable agriculture. In a similar way, residual ash resulting due to metal recovery of biomass incineration can still be used as a construction material or as a soil additive, as long as they are non hazardous, and do not reemplify any of the contaminants (Zulkernain *et al.*, 2023). The water used in the remediation process, especially in constructed wetlands or hydroponic systems, can be treated and reused, minimizing water consumption and discharge (Tanko *et al.*, 2025).

Another possibility of waste valorization is the potential of growing certain species of hydrophytes capable of not only remediating heavy metals but also producing other useful products (Farid *et al.*, 2023). As an example, there are temporal plants that are capable of biofuels, pharmaceuticals, or other biochemicals (Vikraman *et al.*, 2022). The ability of these plants to be able to also remediate the high metals as such could mean that the remediation of those high metals could also be coupled with the production of such high-value goods thus they could be used simultaneously which could help maximize the use of resources and in economic terms. This strategy perfectly suits the point of view of the circular economy that tries to create value out of whatever was previously regarded as waste products and turns the issue of pollution into a source of resource (Amlon Group, 2024).

4.4. Life Cycle Assessment and Sustainability of Hydrophyte-Circular Economy Systems

To determine the sustainability of providing hydrophyte-remediation circular economy implement, an extensive life cycle assessment (LCA) is required. LCA is the process of assessing the environmental costs relating to every stage of the life cycle of a product or a process, viz. acquiring raw materials, processing materials, manufacture, distribution, use, repair, maintenance and disposal or recycling (Srivastav *et al.*, 2023). LCA when applied to hydrophyte-circular economy systems will provide the assessment of their healthy environmental performance with a view of identifying the possible trade-off and the deficiencies that can be improved.

An LCA would take into account energy utilization, greenhouse gas emissions, water use, and even the potential of ecotoxicity, within the overall system (Ahmed *et al.*, 2022). As an example, although hydrophytes present a low-energy remediation alternative to regular measures, one must include its energy costs to harvest, transport, and process the biomass (e.g., as a source of bioenergy or into phytomining) (Kim *et al.*, 2025). Likewise, the emissions during the processing steps (namely, incineration emissions or pyrolysis emissions) need to be given special attention (Madeła & Skuza, 2021). The benefits of resource recovery, such as reduced virgin material extraction and avoided emissions from conventional energy sources, would also be quantified and credited to the system (National Material, 2025).

From a sustainability perspective, the hydrophyte-circular economy systems have several benefits. They also enhances resource efficiency by converting pollutants to valuable resources, which is also the principle of the circular economy (Soo *et al.*, 2024). They have the ability to lessen the dependence on non-renewable resources and limit the waste collection (Iyyappan *et al.*, 2023). Moreover, according to those systems, carbon sequestration in biochar and the generation of renewable energy can combat climate change (Kumar *et al.*, 2024). A detailed LCA can be used so as to ensure that the solutions do not transfer the environmental burden to another stage of the life cycle or another environmental compartment. It gives a strong platform to continuous improvement, which will facilitate the production of more sustainable and genuinely circular systems of heavy metal remediation (Sarah Lee, 2025).

5. WETLANDS AND WASTEWATER

5.1. Constructed Wetlands for Wastewater Treatment

Constructed wetlands (CWs) are anthropogenic structures that resemble natural wetlands in the treatment of different kinds of wastewater, including experimental ones contaminated with heavy metals. Hydrophytes also have a leading position when it comes to CWs because of their assistance with the metal removal process via uptake, adsorption, and rhizospheric precipitation (Xie *et al.*, 2022). For example, planting of the CWs with Typha latifolia (cattail) and Phragmites australis (common reed) which have been already used to treat industrial effluents and acid mine drainage successfully reducing the concentration of metals such as Fe, Mn, Zn, and Cu (Heisi *et al.*, 2023). These emerging plants have strong root systems, which mean they have a high surface area due to the microbial activity and the ability to trap the metals and the biomass of the plants can be harvested after a certain period of time (Rezania *et al.*, 2016). The biomass collected in CWs then can be returned to the bioenergy system or metal extraction processes making the treatment process more sustainable than it appears to be (Khan *et al.*, 2023).

5.1.1. Floating Treatment Wetlands (FTWs)

Artificial floating islands also known as floating treatment wetlands are a derivative of CWs in which vegetation is cultivated on floating mats that float on a contaminated water body (Tanko *et al.*, 2025). Their roots are inserted in the water and come into direct contact with the pollution. Successful deployment of FTWs in lakes, ponds, and even industrial lagoons has been in the removal of heavy metals. Anyway, FTWs of *Eichhornia crassipes* (water hyacinth) and *Pistia stratiotes* (water lettuce) have proved very efficient in eliminating Cd, Pb, and Cu in waters that are polluted (Ali *et al.*, 2020). The strength of FTWs is that they are adjustable to relatively low or high levels of water, and that they cover large surface areas. The biomass harvested at FTWs can be used to convert it into valuable commodities because this material is in direct contact with the water, and possesses a more abundant metal accumulation potential (Farid *et al.*, 2023).

5.1.2. Phytoremediation of Mine Tailings and Industrial Sites

While hydrophytes are mostly adapted to water bodies, they may also be employed with land plants to clean up polluted soils that are saturated with heavy metals such as in mine waste (Gong *et al.*, 2018). An example in this case (high water table areas, areas prone to flood) is that high hydrophytes can grow in high water tables and aid in phytostabilization or phytoextraction (Sharma *et al.*, 2023). The researchers have studied employing different water surface plants to clean up ditches in ports and industrial effluents. These applications illustrate how hydrophytes can be used in other than overtly watery conditions and can be applied to semi-aquatic or wetland sites on land.

5.1.3. Small-Scale and Decentralized Applications

Hydrophyte-based systems are highly adaptable with small-scale and decentralized systems, which means that they can suit rural populations or limited industrial processes (Heisi *et al.*, 2023). The domestic wastewater or localized industrial effluents can also be treated through simple pond systems or even containers with chosen hydrophytes (Rezania *et al.*, 2016). The decentralized method does not require centralized treatment plants and can make local communities responsible in management of their water resources (Tanko *et al.*, 2025). Many hydrophytes do not require much input to grow and maintain making the same systems affordable and readily available to a diverse population (Ali *et al.*, 2020).

5.2. Challenges and Opportunities in Real-World Applications

Despite the significant potential of hydrophyte-based remediation integrated with circular economy principles, several challenges need to be addressed for their widespread adoption and successful implementation in real-world scenarios.

5.2.1. Challenges

4 **Biomass Management and Disposal:** This tons of biomass produced on a mass scale can become logistically cumbersome and despite the valorization options, it may end up being discarded. To process metal-containing biomass, especially the most hazardous metals, safely and efficiently, the situation must therefore be ensured with solid infrastructure and management to avoid secondary pollution (Khan *et al.*, 2023). Metals in the biomass must also be concentrated to an extent that will make recovery economically feasible.

5 Metal Specificity and Bioavailability: Hydrophytes do not have the same capacity to remediate all heavy metals. Efficiency of uptake also depends strongly on the plant species and the particular metal, and the chemical form (speciation) in which that metal occurs in the environment (El Awady *et al.*, 2021). It needs close attention to choice of plant species which are chosen to suit the site contamination profile. In addition to that, bioavailability of metals can be affected by numerous environmental influences, and such remediation is not consistent.

6 Environmental Conditions and Site Heterogeneity: Heterogeneous sites can be found in the real world which will have varying degrees of contamination, soil types, pH and other parameters of the environment. Due to these differences, they can affect the growth of the plant and its capacity to absorb metals, thus a more focused approach cannot be implemented (Gong *et al.*, 2018). Extreme weather type can also put a remediation effort off track, i.e., can cause disruption of a remediation through floods or droughts (Jaishankar et al., 2014).

7 Timeframe for Remediation: Phytoremediation, such as that involving hydrophytes also is slower to occur in comparison to conventional physical or chemical means. Since the process of reducing concentrations of heavy metals in depleted areas can be slow, it might take months and years to reach an efficient level of remediation, depending on the level of contamination at the time of the cleanup and set remediation targets (Ojuederie & Babalola, 2017). Such a longer timeframe may not be appropriate in addressing urgent remediation efforts (Liu *et al.*, 2018).

8 Public Acceptance and Perception: Although it is an eco-friendly strategy, the acceptance and acknowledgement of the population may pose a problem. Uncertainties about the safety associated with the consumption of products produced out of a remediated site or the possibility of metals returning to the environment may impede implementation (Kumar *et al.*, 2018). Achieving communication effectiveness and engagement will help to establish trust and favor (Suman *et al.*, 2018).

5.2.2. Opportunities

- **Cost-Effectiveness:** Hydrophyte based remediation is usually low cost in comparison to the conventional processes, more so in large-scale and long-term projects. When combined with the circular economy's tenets, this boosts economic viability by generating income from recovered materials or bioenergy (Fletcher *et al.*, 2022).

- **Environmental Friendliness:** These systems are environmentally friendly as they have natural operation processes without the use of harsh chemicals and other energy consuming machines. They add biodiversity, clean up water and help restore the ecosystem (Kumar *et al.*, 2018).
- **Versatility and Adaptability:** Hydrophytes may be used in a wide range of aquatic and semi-aquatic environments, including wetlands, contaminated lakes, and in industrial effluent treatment. Their natural malleability enables them to have different design and implementation approaches (Rezania *et al.*, 2016).
- **Resource Recovery and Value Addition:** Phytomining can extract valuable metals and transform contaminated biomass into a resource, bio-converting it into energy and biochar, and this is a dependence on virgin materials, creating a new economic opportunity, and reducing reliance on virgin materials (Madeła & Skuza, 2021).
- **Community Involvement and Capacity Building:** Community involvement can be increased where the hydrophyte systems are not so technical. This interaction encourages skills, local labour as well as environmental consciousness, hence the sustainability development objectives (Tanko *et al.*, 2025).
- **Integration with Other Technologies:** To provide a hybrid solution that is more reliable and effective, hydrophyte-based systems may also be effectively combined with other technologies, such as physical filtering or bioremediation. For instance, pretreatment procedures can reduce the concentration of metals to a level suitable for phytoremediation, or the effluent can be cleaned up either after treatment or before (Farid *et al.*, 2023).

5.3. Economic Viability and Policy Support

An important consideration to the widespread use of the hydrophyte remediation of heavy metal is their economic feasibility, especially when incorporated with the concept of circular economy. Although establishing a system such as the constructed wetland and floating treatment wetlands may be heavy in terms of cost, the cost of maintaining the systems in the long-term may be low compared to other physical and chemical options (Sarah Lee, 2025). The possibility of earning income based on extracted metals, bioenergy or other value products also increases the economic performance of such systems (Srivastav *et al.*, 2023).

For instance, phytomining, although still at the research and development stage in relation to many metals, has a great potential in changing remediation to a profit-generating activity rather than the cost centre (Kumar *et al.*, 2024). This can help cover the cost of remediation and in some cases will yield a profit since the value of the recovered metals can cover the cost of remediation and in some cases make a profit. Likewise, energetic cultivation of the biomass after harvesting can save on energy expenses (or even generate income as a result of energy sales) on the remediation facility (Soo *et al.*, 2024). The fact that the process of biochar production may be sold to the market as a soil enhancer or carbon storage option contributes to other economic gains (Bolan *et al.*, 2014).

Nevertheless, their economic viability is very local and varies depending on the characteristics of metal contamination, the market price of recovered resources and existence of the appropriate land as well as the local energy costs (Zulkernain *et al.*, 2023). As an example, it may not be economical to recover trace quantities of metals, but the higher quantities/valuable metals are recovered it becomes more economic. Even more, the economic value of recycling goes beyond initial revenue generation to cover avoided expenses incurred by the traditional waste management process, less environmental degradation, and better health to the public, yet the costs remain more difficult to calculate monetarily (Ahmed *et al.*, 2022).

Policy support is important to support the economic feasibility and implementation of such polycentric and sustainable solutions to remediation.

6. RECOMMENDATIONS AND FUTURE PERSPECTIVES

Research, policy, and practice must work together to fully realize the promise of the sustainable use of hydrophytes and circular economy principles in heavy metal remediation. This section identifies the future directions to develop this sustainable method and reveals the major lines of advice.

6.1. Policy and Regulatory Frameworks for Circular Economy Integration

To encourage the adoption of these creative solutions, supportive legislative and regulatory frameworks are necessary:

6.3.1 Standards for Clear Remediation: creating precise, scientifically supported concentrations for heavy metal remediation standards that would support environmentally friendly remediation techniques like phytoremediation (Kumar *et al.*, 2024). This involves establishment of tolerable concentration of residual metals in treated waters, soil, and recovered products (Wan *et al.*, 2024).
6.3.2 Incentives for Resource Recovery: Introducing financial incentives (e.g. tax breaks, subsidies, grants) to support the adoption of hydrophyte-based remediation followed by resource recovery by industries and communities (Sarah

Lee, 2025). This may also comprise carbon credits in relation to producing biochar or renewable energy production (Tsai *et al.*, 2021).

6.3.3 Waste-to-Resource Policies: This would involve the creation of policies developed to treat metal-laden biomass which derive from remediation as a resource instead of a waste, simplifying its separation and reuse (Amlon Group, 2024). This necessitates the modification of the current waste management policies to facilitate the access of the circular economy (Srivastav *et al.*, 2023).

6.3.4 Extended Producer Responsibility (EPR): Implementing or strengthening EPR schemes whereby producers are made accountable to all life cycle of their products, including safe handling and recovery of heavy metals from discarded products (Zulkernain *et al.*, 2023). This can establish a monetary reward among industries to make investment in remediation and recycling technologies (National Material, 2025).

6.3.5 Capacity Building: Ensuring education and technical support to the local communities, environmental departments, and industrial figureheads on designing and operationalizing of hydrophyte-based mechanisms and biomass valorization systems (Khan *et al.*, 2023). It develops the local knowledge and creates success of projects in the long term (Ojuederie & Babalola, 2017).

7. CONCLUSION

Heavy metals have become an important environmental problem that prompted the need to have sustainable solutions to remediate the problem without incurring costs or generating wastes through conventional remedial measures. Hydrophytes are grown at low costs and are environmentally protective since they absorb, detoxify and concentrate heavy metals found in polluted water sources. A circular economy approach to hydrophyte-based remediation converts what would otherwise be a metal contaminated organic waste mass into a source of metal recovery, bioenergy, and other by-products such as biochar. This strategy reduces pollution, improves the use of resources and uses economics to create opportunities, which are part of sustainable development goals. Although there were issues such as biomass management, site-specific variability, the practical viability was proven in created and floating treatment wetlands. The continued research into the biology of the hydrophytes, microbial interaction and optimized cultivation as well as of biomass processing is also key to scale these systems. The water-fern technology allows cleaning up the exploitation and making any activity sustainable and efficient in the use of valuable resources.

REFERENCES

Abdelaal, M., Mashaly, I. A., Srour, D. S., Dakhil, M. A., El-Liethy, M. A., El-Keblawy, A., El-Barougy, R. F., Halmy, M. W. A., & El-Sherbeny, G. A. (2021). Phytoremediation perspectives of seven aquatic macrophytes for removal of heavy metals from polluted drains in the Nile Delta of Egypt. *Biology (Basel), 10*(6), 560. DOI: 10.3390/biology10060560 PMID: 34203088

Ahmed, M., Mavukkandy, M. O., Giwa, A., Elektorowicz, M., Katsou, E., Khelifi, O., Naddeo, V., & Hasan, S. W. (2022). Recent developments in hazardous pollutants removal from wastewater and water reuse within a circular economy. *NPJ Clean Water, 5*(1), 12. Advance online publication. DOI: 10.1038/s41545-022-00154-5

Ali, H., Khan, E., & Ilahi, I. (2019). Environmental chemistry and ecotoxicology of hazardous heavy metals: Environmental persistence, toxicity, and bioaccumulation. *Journal of Chemistry, 2019*, 6730305. Advance online publication. DOI: 10.1155/2019/6730305

Ali, S., Abbas, Z., Rizwan, M., Zaheer, I. E., Yavaş, İ., Ünay, A., Abdel-Daim, M. M., Bin-Jumah, M., Hasanuzzaman, M., & Kalderis, D. (2020). Application of floating aquatic plants in phytoremediation of heavy metals polluted water: A review. *Sustainability (Basel), 12*(5), 1927. DOI: 10.3390/su12051927

Amlon Group. (2024). Maximize sustainability and ROI through industrial metals recovery. *Amlon Group.* https://amlongroup.com/news/the-circular-economy-maximizing-sustainability-through-valued-metals-recovery/

Bolan, N., Kunhikrishnan, A., Thangarajan, R., Kumpiene, J., Park, J., Makino, T., Kirkham, M. B., & Scheckel, K. (2014). Remediation of heavy metal(loid)s contaminated soils—To mobilize or to immobilize? *Journal of Hazardous Materials, 266*, 141–166. DOI: 10.1016/j.jhazmat.2013.12.018 PMID: 24394669

El Awady, F. R., Abbas, M. A., Abdelghany, A. M., & El-Amir, Y. A. (2021). Silver modified hydrophytes for heavy metal removal from different water resources. *Biointerface Research in Applied Chemistry, 11*(6), 14555–14563. DOI: 10.33263/BRIAC116.1455514563

Farid, N., Ullah, A., Khan, S., Butt, S., Khan, A. Z., Afsheen, Z., El-Serehy, H. A., Yasmin, H., Ayaz, T., & Ali, Q. (2023). Algae and hydrophytes as potential plants for bioremediation of heavy metals from industrial wastewater. *Water (Basel), 15*(12), 2142. DOI: 10.3390/w15122142

Fletcher, J., Willby, N., Oliver, D. M., & Quilliam, R. S. (2022). Resource recovery and freshwater ecosystem restoration—Prospecting for phytoremediation potential in wild macrophyte stands. *Resources. Environmental Sustainability*, *7*, 100050. Advance online publication. DOI: 10.1016/j.resenv.2022.100050

Gong, Y., Zhao, D., & Wang, Q. (2018). An overview of field-scale studies on remediation of soil contaminated with heavy metals and metalloids: Technical progress over the last decade. *Water Research*, *147*, 440–460. DOI: 10.1016/j.watres.2018.10.024 PMID: 30343201

Heisi, H. D., Awosusi, A. A., Nkuna, R., & Matambo, T. S. (2023). Phytoextraction of anthropogenic heavy metal contamination of the Blesbokspruit wetland: Potential of wetland macrophytes. *Journal of Contaminant Hydrology*, *253*, 104101. Advance online publication. DOI: 10.1016/j.jconhyd.2022.104101 PMID: 36379730

Iyyappan, J., Baskar, G., Deepanraj, B., Vivek Anand, A., Saravanan, R., & Awasthi, M. K. (2023). Promising strategies of circular bioeconomy using heavy metal phytoremediated plants: A critical review. *Chemosphere*, *313*, 137097. Advance online publication. DOI: 10.1016/j.chemosphere.2022.137097 PMID: 36334740

Jaishankar, M., Tseten, T., Anbalagan, N., Mathew, B. B., & Beeregowda, K. N. (2014). Toxicity, mechanism and health effects of some heavy metals. *Interdisciplinary Toxicology*, *7*(2), 60–72. DOI: 10.2478/intox-2014-0009 PMID: 26109881

Kamal, M., Ghaly, A. E., Mahmoud, N., & Côté, R. (2004). Phytoaccumulation of heavy metals by aquatic plants. *Environment International*, *29*(8), 1029–1039. DOI: 10.1016/S0160-4120(03)00091-6 PMID: 14680885

Khan, A. H. A., Kiyani, A., Santiago-Herrera, M., Ibáñez, J., Yousaf, S., Iqbal, M., Martel-Martín, S., & Barros, R. (2023). Sustainability of phytoremediation: Post-harvest stratagems and economic opportunities for the produced metals contaminated biomass. *Journal of Environmental Management, 326*(Pt. B). *Journal of Environmental Management, 116700*, 116700. Advance online publication. DOI: 10.1016/j.jenvman.2022.116700 PMID: 36423411

Khan, S., Kamal, M., Noor, S., & Afzal, S. M. (2024). Assessment of heavy metals and its treatment through phytoremediation in groundwater along River Kabul in District Charsadda. *Frontiers in Environmental Science*, *12*, 1392892. Advance online publication. DOI: 10.3389/fenvs.2024.1392892

Kim, J. Y., Hwang, J.-H., Lee, T., Kim, Y., Bae, S., Kim, D., Lee, J., Lee, J., Tsang, Y. F., & Kwon, E. E. (2025). A review of sustainable metal recovery from e-waste via bioleaching: Current status and strategies for improvement. *Chemical Engineering Journal, 516*, 164149. DOI: 10.1016/j.cej.2025.164149

Kumar, A., Dadhwal, M., Mukherjee, G., Srivastava, A., Gupta, S., & Ahuja, V. (2024). Phytoremediation: Sustainable approach for heavy metal pollution. *Scientifica, 2024*(1), 3909400. Advance online publication. DOI: 10.1155/2024/3909400 PMID: 39430119

Kumar, V., Shahi, S. K., & Singh, S. (2018). Bioremediation: An eco-sustainable approach for restoration of contaminated sites. In Singh, J., Sharma, D., Kumar, G., & Sharma, N. (Eds.), *Microbial bioprospecting for sustainable development* (pp. 123–146). Springer., DOI: 10.1007/978-981-13-0053-0_6

Liu, L., Li, W., Song, W., & Guo, M. (2018). Remediation techniques for heavy metal-contaminated soils: Principles and applicability. *The Science of the Total Environment, 633*, 206–219. DOI: 10.1016/j.scitotenv.2018.03.161 PMID: 29573687

Lone, M. I., He, Z. L., Stoffella, P. J., & Yang, X. E. (2008). Phytoremediation of heavy metal polluted soils and water: Progresses and perspectives. *Journal of Zhejiang University. Science. B., 9*(3), 210–220. DOI: 10.1631/jzus.B0710633 PMID: 18357623

Madeła, M., & Skuza, M. (2021). Towards a circular economy: Analysis of the use of biowaste as biosorbent for the removal of heavy metals. *Energies, 14*(17), 5427. DOI: 10.3390/en14175427

National Material. (2025). *The circular economy of metal processing sustainability.* https://www.nationalmaterial.com/the-circular-economy-of-metal-processing-sustainability/

Ojuederie, O. B., & Babalola, O. O. (2017). Microbial and plant-assisted bioremediation of heavy metal polluted environments: A review. *International Journal of Environmental Research and Public Health, 14*(12), 1504. DOI: 10.3390/ijerph14121504 PMID: 29207531

Rezania, S., Taib, S. M., Md Din, M. F., Dahalan, F. A., & Kamyab, H. (2016). Comprehensive review on phytotechnology: Heavy metals removal by diverse aquatic plant species from wastewater. *Journal of Hazardous Materials, 318*, 587–599. DOI: 10.1016/j.jhazmat.2016.07.053 PMID: 27474848

Sarah Lee. (2025). *Circular economy in environmental remediation.* https://www.numberanalytics.com/blog/circular-economy-environmental-remediation

Sharma, J. K., Kumar, N., Singh, N. P., & Santal, A. R. (2023). Phytoremediation technologies and their mechanism for removal of heavy metals from contaminated soil: An approach for a sustainable environment. *Frontiers in Plant Science, 14*, 1076876. DOI: 10.3389/fpls.2023.1076876 PMID: 36778693

Soo, A., Kim, J., & Shon, H. K. (2024). Technologies for the wastewater circular economy: A review. *Desalination and Water Treatment, 317*, 100205. DOI: 10.1016/j.dwt.2024.100205

Srivastav, A. L., Markandeya, , Patel, N., Pandey, M., Pandey, A. K., Dubey, A. K., Kumar, A., Bhardwaj, A. K., & Chaudhary, V. K. (2023). Concepts of circular economy for sustainable management of electronic wastes: Challenges and management options. *Environmental Science and Pollution Research International, 30*(17), 48654–48675. DOI: 10.1007/s11356-023-26052-y PMID: 36849690

Suman, J., Uhlik, O., Viktorova, J., & Macek, T. (2018). Phytoextraction of heavy metals: A promising tool for clean-up of polluted environment? *Frontiers in Plant Science, 9*, 1476. DOI: 10.3389/fpls.2018.01476 PMID: 30459775

Tanko, W. S., Albalwe, F. M., Bakr, A. A., Ali, A. M., Ibrahim, R. I. H., Abdel-Mageed, A. M., Shahhat, I. M. A. M., Elsheikh, S. Y. S., Hussein, H.-A. A., Qaoud, E.-S. M., Ezzat, S. M., Elshamy, M. M., Abdein, M. A., & Nafea, E. M. (2025). Submerged hydrophytes as a tool for the removal of heavy metals. *Asian Journal of Water, Environment and Pollution, 22*(4), 64–74. DOI: 10.36922/AJWEP025150105

Tsai, C.-H., Shen, Y.-H., & Tsai, W.-T. (2021). Sustainable material management of industrial hazardous waste in Taiwan: Case studies in circular economy. *Sustainability (Basel), 13*(16), 9410. DOI: 10.3390/su13169410

Vikraman Meera, A., Thampatti, K. C. M., John, J., Sudha, B., & Sajeena, A. (2022). Aquatic plants as bioremediators in pollution abatement of heavy metals. In Sharma, P., Pandey, A., Tong, Y. W., & Ngo, H. H. (Eds.), *Current developments in biotechnology and bioengineering* (pp. 223–239). IntechOpen., DOI: 10.5772/intechopen.99627

Wan, Y., Liu, J., Zhuang, Z., Wang, Q., & Li, H. (2024). Heavy metals in agricultural soils: Sources, influencing factors, and remediation strategies. *Toxics, 12*(1), 63. DOI: 10.3390/toxics12010063 PMID: 38251018

Xie, P., Zahoor, F., Iqbal, S. S., Ullah, Z., Noman, M., Din, Z. U., & Yang, W. (2022). Elimination of toxic heavy metals from industrial polluted water by using hydrophytes. *Journal of Cleaner Production, 352*, 131358. DOI: 10.1016/j.jclepro.2022.131358

Yadav, M., Singh, M., Nanda, K., & Singh, S. P. (2022). Genetically engineered plants for phytoremediation of heavy metals. In Sharma, P., Pandey, A., Tong, Y. W., & Ngo, H. H. (Eds.), *Current developments in biotechnology and bioengineering* (pp. 223–239). Elsevier., DOI: 10.1016/B978-0-323-99907-6.00017-7

Zulkernain, N. H., Basant, N., Ng, C. C., Kriti, K., Salari, M., & Mallick, S. (2023). Recovery of precious metals from e-wastes through conventional and phytoremediation treatment methods: A review and prediction. *Journal of Material Cycles and Waste Management*, *25*(6), 2726–2752. DOI: 10.1007/s10163-023-01717-5

Chapter 9

Technological Monitoring and Management in Hydrophyte-Based Heavy Metal Stress Management

Vikanksha Vikanksha
https://orcid.org/0009-0009-0257-4319
Independent Researcher, India

Arun Kumar
Lovely Professional University, India

Jatinder Singh
https://orcid.org/0000-0001-7691-6689
Lovely Professional University, India

ABSTRACT

Remote sensing has emerged as a pivotal technology in the monitoring and management of heavy metal contamination in various environments. By utilizing satellite imagery, aerial photography, and advanced sensor technologies, researchers can effectively assess the spatial distribution and concentration of heavy metals in soil, water, and vegetation. This paper explores the various remote sensing techniques employed in heavy metal monitoring, highlighting their applications in detecting environmental stressors and assessing the health of ecosystems. Through the integration of remote sensing data with geographic information systems (GIS) and machine learning algorithms, it is possible to develop predictive models that facilitate timely interventions and remediation strategies. The findings underscore the importance

DOI: 10.4018/979-8-3373-4037-1.ch009

of remote sensing as a non-invasive and efficient approach to environmental monitoring, providing valuable insights for policymakers and environmental managers in their efforts to combat heavy metal pollution.

1. INTRODUCTION TO HEAVY METAL STRESS

The plant kingdom has a basic role of supporting the large number of diverse forms of life on land and plants are an essential and vital part of the intricate mechanisms of osmotically excluding a variety of solutes (Kosal 2023). These solutes form major ionic, osmotic and gravitational forces which all work together towards improving water quality and water quantity in the ecosystems. The bioavailability of heavy metalloids in the essential rhizosphere can have a significant effect on the solute forces that control the uptake of water and the general effectiveness of plant water relations. In addition, the availability and the bioavailability of these toxic heavy metalloids in the root zone leads to a lethal ambiance likely to cause the breakdown of vital forces of cohesion and adhesion. Directly resulting thereof, the mechanisms of plant water uptake, transport, and transpiration receive seemingly contradictory signals, prompted by the overloading of the soil solution with metalloids, as well as the gravitational forces that act in the surrounding environment and also creates its impact. The fact that this is a problematic situation may become conditionally much worse when subjected to the conditions of water deficiency, which is confirmed by different experiments related to the exposure of plants to heavy metal stress (Goncharuk & Zagoskina, 2023). Consequently, due to the considerable functions that plants serve in ensuring that the soil-plant-atmosphere systems are balanced, there is an increasing need to research further on the influence of excessive heavy metal stress on the water relations of plants. It is also important to explore the physiological processes that cause the detrimental impact on the equilibrium of the solute forces which have become so significant to the plant health. This knowledge, in its turn, can facilitate the way to the creation of proper mitigation and remediation patterns, which is required to maintain the vitality of plants and the stability of their ecosystems (Grifoni et al., 2022). It can be found that heavy metals can be emitted to soil, as well as water resulting the several anthropogenic activities, which encompass mining, different smelting industries, geological weathering, the burning of fossil fuels, the use of fertilizers and pesticides, municipal framework related operations, the operation of thermal power plants, and improper discarding of electronic wastes. These heavy metals are very toxic and difficult to dispose of thus there is a serious environmental concern as the metals are prone to bioaccumulation in organisms. Plants are important components of the Earth eco systems and have been playing a central role in the survival of life on this planet since its very beginning and this is also known as autotrophic vitality. However, heavy metal stresses may also act as an imposing force to the growth and development of plants and it is urgent that they identify the best responses to this kind of environment and also adapt their mechanisms of heavy metal tolerance. Over the last decades, there has been a lot of research, and it is possible to note that various heavy metals produce both positive and negative effects on plants at different concentrations, which is reported in numerous reviews (Asiminicesei et al., 2024). Hypertoxicity of heavy metals may cause diverse adverse effects, such as oxidative stress in plant cells, disturbances in the normal nutritional balance that they need to grow well, changes in the production of phytohormones, cell cycle arrest, and gene expression. Other effects also encompass disruption of root morphology which may disrupt the absorption of important nutrients by a plant. Thus, gaining an understanding of the mechanisms underlying both the toxicity and tolerance of significant plant species in relation to heavy metals can prove to be instrumental in developing and enhancing the biochemical pathways necessary for successful industrial growth and cultivation of plants in contaminated environments.

2. UNDERSTANDING HEAVY METAL CONTAMINATION

Table of contents: 3. Understanding Heavy Metal Contamination The heavy metal contamination of soil and fresh water systems is one of the chief environmental issues which limits the productivity of different crops, and causing various health issues. The impact of heavy metal in agricultural practices has affected the livelihood of many, thus can be treated as a threat to the sustainability of agriculture. Food production is supported by the agricultural industry with an increased global population. However, soil contamination is negatively affecting the performance of the industry (Zahoor & Mushtaq, 2023). Heavy metals are non-biodegradable toxicants, and their immediate discharge in fresh water has further caused aquatic, and terrestrial toxicity. The metal ions upon exceeding their threshold concentrations in soil, and water affect living being and ecological balance, and thus become harmful to the ecosystem and food chain. The aquatic ecosystem is harnessing different mediators, however, the presence of heavy metals in the water ecosystem for long durations and in larger concentrations affects the usefulness of the water bodies restricting their daily utilities by the rural population. To address the effect of heavy metals in water systems, remediation techniques and periodic monitoring schedules need to be implemented. Heavy metal contamination in fresh water systems results in huge economic loss along with extensive damage to ecosystem health. Moreover, currently employed conventional remediation techniques are effective but they are not economical and technically feasible for the large-scale removal of metals. Phytoremediation offers a biologically appealing technique for large-scale metal removal from contaminated water as well as soil systems. Water hyacinth is a model hydrophytic plant known for its unique metal uptake and accumulation capabilities, growing extensively all over the world in heavy metal-contaminated areas with its wide advent of floating, feathery roots (Nwagbara 2023). Through a symprocessing approach, it is possible to remove heavy metals from contaminated water in bulk and it has enhancing the quality of fresh water systems.

3. MECHANISMS OF HEAVY METAL UPTAKE IN HYDROPHYTES

The effects of heavy metals are not confined solely to non-edible plants, as has often been taken for granted, but have also been significantly observed affecting a wide variety of edible plants as well. For instance, the uptake of copper (Cu) by lettuce has been documented, showing a notable translocation of this metal to the leaves and a concerning accumulation in the roots, stems, and leaves. It embraces a great number of interdependent disciplines including geography, biology, geochemistry, geology, physics and ecotoxicology, which provide essential knowledge. By and large, AI-based models have proven to effectively analyze or predict the complex cause-and-effect structure involved in all manner of environmental processes, and hence offer improved frameworks of conceptualization and response to environmental issues. Additionally, it is important to note that the roots of certain plants have the ability to synthesize various organic acids, including citric acid, malic acid, acetic acid, and oxalic acid. These organic acids help in the development of the organo-metal complex, which is water soluble and aids in the absorption of the heavy metals. The lower the pH of the soil, the higher the degree of positive ions of heavy metals that are in the solution. An example is that Pb2+, Cu2+ and Zn2+ ions are more common in such low pH. These metal ions react with exudates of the roots, and more water-soluble forms are formed at alkaline pH levels (Cervantes-Avilés et al., 2021). Moreover, low molecular weight organic acids have been identified to improve the solubility of heavy metals such as lead (Pb), cadmium (Cd) and zinc (Zn) thus, increasing their movement and chances of absorption by plants in the surrounding area of the root surface. Nevertheless, the soluble forms of the heavy metals have first to be diffused to the surface of the root and then absorbed across the cuticular membrane before the plant can adequately absorb it. Besides this, the metal species of the same would be competing with the available sites on the surface, which renders the environmental conditions rather crucial. There has been an observation that the relative high concentration levels of low molecular weight organic acids with low levels of heavy metal ion concentration at a pH of

4.5 provide the most favorable condition favoring optimal heavy metal uptake by the plant roots (Figure 1). Also, metal ions absorption is an important route of entry of essential and non-essential heavy metals into the plant roots. When these metal ions get into the root, they can either enter through the symplastic route via certain ion channels or they can go through the endodermal cells to the xylem system of transportation (Yadav et al., 2021). The heavy metal ions once inside the xylem vessels are then carried to the aerial portions of the plant where they can either be accumulated, translocated or even de-toxified thus affecting the overall well being and growth of the plant.

Figure 1. Mechanism of heavy metal uptake and physiology

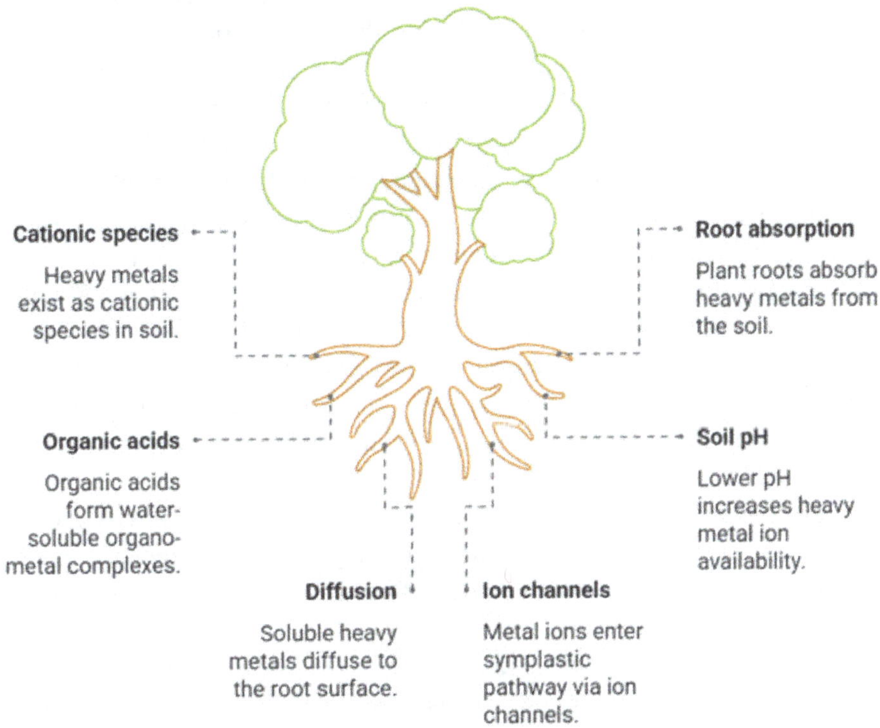

4. ARTIFICIAL INTELLIGENCE IN ENVIRONMENTAL SCIENCE

Artificial Intelligence (AI) is a successful imitation of the human decision-making and learning process as it makes extensive use of complex mathematical and computational frameworks. The AI has the potential to simulate reasoning, learning, representation, perception, social interactions, language comprehension, and even machine emotion, through the use of these models. Due to the fast technological development, AI and its many subdivisions have been applied popularly in a wide range of various industries, proving the huge potential and ability of AI (Thayyib et al., 2023). Environmental science is one of the areas that have been greatly improved through AI. The use of AI and the related methods in the field of environmental science has become especially common in recent years. This influx of use is mainly geared towards addressing the urgent environmental challenges that have been caused by the faults or failures that ensue because of the impreciseness and inherent multi-dimensionality of the traditional methods. Environmental science in itself is holistically concerned with the dynamics of many natural processes as it happens all over the earth and in the atmosphere. Traditionally, environmental systems or processes are modeled by using the physical or mechanistic models that attempt to elucidate the processes that are behind the nature. Together with these models, statistical techniques are regularly used to process the obtained measured data and the complex relationships between different variables (Haslam et al., 2024). The models of statistic which are traditionally used are often assumed to assume that the correlations among the variables are linear in nature. Through the investigations, analysis that is done though, it is often seen that the processes involved are often accompanied with complex non linear correlations. Because of the underlying assumptions of linearity and the complexity of the physical processes involved, it is extremely difficult to formulate extensive physical models to undertake studies in the environmental area. Conversely, the existing databases and information repositories have grown significantly in size in the past few years due to development of technology. Therefore, the interdependences between the many variables are often of high task complexity. Consequently, the traditional models or methodology are no longer effective in responding to or resolving the urgent tasks or complex issues that are common in the field of environmental sciences today. To counter these predicaments, the need to adopt and use contemporary models, which are grounded on the artificial intelligence methods has been noted to be in a growing and increasing demand. This is a tendency that has taken a more sharp turn in recent years in the field of environmental studies and has gained momentum in the last twenty years with an acute demand to find new ways of addressing the complexities and uncertainties that exist when dealing with environmental research and practice (Pan & Zhang, 2023).

5. MACHINE LEARNING TECHNIQUES FOR ENVIRONMENTAL DATA

The Artificial Intelligence that is now synergistically cohesive with the potent and innovative procedures of the use of the Machine Learning tools, has the actually impressive capacity to research, investigate, and process large volumes of data that, in reality, would take only one person an entire lifetime to create and fully understand. Through this huge exposure to high levels of diverse and broad data, AI is also able to create highly complex and sophisticated predictive models, which can be used to accomplish multiple purposes. Over the past several years it has become more and more widespread and effective in the world of remote sensing where it has been used in countless classification and regression processes which can eventually lead to better insights into the workings of the environment. The information provided by a vast number of sophisticated technological sources is now successfully used to perform thorough analysis in the field of the Machine Learning with references to the environmental variables; yet, it is worth noting that there are very few studies that have particularly focused on these environmental factors in the depth and detail (Rane et al., 2024). The task of identifying the most suitable Machine Learning model to be used in certain classification tasks is very much dependent on the complexity in nature of the dataset that is going to be used to train the model, and the set of features to be used in the modeling process. It is necessary to mention that the territory that is assigned classification purposes on the local level is usually very small and restricted. This restriction then reduces the number of samples, which would be used to effectively select the features and also to sufficiently train the classifier to work well. Moreover, environmental data is often intrinsically time or spatio-temporally structured making it even more difficult to analyze. Data in the form of images which have been recorded by passive and active satellites and other sensors with the small footprint have been availed over long durations in several decades. It is in the sheer number of these temporal images that the capacity to train classifiers is greatly enhanced and this increase is bound to cause a rise in the accuracy of performance, especially in those environments which pose difficulties in differentiation. In fact, in conducting a comparative study of both groups of temporal data and solitary images, it is more than abundantly obvious that the latter provide one with a plethora of information about the land surface and the wide range of materials and substances that make it up. Additionally, the exponential growth of Artificial Intelligence and Machine Learning has been higher than ever in the recent decades and is headed towards the only way of growing, now offering a plethora of new opportunities in many fields and areas (Mohamed 2023). The result of this development inevitably is revolutionary developments that have the colossal potential to revolutionize the way we perceive and interact with the complex world around us and create a greater understanding of the environmental systems and processes that control our world. The synergy of these advanced technologies creates an exciting frontier in not only scientific research but also in practical applications that can significantly benefit society.

Since AI has the potential to analyze a huge amount of multi-modal data at high speed, more and more ML models are applied to environmental science. Combined with fast computing technology, ML can be used to go beyond traditional correlation analysis to detect causal relationships between climate-driving factors and climate responses, especially with the big data of recent year. AI and ML are also useful for data assimilation, the goal of data assimilation is to combine observations and a numerical model in an optimal way, and thus to expand the observational record and improve initial conditions used to drive a numerical forecast model (Farchi et al., 2021). Data assimilation also has an advantage of determining model parameters by taking advantage of a large number of observations while the model feedback can prevent drifting.

6. INTEGRATION OF AI IN HYDROPHYTE MANAGEMENT

Hydrophyte-based heavy metal management has proven to be a nature-friendly technology for the reclamation of metal-contaminated lands without disturbing the soil structure. Today, the use of artificial intelligence approaches makes it easier to manage large-scale hydrophyte operations compared to a few years ago when hydrophyte management was largely based on empirical knowledge. A number of machine learning applications have been reported in recent times. However, the literature suggests certain levels of scepticism regarding the reliability of machine learning outputs (Table 1) (Barbierato & Gatti, 2024). Moreover, the flexibility of the hydrophyte-assisted metal/growth inhibition in deciding the necessity of metal dosage/metal type, associated environmental conditions, and other considerations while allowing recovery predictions is inarguably limited. Information route guest-host transfer has emphasized the importance of using appropriate information routes to improve the model performance and reliability in this regard.

Table 1. Overview of AI Integration in Hydrophyte Management

No.	Aspect	Description	Key Benefits
1	AI Technologies	Utilization of algorithms and models to analyze environmental data.	Enhanced data analysis capabilities
2	Machine Learning	Application of ML techniques to predict plant responses to heavy metal stress.	Improved prediction accuracy
3	Data Collection	Gathering data from sensors and IoT devices for real-time monitoring of hydrophyte health.	Timely interventions based on data insights
4	Phytoremediation Techniques	Use of hydrophytes for the removal of heavy metals from contaminated water sources.	Eco-friendly remediation methods
5	Predictive Modeling	Developing models to forecast the effectiveness of different hydrophyte species in remediation.	Optimized species selection for specific sites
6	Real-Time Monitoring	Implementing systems to continuously monitor heavy metal levels and plant health.	Immediate response to environmental changes
7	Decision Support Systems	AI-driven tools that assist in making informed management decisions for hydrophyte cultivation.	Enhanced decision-making processes
8	Case Studies	Examples of successful AI applications in hydrophyte management.	Evidence of effectiveness and best practices
9	Challenges	Identifying barriers to implementing AI in hydrophyte management, such as data quality issues.	Understanding limitations for future research
10	Future Directions	Exploring potential advancements in AI technologies for improved hydrophyte management.	Innovation in environmental management
11	Community Engagement	Involving local communities in AI-driven hydrophyte management initiatives.	Increased awareness and participation
12	Policy Implications	Assessing how AI integration can influence environmental policies related to water management.	Informed policy-making based on data
13	Educational Opportunities	Providing training for stakeholders on the use of AI in hydrophyte management.	Skill development and capacity building
14	Interdisciplinary Approaches	Collaborating across disciplines to enhance AI applications in hydrophyte management.	Comprehensive solutions to complex problems
15	Summary	Recap of the key points discussed regarding the integration of AI in hydrophyte management.	Consolidation of findings for further research

Rather than working independently, machine learning should be incorporated into a more rational decision-based approach to utilize hydrophytes for effective heavy metal phycoremediation. The two-way transfer of information would also help to improve the performance of machine learning outputs, which can be helpful for scientists and engineers working to further improve the adaptability of algorithms (Rolnick et al., 2022). The recent revolution in modeling, structured databases, and low-cost sensor devices being integrated into the artificial intelligence model based on appropriate processing algorithms should help optimize substrate heavy metal inhibition based on the ease of biological joint information processing. Such an integrated device scheme should help to monitor and predict the performance of, and recommend any modifications in, hydrophyte-assisted heavy metal inhibition plans.

7. PREDICTIVE MODELING OF HEAVY METAL STRESS

Models predicting heavy metal (HM) stress in plants may be (i) descriptive, mainly based on simple data fitting, either by linking directly HMs with a model parameter or one or more model parameters to one or multiple synonyms of stress; or (ii) mechanism-based, based on the geometric form of fundamental equations re-arranged into fitting forms linked to HM stress. The main descriptive models are moisture status evaluated through empirical models or models linking transpiration, temperature, and vapour pressure; or models relating relative water deficit to various meteorological parameters (Nandgude et al., 2023). These have been successfully applied for describing the influence of HMs on plant water status in several angiosperm species. Support Vector Machine (SVM) model parameterizations, which specifically describe leaf succulence or the reflectance variables derived from satellite observations, strongly indicate that heavy metals (HMs) play a significant role in inducing stomatal closure. This physiological response is largely achieved by reducing transpiration rates or by having a direct negative impact on the functioning of epidermal cells. Even more interesting is the fact that the impact of HMs do not limit to the affected plants but are transferred to other plants who may not be directly in contact with these toxic elements, indicating increased ecological consequences (Figure 2). Recent studies have introduced the mechanism-based modeling and the parameterization of these models and demonstrated the unexplored potential that these models have in quantitatively tackling stresses caused by heavy metals. More sophisticated methods of models are actively under development, namely the integration of mechanistically-driven eco-physiological models with gas exchange models. Nevertheless, recent uses have chosen to use a quite simple stomatal conductance model instead of using a full photosynthesis model. These types of models prove to be well-suited for describing the various HM-induced stresses observed across different plant species and the multitude of stress symptoms that may arise as a result; indeed, research has successfully parameterized the relationship that exists between the reduction of the maximum leaf photosynthetic rate and the average foliar concentrations of heavy metals such as cadmium, lead, and mercury, thereby considering both single and mixed contaminations (Ikkonen & Kaznina, 2022).

Figure 2 Analyzing heavy metal stress in plants

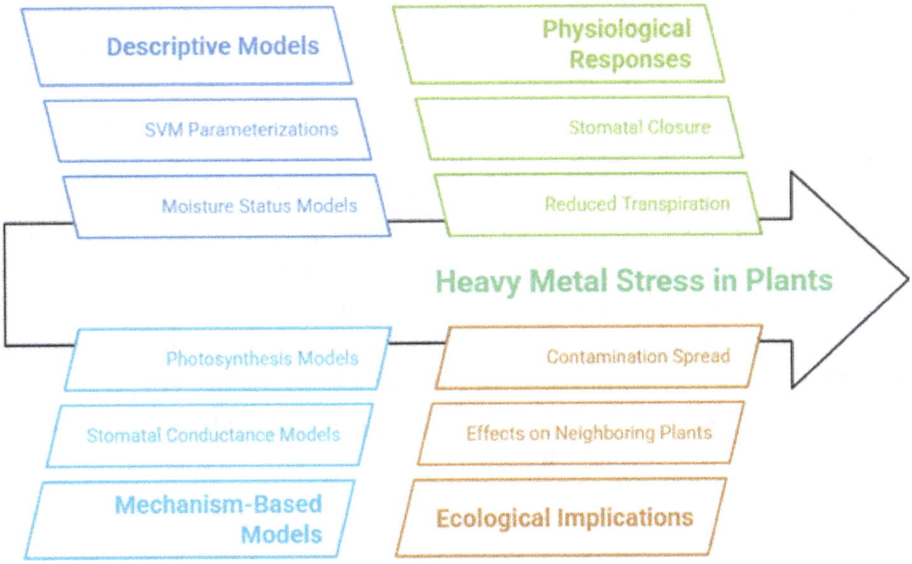

8. DATA COLLECTION METHODS FOR HYDROPHYTE STUDIES

Data collection is an exceptionally crucial step in any scientific endeavor, and studies investigating the phytoremediation potential of hydrophytes necessitate the same three core data types that are relevant to other environmental fields. These are the location data that make the studies reproducible, permit the in-depth spatial study of interest, or enable the association with the other relevant spatial data; observation data that are data that are indicative of the outcomes, process, or mechanism of interest that are described in a clear and concise way; and finally, the attribution data that offer the necessary details about the observed sites or organism under study (Blampey et al., 2024). With phytoremediation, a variety of six sampling methods can be considered a source of data, all of which differ in their features like the number and level of experience of the individuals participating in the sampling, their training, logistics, cost implications, time taken to complete the task, repeatability, processes, and the final size of the data footprint. The examination of remediating plants which exist in a certain area of interest is the first and sometimes the only way of collecting data. This often happens when a person working in a locality known to contain anomalies is casually in a position to note and in a few instances collects the plants which can be remedied. The second approach to the collection of data is the designed sampling, which involves the predetermined investigation and evaluation of the possible remediators. The most commonly used method of this is by the application of hedonic methods where the viewer scores the phytoremediation potential of an organism on a number of criteria that are pertinent to a particular contaminant of interest and response on the results of their observations. The third approach is by collecting the plants of their natural habitats and letting them go through cleaning procedures in well controlled laboratory conditions, that effectively recreates an event that may happen when trying to remediate a situation in the field. The fourth data collection technique presents the plants with the contaminants in a regulated fashion to permit the researchers to thoroughly manage the source and the concentration(s) of the contaminant(s) that were introduced during the cleanup effort (Liu et al., 2024). It is however worth noting that these testing processes may on certain occasions be inconsistent with the basic tenets of green chemistry and as such should be done with care and one only when using plants with a scientifically developed phytoremediation potential.

9. ROLE OF REMOTE SENSING IN HEAVY METAL MONITORING

Remote sensing is a given procedure, which entails the quantification of something or a phenomenon without necessarily having to come into physical contact with it in any manner. The application of different remote sensing methods can greatly increase the spatial and temporal transferability of data of analytes. Such a method is associated with a variety of various approaches that can be applied to obtain information that pertains to an object without necessarily touching the material that is being measured or examined. Notably, the remote sensing is a non-destructive technique of collecting data, which enables scholars to examine large regions and repeat the process on many occasions. Such a methodological benefit renders it especially helpful in keeping track of metal concentrations in the framework of the phytoremediation research. Remote sensing is a process that assumes the use of sensors that may be mounted on space stations or in the air (Table 2). Such sensors are uniquely engineered to respond to the electromagnetic spectrum as radar sensors are sensitive to emitted or backscattered radiations of the different materials within the scene under observation (Goddijn-Murphy et al., 2024). This technology requires certain sensors, which are mounted on airborne or satellite platform since these sensors are dedicated to measuring samples that are covering great distance. Through the interactions of the electromagnetic radiations with the carriers of such radiations and the vegetation scenes, the remote sensing techniques allow the evaluation of vegetative scenes. For instance, they can determine whether the vegetation appears to be green, alive, and functioning properly. Recent advancements in the techniques related to remote sensing have significantly improved our ability to detect alterations in the chemical composition of vegetation over time, thus making these technologies even more relevant and useful in ecological and environmental studies.

Table 2. Role of Remote Sensing in Heavy Metal Monitoring

No.	Remote Sensing Technique	Description	Applications in Heavy Metal Monitoring	Benefits
1	Satellite Imagery	Utilizes satellite data to capture large-scale environmental conditions.	Monitoring heavy metal contamination in soil and water.	Provides broad spatial coverage
2	Aerial Photography	Captures images from aircraft to analyze environmental conditions at a finer scale.	Assessing vegetation health in contaminated areas.	Higher resolution data for localized monitoring
3	Hyperspectral Imaging	Captures a wide spectrum of light to identify specific materials based on their spectral signatures.	Detecting heavy metal presence in plants and soil.	Increased sensitivity to metal concentrations
4	LiDAR (Light Detection and Ranging)	Uses laser pulses to measure distances and create 3D models of landscapes.	Mapping terrain and vegetation that may be affected by metals.	Detailed topographic data
5	Thermal Imaging	Measures surface temperature variations to identify stressed vegetation.	Identifying plants stressed by heavy metal exposure.	Non-invasive assessment of plant health
6	Ground Penetrating Radar (GPR)	Uses radar pulses to image the subsurface and detect contaminants.	Locating heavy metal hotspots in soil layers.	Provides subsurface information
7	UAV (Unmanned Aerial Vehicle)	Drones equipped with sensors for targeted monitoring of specific sites.	Real-time monitoring of contaminated sites.	Flexibility and cost-effectiveness
8	GIS (Geographic Information Systems)	Integrates spatial data for analysis and visualization.	Mapping heavy metal distribution and sources.	Enhanced decision-making through data analysis
9	Multi-spectral Sensors	Captures data across multiple wavelengths to assess environmental conditions.	Monitoring changes in vegetation related to metal uptake.	Cost-effective and versatile
10	Image Processing Techniques	Analyzes remote sensing data to extract meaningful information.	Identifying and quantifying heavy metal contamination.	Improved accuracy in detection
11	Data Fusion	Combines data from multiple sources for comprehensive analysis.	Integrating satellite and ground data for better insights.	Holistic view of contamination issues
12	Spectral Reflectance Analysis	Evaluates how different materials reflect light to identify contamination.	Assessing the presence of heavy metals in vegetation.	Precise identification of contaminants

continued on following page

Table 2. Continued

No.	Remote Sensing Technique	Description	Applications in Heavy Metal Monitoring	Benefits
13	Change Detection	Analyzes temporal data to identify changes in environmental conditions over time.	Tracking the impact of remediation efforts.	Monitoring effectiveness of interventions
14	Environmental Modeling	Uses remote sensing data to create models predicting heavy metal behavior in ecosystems.	Forecasting future contamination scenarios.	Informs proactive management strategies
15	Community Engagement	Involves local communities in monitoring efforts using remote sensing data.	Educating communities about heavy metal risks.	Builds trust and collaboration

Heavy metals and metalloids, like Cadmium and Lead, are strongly absorbed on plant surface and plant metabolism processes are not heavily influenced by the existence of these elements in plant surface, thus in this stage remote sensing can be proved to be efficient in detecting these elements and quantifying their content. Remote sensing is a powerful tool in the study of heavy metal monitoring in leaf surfaces of sensitive plants. In hypothesis of having enough sensitivity in the involved processes, the LIDAR method appears as the most promising technique in predicting the surface content (Gharineiat et al., 2022). Moreover in detecting these elements during the first experienced stages of stress induced by heavy metal pollution, Trichoderma like fungi play an important role.

10. AI-DRIVEN DECISION SUPPORT SYSTEMS FOR ENVIRONMENTAL MANAGEMENT

AI-based decision support systems and artificial intelligence (AI) technologies have become popular in the last few years in environmental management. The fusion of the artificial intelligence and the conventional sciences gives us the integrated and holistic progress in solving complex problems. The two issues that are of significant concern to the whole planet are climate change and environmental sustainability. AI-based solutions may facilitate the improved perception of the reasons behind climate change and further elaborate potential solutions. Applications in air pollution forecasting, forest fire detection, wildlife habitat protection, detecting plant and animal diseases, soil management, biodiversity monitoring, climate adaptation and mitigation, flooding management, and renewable energy forecasting are only but a few applications (Ahmad, 2024). In fact, AI technologies complement traditional methods of monitoring, prediction, and data assimilation that remain, for the most part, the foundation of any climate action.

Appropriate information is essential to the success of climate actions. Climate policies rely on climate information at different scales to support decision-making. Climate information is also required to assist local populations in their day-to-day life when confronted with weather events such as flooding or drought. However, the production of climate information and, more importantly, its dissemination remain an important challenge. uncertainties relate both to the dissemination of climate information to end-users and to issues of perception of, and sensitivity to, climate change (Kumar et al., 2021). Interactive information visualization can help provide clear and easily understandable climate information, and the use of climate analogues is another promising approach.

11. CHALLENGES IN IMPLEMENTING AI IN ENVIRONMENTAL RESEARCH

AI and ML models are mostly focused on understanding the surface-level process in a particular implementation. Despite their apparent flexibility and concerned specialized implementation, the rich cross-scale dynamics in natural ecosystems with all their unpredictability, being typically unique (in terms of social, geographical, and environmental drivers), questions general solutions based on large datasets possibly irrelevant to any local implementation. Simultaneously, the parameterization of AI and ML models is typically computationally heavy, limiting their applicability for predictive modeling in regions with low population levels and no automated data gathering.

It is evident that automation of the data gathering process is key to the successful implementation of AI in environmental research. Remote sensing has become a paramount solution to gather independent data with good spatial and temporal coverage over densely populated areas. However, the shortage of appropriate multi-temporal data covering the entire state of the monitored ecosystem (with the dynamical processes being especially important for unstable natural zones) limits the AI implementation in this area. The challenge here is to propose relevant implementations (in terms of variables and scales) on which the larger dataset will provide reasonable educated AI-based responses, which will rely on single observation data (Whang et al., 2023). Here, the recent development in the area of nonlinear process data mining offers new perspectives for the development of robust AI and ML models applicable to particular implementations of prediction-based human-environment interaction models. Other challenges include (i) development of AI-based tools fitted for scientists having limited ML background but crucial local environmental knowledge; (ii) ergonomics of the interaction and data exchange processes in mixed expert AI systems; (iii) providing validity analysis for AI, and most essential, (iv) avoiding algorithmic distraction and overloading the consideration process with unsuitable AI-implemented responses and turf priority bias in AI data selection and model building.

12. TECHNOLOGICAL INNOVATIONS IN PHYTOREMEDIATION

Recent discoveries have revealed that higher plants possess an innate immunity against low doses of toxins like metals. At higher concentrations, these environmental stressors can disrupt the regular activities of plants like photosynthesis, respiration, and nitrogen fixation, leading to detrimental effects on economic activities on which human beings thrive. However, low concentrations of such environmental stressors can help plants clear structural blocks, forming defensive envelopes and valves as a hindrance to such invading stressors. Experiments carried out using proteomic and transcriptomic techniques on plants have led to the identification of genes that play a crucial role in the process of duct formation around invading environmental stressors (Satrio et al., 2024). Such scientific studies on several dicot and monocot plants, along with their respective algal and fungal associates, need to be carried out to enhance the current available technological methods to assist plants in clearing such structural blocks.

The use of phytoremediation to restore metal-contaminated sites has grown in popularity over the past 25 years. Vegetation restoration is a sustainable remedial approach that is typically cost-effective and aesthetically pleasing. The applications of phytoremediation vary in specific circumstances using different plant species. Most commercial-scale phytoremediation systems use plants in aboveground structures operated under natural conditions, unaffected or unsupported greenhouse formats. Various research teams have carried out phytoremediation experimentation in controlled environments such as greenhouses or growth chambers to exploit genetic engineering or bioengineering for expected outcomes. It is possible to scale up beneficial bioremediation technology and expedite the commercialization of processes once they have been characterized and optimized. Sometimes crops are contemplated as potential study materials in phytoremediation, which can also be applicable to alleviate contamination in soils around agricultural sites (Moreira et al., 2021). However, this can lead to hazardous contamination of food supply chains if such methods are not properly employed. Therefore, caution should be exercised while working with these agricultural crops, and several precautionary measures should be adopted to ensure that the food supply chains are not endangered.

13. MONITORING AND EVALUATION OF PHYTOREMEDIATION EFFORTS

Artificial Intelligence (AI) and Machine Learning (ML) technologies can generate or optimize existing monitoring schemes that can help in their long-term effective functioning. Majority of these present recovery evaluation schemes rely only on simple micro-macro plant growth and performance indicators that lack specificity and precision with a limited range. Continually assessing these data will ensure timely recovery moments/pathways revealing strong/weak plant-metal interactions. Determining selection pressures and how plants develop strategies in facing and subverting these pressures in long-term studies will allow researchers to decide whether to establish a new intervention strategy or if the existing approach is yielding beneficial return (Russ et al., 2023).

Taking into consideration the beneficial potential and environmental risks given by the utilization of transgenic plants, the prudent use of the framework presented might allow, when appropriate, a careful monitoring of transgenic plant performance/realization in HM polluted field scenarios while avoiding risks linked to TM release. Remote and automatized monitoring of selected plant parameters like pigment composition or growth can be realized, providing for a large amount of data generation while reducing the risk of human error and providing data at a much higher resolution in space and time. This strategy can also be coupled with systems allowing to optimize transect selections, either spatially in relation to the plants, or temporally in relation to the action of their plant-cutting systems. These monitoring systems will also produce data relative to the concentrations of specific HM related parameters, allowing predictive modeling of the pollution un-pollution scenario.

14. ECONOMIC ASPECTS OF AI IN ENVIRONMENTAL SOLUTIONS

Precision agriculture, which applies sophisticated techniques for monitoring a variety of environmental conditions, ecosystem and crop health, and farming practices, also to allocate resources more effectively and reduce application costs, could benefit enormously from the use of AI capabilities. For example, currently, companies are assisting farmers in risk assessments and predictive analyses by using optical, infrared, and thermal camera technologies to monitor crop health and ecosystem conditions (Wu et al., 2023). These companies exclusively rely on algorithms developed to assess these data. However, the successful adoption of these technologies by farmers will only occur if field studies prove substantial financial advantages to them. Without question, AI-enabled monitoring systems can evaluate crop and environmental health much faster and more accurately than sensors and field workers; however, sensors typically require relatively low investments and sample data over large areas but relatively long time frames, while AI can reduce the need for field workers but represents a much larger initial investment. What's more, no AI service provider is currently replacing the worker monitoring methane emissions from cattle ranches, the sensitive network of drones assessing the deployed land reform program, or the satellite imagery provider mapping land use.

The AI-based solutions supporting decision-making in environmental agencies can detect irregularities in monitoring and fine, among others, large-scale deforestation and emissions of the cattle ranching sector. Brazil is using AI-enabled satellite data together with environmental legislation as decision rules to assess the flexibility with which agency staff can grant approvals, particularly for private investments (Raghuvanshi et al., 2024). Nonetheless, AI companies must cooperate closely with agencies and actively promote positive results as perceived by agency staff.

15. COMPARATIVE ANALYSIS OF TRADITIONAL VS. AI APPROACHES

Plant-based bioremediation entails a series of actions implemented by plants to reduce the concentrations and toxicity of metals in contaminated sites, thereby restoring ecological integrity and improving quality of life. An advantage of using hydrophytes in bioremediation is the reduced expense compared to conventional technologies. However, the process requires an understanding of the molecular responses of hydrophytes to abiotic stress, including the possible toxic effects of high metal concentrations on the plant and the underlying mechanisms of tolerance and the accumulation of metals in tissues and organs (root, stem, leaf, flower, and fruit). Furthermore, a low tolerance to heavy metal concentrations can slow down the bioremediation process. Moreover, not all plants have the same effectiveness or speed in mobilizing heavy metals or restoring the ecological integrity of the contaminated site (Moreiraet al., 2021). In this context, AI-based models can enhance the efficiency of bioremediation strategies via hydrophytes.

Notably, artificial intelligence approaches provide comparisons between multiple and/or complex levels of observations (geographical, pedological, climatological, biogeochemical, and biological) and assist in estimating or predicting the outcomes of future observations to assess the potential of specific species for in situ bioremediation strategies. The scientific community, policymakers, and society in general have turned their attention to AI-based and machine learning approaches because these technologies and methods surpass traditional techniques in several fields (Arsalan et al., 2025). By reviewing specific studies where traditional experimental design was transformed to use machine learning methods to estimate bioremediation potential, we describe the advantages of AI models. Optimize basic experimental design of plant growth conditions prior to experimental determination of metal tolerance and/or hyperaccumulation; Explore and model strong relationships between plant growth-promoting factors and hyperaccumulation potential of cadmium, lead, or zinc in plants; Estimate or predict heavy metal concentrations in substrates.

16. ROLE OF BIG DATA IN ENVIRONMENTAL MONITORING

It has been estimated that by 2050, the global population will reach 200 billion. With these many people, food is going to be essential. However, with technological advancement, soil contamination with heavy metals is of growing concern. The adaptability and ability to thrive in adverse conditions exhibited by certain aquatic plants can act as an effective trigger in the mitigation of heavy metal-stressed plants. Artificial intelligence tools along with machine learning approaches have recently emerged as powerful tools in the study of phytoremediation, providing accurate prediction and identification of the various challenges posed when utilizing aquatic plants as potential agents in heavy metal-translocated areas. In this manuscript, we summarize the significance of utilizing these technologies and methods in dealing with the issues posed by aquatic plants to amplify and ameliorate heavy metal phytoremediation (Sharma et al., 2023).

To assist data-driven environmental decision-making, various information and communication technologies are rapidly approaching a data-centric approach. Environmental data is increasingly generated through citizen input and open data initiatives. Such data has great potential contribution to environmental management. However, cheap and easy data generation technologies also make context-specific data constrained, which makes it difficult to rely solely on the data-driven approach, worthiness of data for the specified objectives and multiple levels increasing environment monitoring applications calls for positioning big data utilization in a comparative analytical framework, hierarchically organized by combinations of objectives, techniques, and types of value creation, guiding the integration with conventional data collection. Despite the clear potential of big data, it does not supersede traditional means entirely. The future challenge will be to bring big data and traditional data into a fruitful coexistence (Bonacchi 2022).

17. CONCLUSION

We create reports on the use of Artificial Intelligence and Machine Learning in hydrophyte-based techniques. Heavy metal pollution has become a 21st-century curse on humanity. The natural beauty, landscape, and aesthetics of wetlands and rivers are being adversely affected by the continuous release of heavy metals into the environment. Hydrophytes have recently come to light as potential agents for the cleaning of heavy metal-contaminated water, especially if the cleaning process is implemented using the latest advances in Artificial Intelligence and Machine Learning. Artificial Intelligence and Machine Learning have the potential to accelerate the Hydrophyte-assisted Heavy Metal Contaminated Water Cleaning Process. Not only do they increase the rate of heavy metal removal by hydrophytes, their implementation ensures cost-effective scaling of the pollutant-cleansing process.

The world, and indeed the entire Universe, operates as a remarkably closed system, and it is crucial that we treat it with the utmost respect and care in this capacity. The potential destruction of Earth will inevitably influence every object and entity that is intricately embedded within her environment, making it unreasonable to expect that their delicate balance would remain stable and unchanged. Our understanding of the most minute levels of basic subatomic systems reveals that these cannot be disturbed in isolation; rather, any alterations will inevitably trigger significant reactions at the outer levels, reverberating throughout the larger system. These outer levels encompass the intricate biochemical systems that involve bacteria, plants, and animals that call Earth home. The intelligent awareness and critical thinking of human beings have progressed to such an advanced stage that we have taken on the role of manipulators of the elements, harnessing the power of fissioning atoms to produce enormous bursts of energy, all in the pursuit of enhancing our comfortable lifestyles and conveniences. However, is it, therefore, unreasonable to consider that there may be profound disadvantages regarding nature's energy rebounds at the outer levels? These disadvantages could potentially manifest in the form of disruptions within the food chain effect, a consequence stemming from the Environmental Impact caused by numerous industries. It is vital to remember that Earth is not merely a machine functioning with crisp, mechanical precision; rather, it is a complex, living system evolving over billions of years, endowed with the remarkable ability to repair itself within certain limitations. Unfortunately, these limitations in the flow of energy throughout the Earth system have been strained, as we have long surpassed the threshold of burdening the system with pollutants and deleterious substances. This excessive pollution has significantly disrupted and compromised our access to the bountiful resources that nature has generously provided, leading to lasting repercussions for both humanity and the planet.

REFERENCES

Ahmad, M. (2024). Mapping Green Infrastructure Harnessing OSM Data for Sustainable Development. In *Marketing and Resource Management for Green Transitions in Economies* (pp. 105–125). IGI Global. DOI: 10.4018/979-8-3693-3439-3.ch005

Arsalan, M. H., Mubin, O., Al Mahmud, A., Khan, I. A., & Hassan, A. J. (2025, April). Mapping Data-Driven Research Impact Science: The Role of Machine Learning and Artificial Intelligence. In *Metrics* (Vol. 2, No. 2, p. 5). MDPI.

Asiminicesei, D. M., Fertu, D. I., & Gavrilescu, M. (2024). Impact of heavy metal pollution in the environment on the metabolic profile of medicinal plants and their therapeutic potential. *Plants, 13*(6), 913. DOI: 10.3390/plants13060913 PMID: 38592933

Barbierato, E., & Gatti, A. (2024). The challenges of machine learning: A critical review. *Electronics (Basel), 13*(2), 416. DOI: 10.3390/electronics13020416

Blampey, Q., Mulder, K., Gardet, M., Christodoulidis, S., Dutertre, C. A., André, F., Ginhoux, F., & Cournède, P. H. (2024). Sopa: A technology-invariant pipeline for analyses of image-based spatial omics. *Nature Communications, 15*(1), 4981. DOI: 10.1038/s41467-024-48981-z PMID: 38862483

. Bonacchi, C. (2022). *Heritage and Nationalism*

Cervantes-Avilés, P., Huang, X., & Keller, A. A. (2021). Dissolution and aggregation of metal oxide nanoparticles in root exudates and soil leachate: Implications for nanoagrochemical application. *Environmental Science & Technology, 55*(20), 13443–13451. DOI: 10.1021/acs.est.1c00767 PMID: 34029070

Farchi, A., Laloyaux, P., Bonavita, M., & Bocquet, M. (2021). Using machine learning to correct model error in data assimilation and forecast applications. *Quarterly Journal of the Royal Meteorological Society, 147*(739), 3067–3084. DOI: 10.1002/qj.4116

Gharineiat, Z., Tarsha Kurdi, F., & Campbell, G. (2022). Review of automatic processing of topography and surface feature identification LiDAR data using machine learning techniques. *Remote Sensing (Basel), 14*(19), 4685. DOI: 10.3390/rs14194685

Goddijn-Murphy, L., Martínez-Vicente, V., Dierssen, H. M., Raimondi, V., Gandini, E., Foster, R., & Chirayath, V. (2024). Emerging technologies for remote sensing of floating and submerged plastic litter. *Remote Sensing (Basel), 16*(10), 1770. DOI: 10.3390/rs16101770

Goncharuk, E. A., & Zagoskina, N. V. (2023). Heavy metals, their phytotoxicity, and the role of phenolic antioxidants in plant stress responses with focus on cadmium. *Molecules (Basel, Switzerland), 28*(9), 3921. DOI: 10.3390/molecules28093921 PMID: 37175331

Grifoni, M., Franchi, E., Fusini, D., Vocciante, M., Barbafieri, M., Pedron, F., Rosellini, I., & Petruzzelli, G. (2022). Soil remediation: Towards a resilient and adaptive approach to deal with the ever-changing environmental challenges. *Environments (Basel, Switzerland), 9*(2), 18. DOI: 10.3390/environments9020018

. Haslam, S. A., Steffens, N. K., Cruwys, T., & McGarty, C. (2024). Research methods and statistics in psychology.

Ikkonen, E., & Kaznina, N. (2022). Physiological responses of lettuce (Lactuca sativa L.) to soil contamination with Pb. *Horticulturae, 8*(10), 951. DOI: 10.3390/horticulturae8100951

Kosal, E. (2023). *Plant Kingdom. Introductory Biology: Ecology*. Evolution, and Biodiversity.

Kumar, U., Werners, S. E., Paparrizos, S., Datta, D. K., & Ludwig, F. (2021). Co-producing climate information services with smallholder farmers in the Lower Bengal Delta: How forecast visualization and communication support farmers' decision-making. *Climate Risk Management, 33*, 100346. DOI: 10.1016/j.crm.2021.100346

Liu, X., Sathishkumar, K., Zhang, H., Saxena, K. K., Zhang, F., Naraginiti, S., & Guo, X. (2024). Frontiers in environmental cleanup: Recent advances in remediation of emerging pollutants from soil and water. *Journal of Hazardous Materials Advances, 16*, 100461. DOI: 10.1016/j.hazadv.2024.100461

Mohamed, M. (2023). Agricultural sustainability in the age of deep learning: Current trends, challenges, and future trajectories. *Sustainable Machine Intelligence Journal, 4*, 2–1. DOI: 10.61185/SMIJ.2023.44102

Moreira, H., Pereira, S. I., Mench, M., Garbisu, C., Kidd, P., & Castro, P. M. (2021). Phytomanagement of metal (loid)-contaminated soils: Options, efficiency and value. *Frontiers in Environmental Science, 9*, 661423. DOI: 10.3389/fenvs.2021.661423

Nandgude, N., Singh, T. P., Nandgude, S., & Tiwari, M. (2023). Drought prediction: A comprehensive review of different drought prediction models and adopted technologies. *Sustainability (Basel), 15*(15), 11684. DOI: 10.3390/su151511684

Nwagbara, V. U. (2023). *Comparative Analysis of Water Hyacinth Efficiency as Biosorbent and Phyto remediating Plant for Removal of Lead (Pb)*. Water Contaminants.

Pan, Y., & Zhang, L. (2023). Integrating BIM and AI for smart construction management: Current status and future directions. *Archives of Computational Methods in Engineering*, *30*(2), 1081–1110. DOI: 10.1007/s11831-022-09830-8

Raghuvanshi, R., Sharma, T., Rena, R., & Rai, R. (2024). Artificial Intelligence for Smart City Development in BRICS Economies: Opportunities and Challenges. *Handbook of Artificial Intelligence for Smart City Development*, 173-208.

Rane, N. L., Paramesha, M., Choudhary, S. P., & Rane, J. (2024). Machine learning and deep learning for big data analytics: A review of methods and applications. *SSRN*, *2*(3), 172–197. DOI: 10.2139/ssrn.4835655

Rolnick, D., Donti, P. L., Kaack, L. H., Kochanski, K., Lacoste, A., Sankaran, K., Ross, A. S., Milojevic-Dupont, N., Jaques, N., Waldman-Brown, A., Luccioni, A. S., Maharaj, T., Sherwin, E. D., Mukkavilli, S. K., Kording, K. P., Gomes, C. P., Ng, A. Y., Hassabis, D., Platt, J. C., & Bengio, Y. (2022). Tackling climate change with machine learning. *ACM Computing Surveys*, *55*(2), 1–96. DOI: 10.1145/3485128

Russ, D., Fitzpatrick, C. R., Teixeira, P. J., & Dangl, J. L. (2023). Deep discovery informs difficult deployment in plant microbiome science. *Cell*, *186*(21), 4496–4513. DOI: 10.1016/j.cell.2023.08.035 PMID: 37832524

Satrio, R. D., Fendiyanto, M. H., & Miftahudin, M. (2024). Tools and techniques used at global scale through genomics, transcriptomics, proteomics, and metabolomics to investigate plant stress responses at the molecular level. In *Molecular Dynamics of Plant Stress and its Management* (pp. 555–607). Springer Nature Singapore. DOI: 10.1007/978-981-97-1699-9_25

Sharma, J. K., Kumar, N., Singh, N. P., & Santal, A. R. (2023). Phytoremediation technologies and their mechanism for removal of heavy metal from contaminated soil: An approach for a sustainable environment. *Frontiers in Plant Science*, *14*, 1076876. DOI: 10.3389/fpls.2023.1076876 PMID: 36778693

Thayyib, P. V., Mamilla, R., Khan, M., Fatima, H., Asim, M., Anwar, I., Shamsudheen, M. K., & Khan, M. A. (2023). State-of-the-art of artificial intelligence and big data analytics reviews in five different domains: A bibliometric summary. *Sustainability (Basel)*, *15*(5), 4026. DOI: 10.3390/su15054026

Whang, S. E., Roh, Y., Song, H., & Lee, J. G. (2023). Data collection and quality challenges in deep learning: A data-centric ai perspective. *The VLDB Journal*, *32*(4), 791–813. DOI: 10.1007/s00778-022-00775-9

Wu, B., Zhang, M., Zeng, H., Tian, F., Potgieter, A. B., Qin, X., Yan, N., Chang, S., Zhao, Y., Dong, Q., Boken, V., Plotnikov, D., Guo, H., Wu, F., Zhao, H., Deronde, B., Tits, L., & Loupian, E. (2023). Challenges and opportunities in remote sensing-based crop monitoring: A review. *National Science Review*, *10*(4), nwac290. DOI: 10.1093/nsr/nwac290 PMID: 36960224

Yadav, V., Arif, N., Kováč, J., Singh, V. P., Tripathi, D. K., Chauhan, D. K., & Vaculík, M. (2021). Structural modifications of plant organs and tissues by metals and metalloids in the environment: A review. *Plant Physiology and Biochemistry*, *159*, 100–112. DOI: 10.1016/j.plaphy.2020.11.047 PMID: 33359959

Zahoor, I., & Mushtaq, A. (2023). Water pollution from agricultural activities: A critical global review. *International Journal of Chemical and Biochemical Science*, *23*(1), 164–176.

Chapter 10
Synergistic Role of Artificial Intelligence and Machine Learning in Enhancing Hydrophyte-Based Heavy Metal Stress Management

Pushkar Singh Rawat
https://orcid.org/0000-0002-8679-5603
University of Lucknow, India

Shalini Singh
https://orcid.org/0000-0003-3351-5732
Shri Ramswaroop Memorial University, India

ABSTRACT

Heavy metal pollution threatens ecosystems and human health. Hydrophytes, with their natural ability to absorb and detoxify metals, are valuable for phytoremediation. However, their efficiency varies due to species differences and environmental factors. Integrating Artificial Intelligence (AI) and Machine Learning (ML) enhances this process by enabling predictive modeling, stress detection, and real-time monitoring. AI-driven tools optimize plant selection, growth conditions, and risk assessment while improving scalability and sustainability. Despite challenges like data quality, costs, and ethical concerns, the synergy of hydrophytes with AI/ML offers a promising, eco-friendly solution for restoring aquatic ecosystems.

DOI: 10.4018/979-8-3373-4037-1.ch010

1. INTRODUCTION

Heavy metal pollution in aquatic ecosystems is a major global concern, primarily driven by industrialization, urbanization, and climate change. The primary sources are mining wastes, landfill leachates, municipal and industrial effluents, urban runoffs and natural processes such as volcanic eruptions and the erosion of rocks. Toxic metals include cadmium (Cd), lead (Pb), mercury (Hg), copper (Cu), zinc (Zn), nickel (Ni), manganese (Mn), arsenic (As) and chromium (Cr) not only toxic but also highly persistent and non-biodegradable. This implies that they will accumulate with time in our waters, sediments, aquatic plants and animals. Such heavy metals may pile up in aquatic organisms and multiply via the food chain posing serious threats to the well-being of the aquatic ecosystems as well as the health of humans who consume the aquatic organisms (Hama Aziz et al., 2023).

The heavy metals interfere with the normal physiological and biochemical processes of the aquatic organisms, damaging DNA, disrupting growth, affecting reproduction, respiratory disturbance, and death of fish, invertebrates, and other living beings (Jeong et al., 2023). These effects decrease biodiversity, impair the water quality, and destabilize food webs. Even low doses of contaminated water or seafood can cause organ damage, neurological disorders and risks of cancer. In addition, their toxicity can be modified depending on exposure to pollutants such as microplastic or persistent organic compounds (Abubakar et al., 2024). Using such monitoring tools as the Heavy Metal Pollution Index and Degree of Contamination can be considered essential information on the state of pollution. Remediation measures can include traditional processes such as chemical precipitation and adsorption or other modern techniques such as bioremediation, phytoremediation and oxidation processes. Nevertheless, there are still the problems of cost and efficiency (M. Sharma et al., 2025). To combat this problem, governments, industries, and communities should collaborate to minimize discharges, enhance the level of wastewater management, and safeguard aquatic ecosystems to be sustainable.

1.1. Overview of Hydrophytes and Their Role in Phytoremediation

Aquatic plants or hydrophytes are macrophytes which adapt to thrive in waterlogged habitats and are classified into floating, submerged and emergent species. They play an important role in aquatic ecosystems as they stabilize the sediments, provide habitats, and control nutrient cycles. In the recent times, they have attracted much interest towards their prospects in phytoremediation of water, sediments and soils using plants to remediate the pollutants (Thakur et al., 2023). Their direct contact with water, high production of biomass and rapid growth allow them to absorb and store contaminants due to their high ability to absorb, especially heavy metals and nutrients. Many phytoremediation strategies have been employed by hydrophytes among which they are phytoextraction, phytodegradation, rhizofiltration, phytostabilization, and phytovolatilization (Rahman & Hasegawa, 2011). Some of the common species used are water hyacinth (*Eichhornia crassipes*), duckweed (*Lemna spp.),* water lettuce (*Pistia stratiotes*), and several other emergent species, such as *Iris pseudacorus* and *Acorus gramineus* (Soliman et al., 2019). These plants eliminate various pollutants like heavy metals nutrients (nitrogen, phosphorus), organic contaminants (e.g., atrazine) and pathogens of domestic, agricultural and industrial wastewater.

Wetlands that have been built that combine microbes and hydrophytes are considered to be eco-friendly, cost-effective, and sustainable in treating wastewater and recovering resources. Remediation efficiency however is dependent on plant species, the type of pollutant, physiology and environmental conditions. As an example, emergent hydrophytes tend to accumulate heavy metals as compared to floating or submerged counterparts and can develop antioxidative systems that make them more tolerant (Soliman et al., 2019). In addition to remediation, hydrophytes minimize greenhouse gases, recycle nutrients and aid in restoration of ecosystems. They are an alternative to traditional treatment that promote biodiversity, recovery of resources, and circular bioeconomies that are green. Further studies are required to streamline the process of species selection, system design, and biomass management to apply it on a sustainable basis (Thakur et al., 2023).

Table 1. Hydrophyte Remediation Mechanisms

Mechanism	Description
Phytoextraction	The uptake and accumulation of contaminants into plant tissues.
Phytodegradation	The breakdown of pollutants via plant enzymes.
Rhizofiltration	Adsorption or absorption of pollutants by roots.
Phytostabilization	The immobilization of contaminants in sediments.
Phytovolatilization	The conversion of pollutants into volatile forms that are released into the atmosphere.

1.2. Need for Integrating Artificial Intelligence (AI) and Machine Learning (ML) in Heavy Metals Stress Management

AI and ML in the management of stress in heavy metals are becoming a critical tool in the detection, monitoring, and mitigation of stress. The traditional approaches to the identification and elimination of heavy metals tend to be rather expensive, time-consuming, and ineffective when working under the intricate environmental conditions or when monitoring a large number of sites (Lakshmi et al., 2021). Conversely, AI and ML provide potent remedies because they allow working with massive volumes of data, improve predictive modelling, and make informed decision-making in environmental management (Wei et al., 2024). ML and hyperspectral imaging have demonstrated excellent accuracy in the detection of plant-stressed out by heavy metals, even at low traces in the case of plant stress detection. As an illustration, spectral-based non-invasive support vector machine (SVM) models have 100 percent accuracy to detect mercury-stressed tobacco plants, demonstrating the potential of non-invasive, early detection tools (Yu et al., 2021). Artificial neural networks (ANN) and SVM may be used to simulate and predict heavy metal levels in water using physicochemical parameters, albeit with a small amount of data, to be used as an environmental monitoring tool. Such models are also used to determine the key determinants of heavy metal behaviour and enhance monitoring and management control plans (Azizi et al., 2022).

Remediation is also being revolutionised by AI and ML. By examining feedstock attributes, pyrolysis parameters, and the adsorption rate dynamics, the ML models can optimize the biochar synthesis to heavy metal adsorption, and thus render the remediation process cheaper and more sustainable (Lakshmi et al., 2021). In addition, the models based on ML can anticipate heavy metal distribution in soils and sediments, which can be applied to large-scale risk measurements and intervention (Azizi et al., 2022). Despite the persistence of issues in data quality, algorithm creation, and general interdisciplinarity, AI and ML have a significant potential to provide smarter, faster, and more adaptive solutions to the heavy metal stress management.

1.3. Scope and Objectives of the Review

The AI + ML in addressing stresses of heavy metals in hydrophytes are a smarter data-driven solution to environmental management. Conventional monitoring and remediation techniques usually have problems such as slowness, inaccessibility, and the complexity of the interaction with the plant environment. AI and ML intervene to address these gaps with the help of sorting through large and complex datasets to find trends that could be used to inform effective phytoremediation efforts. This combination can be used to detect real time stress levels in hydrophytes with hyperspectral imaging with ML algorithms, forecasts distribution of heavy metals in water bodies, and optimizes phytoremediation methods with predictive models. The aims of this strategy are varied. First, it enhances speed and accuracy in the process of determining heavy metal stress in the plants and thus taking appropriate measures on time that are likely to reduce ecological damage. Second, AI and ML models will have the ability to predict the contaminating trends across time and space, which will help in strategically positioning hydrophytes in order to effectively remove the pollutants. In addition, the technologies improve the selection and management of hydrophyte species by assessing physiological characteristics, environmental features, and operation parameters that subsequently increase pollutant uptake and resistance. They also facilitate massive monitoring and management and thus the phytoremediation becomes feasible and cost effective. Notably, AI and ML systems are flexible, undergoing constant updates to the models as new information appears in order to become a better decision maker over time.

2. Heavy Metal Stress in Aquatic Systems

2.1. Physiological and Biochemical Impact on Aquatic Plants

The impact of heavy metal stress on the aquatic environment is strong and affects the physiology and biochemistry of the aquatic plants causing impaired growth, metabolism, and health. Exposing the plants to such metals as Cu, Pb, and Cd harms their cell structures and photosynthetic systems thereby reducing their biomass and chlorophyll content. This type of stress also leads to oxidative injury, which is reflected by the elevated concentrations of malondialdehyde (MDA), which is the predictor of lipid peroxidation (LPO). Plants in response protect themselves by using antioxidant enzymes including superoxide dismutase (SOD), peroxidase, and catalase. Nonetheless, they are not very effective when it comes to the oxidative stress resulting from the exposure to heavy metals (Nguyen et al., 2021; Wani et al., 2020).

The presence of heavy metals disturbs the normal functioning of the body such as uptake of nutrients, water and photosynthesis, which eventually reduces the productivity of the plants. The destruction of the photosynthetic equipment implies the reduced energy generation, whereas the destruction of membranes and organelles also impairs the functioning of cells. Moreover, due to heavy metal contamination, the composition of the microbial community related to plant biofilm, nutrient cycling, and plant microbe interactions can be altered (S. Huang et al., 2020). The accumulation of persistent metal (Hg, Pb, Cd, As, and Cr) is particularly harmful, disrupting the work of cells and hormones, and causing developmental disorders in plants and other organisms in the long run. Aquatic vegetation combats toxicity through the following mechanisms: compartmentalization of metals by tissue masses, the formation of metal binding proteins, and the regulation of phytohormone concentrations to allow responses to stress (Nguyen et al., 2021). Nevertheless, prolonged or excessive exposure may overwhelm these defenses and lead to impaired and slow growth, reduced reproductive ability and exposure to disease. These physiological and biochemical disturbances do not only occur to individual plants, but they are a threat to ecosystem stability and aquatic food web dynamics.

2.2. Mechanisms of Heavy Metal Uptake, Translocation, and Detoxification in Hydrophytes

Hydrophytes have evolved special mechanisms of dealing with heavy metal stress in their water environments. This includes absorption, transport and de-toxification of harmful metals. They do so by absorbing, transporting and detoxification of toxic metals. Although most of the metals are absorbed by the roots, in some cases, they are also absorbed by the leaves (Shahid et al., 2017). It all begins in the rhizosphere, where root exudates and microbes interactions make the metals more soluble and available (Wagh et al., 2024). Critical transporter proteins such as ATP-binding cassette (ABC), heavy metal ATPases (HMA), metal tolerance proteins (MTP), and natural resistance-associated macrophage proteins (NRAMP), play a vital role in regulating the entry of ions into root cells, which is usually influenced by plant hormones and stress signals (Ghuge et al., 2023). When the metals are in the roots they may be stored in the roots or may be conducted up to the shoots and the leaves via the xylem. Most hydrophytes, including *Scirpus validus*, *Typha orientalis*, and *Phragmites australis*, are more likely to store more metals in the root of the plant. Conversely, hyper accumulator species can easily transport metals to their aerial locations to store or detoxify. Also, some beneficial microbes, such as *Bacillus xiamenensis*, can promote this uptake and translocation by improving growth and resistance to stress (Wagh et al., 2024).

Detoxification occurs by keeping the metals immobile in the cell walls or they can be bound by other substances such as phytochelatins, metallothionein, and organic acids. The metals are isolated into vacuoles to keep them out of all key processes, and antioxidant enzymes are used to fight against the reactive oxygen species (ROS) (Yang et al., 2022). Genetically, stress also induces tolerance-related genes, microRNAs, and proteins (Ghuge et al., 2023). Such strategies will not only ensure the health of plants, but also contribute to phytoremediation, which improves water quality and microbial diversity. As molecular genetics and omics continue to improve hydrophytes remediation capability is only improving, such as the development of transgenic and gene edited plants.

3. Hydrophyte-Based Remediation Strategies

3.1. Classification and Selection of Hydrophyte Species

The concept of hydrophyte based remediation is based on the utilization of aquatic vegetation to treat contaminated water through the elimination of heavy metals and contaminants. It is very affordable and environmentally friendly and so it is a good substitute to the conventional methods of treatment. It is possible to classify these hydrophytes into various categories: emergent plants (*Typha latifolia* and *Phragmites australis*), floating plants (*Eichhornia crassipes* and *Pistia stratiotes*), submerged (*Hydrilla verticillata* and *Potamogeton crispus*), and free floating algae (*Zygnema pectiantum* and the species of *Spyrogyra)* (Ayaz et al., 2020). The uptake of pollutants differs with each group and it is just how important it is to select the correct species to a particular contaminant and environment.

In the choice of plants, such factors as the efficiency of the removal of pollutants, adaptability, the growth rate, biomass yield, and tolerance to contaminants are involved. As an example, water cabbage (*Pistia stratiotes*) is very effective in cadmium removal, water hyacinth (*Eichhornia crassipes*) is very effective in lead removal, and duckweed (*Lemna gibba*) is well known in copper removal (Xie et al., 2022). Equally, *Typha latifolia* (cattail) and *Iris pseudacorum* have the ability to absorb various heavy metals and nutrients in wastewater (Farid et al., 2023). More intriguingly, a number of these species are likely to gain greater amounts of metals in their roots than in their shoots, which should be taken into account when biomass is safely disposed of (Xie et al., 2022).

Constructed wetlands are usually made with a combination of various species of hydrophytes to enhance the capacity to eliminate pollutants and enhance resilience of the ecosystem. These hybrid systems can be useful in promoting the uptake of heavy metal, decreasing the biological and chemical oxygen demand, and decreasing the total dissolved solids (Farid et al., 2023). In addition to this, the introduction of ornamental hydrophytes and plant-microbe complexes can also be used to increase the effectiveness of remediation and support biodiversity (Thakur et al., 2023).

Table 2. Different types of hydrophytes have varying efficiencies and are effective at removing specific pollutants.

Hydrophyte Classification	Examples of Species	Pollutants Removed	Key Advantage
Emergent	*Typha latifolia* (cattail), *Phragmites australis*, *Iris pseudacorus*, *Typha orientalis*, *Pontederia cordata*, *Persicaria salicifolia*	Heavy metals like Cd, Pb, Cr, Ni, Mn, Zn, Cu and nutrients	Superior at heavy metal accumulation and long-term remediation
Floating	*Eichhornia crassipes* (water hyacinth), *Pistia stratiotes* (water lettuce), *Lemna* spp. (duckweed)	Cadmium, lead, copper, organic pollutants, nutrients	Useful for rapid surface water treatment and easy to harvest
Submerged	*Hydrilla verticillata*, *Potamogeton crispus*, *Myriophyllum aquaticum*	Heavy metals like Cu, Cr, As, Pb, organic pollutants like erythromycin	Superior removal efficiency for certain pollutants due to greater surface area in contact with water

3.2. Limitations and Challenges of Conventional Hydrophyte-Based Remediation

Hydrophyte-based remediation is a relatively inexpensive and eco-friendly method of removing pollutants in water bodies; however, it has its own range of challenges that restrict its larger application. The inconsistency in the effectiveness of various hydrophyte species is one of the primary obstacles, and even the same species may also have different levels of effectiveness based on the environmental conditions. The nature of the sediment, the chemistry of the water, the presence of herbivores, algal competition, and special ecological interactions of a location can all make a significant contribution to the extent of the plants establishing and absorbing pollutants (Rodrigo, 2021). The performance of hydrophytes too depends on the nature and quantities of pollutants and physiological and antioxidative capacities of the plant. The other problem is the fact that remediation process is usually slower than the traditional physico-chemical methods. Hydrophyte systems take extensive times to produce a discernible decrease in pollutants thus are not suitable in areas that require quick clean-ups (Hu et al., 2020). Besides, the accumulation of these heavy metals and other pollutants in the biomass of the plants poses a question of disposal and management of the products since improper handling may result in secondary pollution (Ayaz et al., 2020). The use of large-scale and long-term systems is also not common because of the difficulties of maintaining the plants healthy, continued existence of the species, and long-term ecological balance. The results of the remediation may also be complicated by environmental factors such as climate changes, availability of water, and interaction of the pollutants. There are even cases of contaminants when they are too high to the extent that they inhibit plant growth or cause toxicity (Hu et al., 2020). Also, we still have unanswered questions in biology, ecology, and the mode of propagation of the hydrophytes, which complicates the restoration process. Finally, the policy and regulatory frameworks, however, tend to lag behind the scientific developments and it is, therefore, hard to apply these approaches to a greater scale. To deal with them, we will require interdisciplinary studies, improved selection of species, improved management protocols, and enabling regulations to make full use of the potential of hydrophyte remediation.

Table 3. Challenges of Conventional Hydrophyte Remediation

Challenge	Description
Variability in Efficiency	Pollutant removal efficiency can vary significantly between species and under different environmental conditions.
Slow Rate of Remediation	Phytoremediation may require long periods to achieve significant contaminant reduction, making it unsuitable for sites needing rapid cleanup.
Biomass Management	The safe disposal of contaminated plant material is necessary to prevent secondary pollution.
Environmental Factors	The presence of multiple contaminants or climate variability can complicate remediation efforts.
Lack of Research	There is a need for more research on species biology, propagation, and revegetation techniques.
Policy and Regulatory Frameworks	Regulations often lag behind scientific advancements, limiting the integration of hydrophyte-based remediation into water management strategies.

Conventional methods face several limitations that can hinder their effectiveness and scalability

4. ARTIFICIAL INTELLIGENCE AND MACHINE LEARNING: PRINCIPLES AND RELEVANCE

4.1. Overview of AI and ML Techniques

AI and ML are revolutionary fields in computer science, involved in the construction of systems capable of performing the tasks traditionally requiring human intelligence, such as learning, reasoning, and decision making. AI as a general term refers to the umbrella aspect whereas ML is a vital component of AI that enables systems to learn based on information and improves as time advances with the absence of explicit programming. In ML, deep learning (DL) has actually asserted itself as a leading figure in its capability to utilize neural networks to examine and examine sophisticated forms of information like pictures, text and audio, frequently surpassing or even surpassing human capabilities (Mian et al., 2024).

There are three general categories of ML techniques: supervised learning (the models learn using labelled data); unsupervised learning (the models discover existing patterns in untitled data); and reinforcement learning (the agents discover the most appropriate actions via trial and error). There are such simple algorithms as SVM, decision trees, naive Bayes classifiers, ANN and more powerful deep learning systems such as convolutional and recurrent neural networks (Górriz et al., 2023). Such practices have been successful in various spheres. AI/ML is applied in healthcare in the fields of diagnostics, imaging, and predictive analytics. It facilitates design optimization and damage detection in structural engineering, and property prediction and the identification of new compounds in materials science. Drug discovery and better clinical trials have also been realized in the pharmaceutical industry (Cao et al., 2024). The recent developments, caused by big data and the power of the computing engine, have accelerated the development in numerous fields, life sciences, and sustainability studies. Nevertheless, it has not yet passed all the hurdles, including explainable AI, ethical concerns, and making sure that AI can be used reliably in the real world (Górriz et al., 2023).

4.2. Data Requirements and Sources in Aquatic Pollution Studies

AI and ML are sweeping across the aquatic pollution research sector, however, the success really depends on the quality, quantity, and diversity of data we possess at our disposal. In order to develop robust models, we require good datasets, which effectively indicate the pollution level like heavy metals, pesticides, and organic contaminants and key water quality parameters such as pH, turbidity, and dissolved oxygen. There is also a strong need to have hydrological data, i.e. precipitations and stream flow, and metadata of where and when samples were collected and the environment conditions at that time (Connors et al., 2019). To make our results reproducible and reliable, we need to make sure that the datasets are standardized, harmonized, and linked to their source. The most significant data sources include the national and international water quality monitoring programs. In the example of the Water Quality Portal (WQP), more than 290 million records are united by 2.7 million locations. Then there is the EnviroTox database which has over 91,000 aquatic toxicity records in 1563 species and 4016 chemicals (Connors et al., 2019). Remote sensing, satellite and high-throughput sequencing, and government or institutional reports are also valuable sources of information to us. There are especially long-term monitoring studies, which are essential to us because they aid in monitoring the trends of pollution and how effective the mitigation measures are, though they do demand regular methodologies and elaborate metadata (Chow et al., 2020).

AI and ML models excel with integrated datasets that include chemical, physical, biological, and land use data to locate the sources, predict the risks and trends. With the help of connecting the data on the quality of water with the data on the industrial, agricultural and weather data, we would be able to build a even better prediction. Nonetheless we encounter such problems as variable formats, space and time gaps between data and lack of sharing data. Addressing these barriers requires continued work on standardization and harmonization and the development of open-access systems to support AI-based research in water pollution (Connors et al., 2019).

4.3. Predictive Modeling and Decision Support Systems

AI and ML have actually revolutionized the world of predictive modeling and decision support systems (DSS). They enable organizations to make knowledgeable, data-based decisions, optimize their functions and even foresee future trends. Predictive modeling uses both past and present information to predict an event and DSS uses the information and combines it with analytical tools to aid in making more intricate decisions. However, recently, modern DSS are becoming more and more merged with predictive and prescriptive analytics, that is, they do not just provide the forecasts; they also offer the recommendations on how to obtain the best results (Ribeiro et al., 2023). Among the most notable innovations in this area, it is possible to mention explainable AI, which fosters trust because it sheds light on the process of making predictions. As an example, the frameworks based on the application of Shapley Values in game theory can be used to elucidate the logic underlying predictions, which increases the confidence of stakeholders (Galanti et al., 2023). Multivariate DSS are used in the healthcare industry, especially in radiation oncology, to forecast treatment outcomes, recurrence and toxicity using clinical, imaging, and molecular data. This practice will promote customized medicine and constantly upgrade models with new information (Lambin et al., 2013).

AI-based DSS are shaking the industry in terms of predictive maintenance, process optimization, and quality control in different industries. These systems can use sensor data, extract features and ML models, such as decision trees and neural networks, to predict failure, optimize maintenance schedules, and reduce costs. Combined with the IoT and cloud-based models, they enhance the scalability and responsiveness in real-time significantly (Rosati et al., 2023). Quantitative models such as simulation, optimization and decision analysis are also used in model driven DSS to assess various situations and inform managers or policymakers. Indicatively, the agent based modelling may play a crucial role in the evaluation of the methods of the public health and the provision of resources (Johnson et al.,

2024). Convergence of both predictive and prescriptive analytics, usually boosted by neural networks, fuzzy logic, and genetic algorithms, is emerging as an effective approach to addressing complex and uncertain decisions in a wide range of areas.

5. INTEGRATION OF AI/ML IN HYDROPHYTE-BASED HEAVY METAL STRESS MANAGEMENT

5.1. Predictive Modeling for Heavy Metal Uptake and Accumulation

The manner of incorporating AI and ML into the process of controlling heavy metal stress in aquatic plants is indeed revolutionizing the field of study. The scientists have been able to predict, monitor and refine the amount of heavy metals these plants absorb and store. The sifting of complicated environmental data such as physicochemical parameters and heavy metals are being performed using tools such as the one available on the WEKA platform to identify trends and optimize phytoremediation plans. This new method also streamlines the operation and enables it to be more scalable and cost-effective in cleaning up contaminants such as Pb, Zn and Ni in our water sources, which are demonstrated on contaminated rivers (Raghavendra & Kumar, 2023). With the help of data mining and AI, scientists are discovering the influence of environmental conditions on plant uptake, which can be used to create more sustainable and intelligent cleanup.

ML algorithms combined with high throughput phenotyping technologies, e.g., visible/near-infrared (V/NIR) spectroscopy and hyperspectral imaging (HSI) allow fast and non-invasive assessments of the plant stress and heavy metal concentration. The methods are able to distinguish the stressful and healthy plants, identify the type and extent of stress, and determine the content of heavy metals correctly. SVM and discriminant analysis, which are examples of ML models, have also demonstrated high potential in the classification of plant stress and the measurement of heavy metal uptake, but their effectiveness can also depend on several variables, such as metal concentration and plant structure (Zhai et al., 2023). As an example, LS-SVM models concentrating on the particular spectral features demonstrated 100 percent accuracy in detecting Hg stressed tobacco plants over controls, which is a phenomenal demonstration of what AI and ML can do in the detection of precise stress (Yu et al., 2021).

In recent times we have been observing some truly exciting advancement in AI aided platforms that can monitor heavy metals in environmental samples in real time and in high throughput. These novel systems employ high-tech enhanced methods, such as deep eutectic solvents, coupled with AI facilitated image analysis. Quick, portable and sensitive detection of such metals as Hg in water, soil and industrial waste can be achieved using this dynamic combination. They are capable of simultaneously analyzing a variety of samples and are able to give real-time feedback, which can enhance environmental monitoring and decision-making (Peng et al., 2025). Beyond that, ML models are rising to the occasion and predict bioaccumulation, and ecological impacts of heavy metals in different environmental and climatic conditions. Considering various species and environmental situations, the models are capable of predicting not only the absorption and accumulation of metals in aquatic flora but also the overall histopathology of aquatic flora. Such knowledge is critical to the development of more informed management and conservation practices (Obadimu et al., 2024).

5.2. Optimization of Plant Species and Growth Conditions

The integration of AI and ML in addressing heavy metal stress with the help of hydrophytes is proving to be a truly promising opportunity in the process of choosing a suitable plant species and optimizing their living conditions to perform a productive phytoremediation. Use the example of such ML models as a random forest, extreme gradient boosting (XGBoost), and multilayer perceptron (MLP). These models have been fairly useful in forecasting and enhancing the uptake and elimination of the heavy metals including cadmium by the aquatic plants such as *Ceratophyllum demersum*. They are able to predict the bioconcentration factors and removal efficiencies in diverse exposure time and concentration of contaminants and this allows the researcher to determine the most suitable plant species and conditions under which maximum remediation efficacy will be achieved (Aasim et al., 2023). On the same note, the response and sensitivity of various genotypes to Cd stress in plant species such as goji berry has been examined by applying ML techniques, which have provided intricate interactions between heavy metal exposure and other growth parameters. This information can be used to select more tolerant and accumulative genotypes, which will further narrow the selection of species to be used in remediation work (Isak et al., 2024). In addition, AI and ML driven systems are being used to optimize the conditions of growth in real-time. As an example, the use of the IoT to monitor nutrient and heavy metal in hydroponic or aquaponic systems can be continuously monitored by using IoT-based monitoring platforms applicable with the implementation of ML algorithms. Such platforms work with models such as linear SVM to identify the highest possible concentrations of metals, including Fe, Cu and Zn, and automatically adjust the dosage of nutrients to achieve the optimal plant growth and remediation performance (Dhal et al., 2023). Also, high throughput phenotyping methods, such as hyperspectral imaging with ML, can be used to carry out rapid, non-invasive measurements of plant stress and metal accumulation to optimise the environmental conditions of different hydrophyte species (Yu et al., 2021).

The AI and ML applications within this area is not only the fastest means of identifying the most successful kinds of plants and genotypes to soak up heavy metals, but also makes it possible to do real-time adjustments to the environmental parameters such as nutrient content, duration of exposure, and concentrations of contaminants. This increases the effectiveness of phytoremediation program. These innovations are also rendering plant-based cleanup processes more focused, scaled and tailored to particular location problems (Isak et al., 2024).

5.3. AI-Driven Monitoring of Water Quality and Plant Health

The application of AI and ML in the control of heavy metal stress in aquatic plants is indeed picking up steam particularly in real time control of water quality and plant health. These AI-driven systems are based on the network of IoT sensors to monitor the necessary environmental variables such as water temperature, pH, and dissolved oxygen, and contaminants. They also keep track of plant specific indicators like moisture content, nutrients, and disease indicators (Ponnuru et al., 2025). These systems have the capability to filter large volumes of sensor and image data using the state of the art algorithms in ML, such as ANN, SVM, decision trees, and random forests. They can identify water pollution, stress, and diseases at an earlier stage and with high precision and efficiency (Rajitha et al., 2024). As an illustration, AI-based water quality monitoring platforms have proven to forecast pollution rates with 95% accuracy, cut the number of mistakes by half relative to the traditional methods and cut the efficiency of real-time monitoring by 40 percent (Durgun, 2024).

AI-powered robots and sensor systems are indeed changing things in the context of plant health monitoring. They offer an invasive-free method to check the performance of plants by detecting the slightest variations in their physiological indicators. Such systems can evaluate the disease or stress in plants using high-level deep learning algorithms, such as convolutional neural networks (Sahoo et al., 2024). Automation of irrigation and detection of pests and diseases and control of nutrients are all possible with these innovative technologies, aiming at enhancing plant growth and increasing heavy metal uptake in hydrophytes-based systems (M. A. H. Ali et al., 2024). With the integration of AI, the IoT, and robotics, it is possible to drive the independence of the decisions made, such as changing irrigation times or turning on the pumps depending on the real-time information, which would improve the efficiency and sustainability of the resources (Sahoo et al., 2024). Although these improvements have enormous potential in managing the environment and sustainable agriculture, issues that have to be addressed include scaling of these systems, data security, and system adaptation to regions with minimal resources (Ponnuru et al., 2025).

5.4. Image-Based Hydrophyte Health Assessment Using Deep Learning

Deep learning has been very useful in changing how we determine the health of hydrophytes, or aquatic plants. It offers rapid, precise and computerized assessments of the health and stress of plants in the water bodies. Convolutional neural networks (CNNs) and other advanced models like ResNet, DenseNet, and InceptionV3 have been found to be very useful in the classification of aquatic vegetation, detection of invasive species, and health status based on different image samples, including aerial RGB, hydroacoustic, and microscope images (M. Ali et al., 2025). As an example, deep learning models that are trained on multi-sensed data can be used to detect and recognize submerged aquatic plants such as Hydrilla verticillata, even in adverse conditions with varying depths, murkiness and lighting, with classification accuracies of more than 80 percent (Perrin et al., 2022). Similarly, CNN-based models have been applied to categorize both microscopic algae species and harmful algal blooms, with customized models reaching an accuracy rate of 96.7 percent, outperforming popular transfer learning models, such as ResNet50 and VGG16 (M. Ali et al., 2025).

The explainable AI methods, such as Grad-CAM, and its application are one of the biggest improvements in this area, as they generate visual heatmaps that highlight which regions of an image are the most influential to the model. This method is a way to not only increase the transparency and develop trust in automated assessments but also help the researchers learn about the visual clues that are linked to the plant health or the stress (Milosavljević et al., 2023). People are finding it easier to use deep learning to classify images by using open-source software like Imagelytics, which enables them to create their own models to perform bio assessment tasks and visualize the decision-making process, especially when detecting aquatic bio indicators (Milosavljević et al., 2023).

Accurate classification can be enhanced by adding contextual metadata, such as geotemporal and hydrographic data and image data. As an illustration, when meta-data was added, accuracy increased to more than 92 percent in a study that sought to classify the images of planktons (Ellen et al., 2019). They are extensible and utilize data augmentation and more complex network architectures and are thus well suited to various aquatic monitoring problems, including measuring the health of coral reefs, and finding invasive or stressed aquatic vegetation (M. Ali et al., 2025).

Table 4. AI and ML Applications in Hydrophyte-Based Heavy Metal Stress Management

AI/ML Technique	Application	Benefits
Machine Learning Models	Prediction of metal uptake efficiency	Improved species selection
Deep Learning (CNN, RNN)	Image-based monitoring of hydrophyte growth	Real-time assessment
Big Data Analytics	Analysis of environmental datasets	Pattern recognition, forecasting
IoT + AI Integration	Remote sensing and water quality monitoring	Adaptive site management

5.5. Case Studies and Successful Implementations

The last few years have seen AI and ML actually grab the headlines in terms of addressing the issue of heavy metal stress in our environmental systems. They demonstrate their efficiency in predicting, monitoring, and fine-tuning remediation procedures in a bunch of case studies. Although much of the work has focused on biochar and soil systems, much of the research can be used in hydrophyte-based (aquatic plant) remediation as well.

A remarkable study involved 20 various AI models such as SVM, random forests (RF), ANN and hybrid ensemble models as well in predicting the extent to which biochar absorbs heavy metals in varied environmental conditions. The SVM-ANN ensemble model was the most precise and offered credible early warnings on pollution and assisted in forming effective remediation strategies (Ke et al., 2021). Similarly, the ensemble ML models such as XGBoost, LightGBM, and Random Forest Regressor have been used to predict the efficiency of metals as an adsorbent such as Pb, Cd, Ni, Cu, and Zn, with XGBoost reporting a high determination coefficient ($R^2 = 0.92$). These models identified very important parameters including the initial metal biochar ratio and pH and can provide beneficial information to comply with optimization of remediation activity (Yaseen & Alhalimi, 2025).

Multilayer perceptron (MLP) neural networks have also been successfully applied in another application to estimate the amounts of the heavy metals in both plants and soil. This technique has greatly reduced the requirement of expensive and time-consuming lab analyses, which has made environmental monitoring faster and more efficient remediation programs (Sari et al., 2023). Also, it has been demonstrated that ML models paired with genetic algorithms can be useful to improve the immobilization of heavy metals in the process of composting. Gradient boosting regression has also been very precise in its ability to predict bioavailability and also identify the important contributing factors such as initial metal levels and the composting time. The ability of machine learning to lower the risks related to waste management is evidenced by impressive immobilization rates of metals like Cu, Zn, Cd, As, and Cr, which have been optimized (Guo et al., 2022).

Heavy metal contamination in the soil has also been predicted and mapped by use of AI and ML. Models that are automated with hyperparameters and random forest and XGBoost have been useful in mapping the concentrations across agricultural regions. Combined with explainable AI methods such as SHAP, these models have managed to reveal the most important environmental influences, which opens the way to specific defensive actions (Song et al., 2023). Support vector regression (SVR) and ANN are other techniques that have been used to predict breakthrough curves of multi metal bioadsorbents, which has given better results in comparison with conventional regression techniques (Aftab et al., 2022). The data regarding the use of hydrophyte based systems is less numerous, however, the success of AI and ML in similar fields confirms that they can be used in the remediation of aquatic plants. These technologies are a ray of hope in the achievement of more efficient, scalable, and sustainable phytoremediation solutions by enabling precise predictions, real time control, and optimized plans (Guo et al., 2022).

6. HYBRID APPROACHES AND MULTI-OMICS INTEGRATION

6.1. AI in Genomic and Transcriptomic Data Analysis of Hydrophytes

The manner in which AI and hybrid multiomics methods are converging is truly transforming the game in the manner in which genomic and transcriptomic data are analyzed in hydrophytes. Although much of the exciting breakthrough is now being emphasized in more of the medical and biomedical sciences, the possibilities are enormous. AI, and in particular DL, is striding with great success in functional genomics which covers genomics, transcriptomics, epigenomics, proteomics, and metabolomics. These technologically advanced approaches can reveal detailed patterns of large, high throughput data sets, and in that way, it becomes far simpler to determine the functions of genes along with regulatory networks and stress response pathways that will be pivotal in comprehending how vegetation endures and accumulates heavy metals (Caudai et al., 2021). With the combination of various layers of omics such as genomics and transcriptomics, we can understand the hydrophyte biology and their responses to environmental stress better.

An important factor to successful AI driven omics analysis is the maximization of data pre-processing. It has been discovered that the processes of normalization and scaling that we choose can greatly influence the results of AI models in transcriptomic data analyzing. Pre-processing pipelines, encompassing several normalization methods (such as TPM and RPKM) and scaling, should be evaluated in a systematic manner to come up with powerful predictive models. It is especially critical in a case of complex datasets as personalized pre-processing may significantly improve disease or stress diagnosis accuracy (Lee et al., 2025). Specifically, AI and ML models are currently developed to overcome the special demands of plant transcriptomics, including high biological variability and technical noise (Ahmed, 2025). Conservation genomics is another area where AI is storming; ML models are used to assess the threat of extinction and apply the data to the management of biodiversity. Such new methods indicate that AI can enhance and speed up the conservation assessment, which could also be adapted to hydrophyte species in the environment monitoring and management (van Oosterhout, 2024). To make things worse, new AI ideas such as Genetic AI apply evolutionary simulations to read through omics data and identify new correlations and patterns without relying on predefined rules or training data (Wissgott, 2025). Nevertheless, several issues require solving, among which are the necessity of user friendly, reproducible AI/ML systems and the challenge of multiomics integration of biological system data to have a comprehensive perspective. The current trend is to create FAIR (Findable, Accessible, Intelligible, and Reproducible) AI solutions to make these tools more accessible and guarantee that the results are sound and can be perceived by any person, not only experts in the sphere of computation (Ahmed, 2025). With the continued development of AI and multi omics integration, their application in hydrophyte genomics and transcriptomics will allow new opportunities to be explored and learn to be more resilient to heavy metal stress, and eventually contribute to not only cleaning up the environment but also conserving it (Caudai et al., 2021).

6.2. Combining Physiological, Environmental, and Remote Sensing Data

Integration of physiological, environmental, and remote sensing data, also known as hybrid or multiomics strategies have become an essential component of current environmental surveillance and ecosystem research. Using physiological data (such as plant health metrics), environmental (temperature, rainfall, and the level of pollution) and remote sensing data (collected by satellites, drones, or ground sensors) allows researchers to create a complex view of ecosystem dynamics and their responses to stress. Multispectral and radar imagery are remote sensing technologies that offered a lot of high resolution information on the land cover, vegetation health, water quality, and surface temperature. It is possible to combine this information with ground based physiological and environmental data in order to analyze it thoroughly (Aakash et al., 2023).

Machine learning and data fusion are important in the integration of these heterogeneous streams of data. The latest developments in ML, including DL and ensemble models, allow smart feature extraction and predictive modelling that is critical in such applications as environmental risk assessment, early warning systems, and real-time decision-making (Chang & Bai, 2018). As an example, the integration of satellite based land surface temperature and PM data with ground based environmental and health data has played a major role in research on the effect of environmental risk factors on the result in the form of public health (Al-Hamdan et al., 2014). Place based curriculum connecting remote sensing and environmental science with local physiological measurements in educational and research environments can help learners to have a more in-depth view of watershed and ecosystem processes. The drought, land cover changes, and water quality can be analyzed through different parameters with the help of this method (Ignatius & Flood, 2021).

The combination of remote sensing and GIS in fact improves our spatial data analysis capacity. It assists us charting and tracking environmental changes of different magnitudes. With further development of sensor technologies and data processing options, the integration of physiological, environmental, and remote sensing data in particular with the help of AI will enhance the precision, scalability, and our general perception of the way we view and manage ecosystems (Aakash et al., 2023).

6.3. Digital Twins for Phytoremediation Systems

Phytoremediation systems are changing fast due to the use of digital twins. They are virtual models of real plants and ecosystems allowing real time observation, modeling as well as fining tuning of remediation processes. Digital twins find applications in both agriculture and environmental management; they combine the riches of multiomics data think genomic, transcriptomic, and proteomic findings and physiological measurements, environmental sensor data and remote sensing imagery. This combination is very rich and assists in creating dynamic and practical models of plant systems. Such models are capable of simulating plant growth, estimating the way plants would respond to stresses of heavy metal, and optimizing other interventions such as irrigation, fertilization and remediation of pollutants strategies (Verdouw et al., 2021).

In this area, 3D functional plant modelling frameworks have been developed which is a significant improvement. These models not only represent the architecture but also the biochemical processes of plants in digital twins, resulting in reduced prediction errors of plant growth, productivity, and stress tolerance. They are also useful in practical applications, including disease forecasting, maximization of pruning, and resource management (Mitsanis et al., 2024). Digital twins have already been useful in agriculture in the field of precision farming, where it is possible to make decisions based on data and improve the sustainability, use of resources, and crop health. It is achieved through combining real time information of soil, plants and environmental sensors with high resolution image (Peladarinos et al., 2023). These systems allow the remote control and simulation of the interventions and give the stakeholders the power to predict the consequences of the changes in the environment or the management action and implement it in the field (Verdouw et al., 2021).

The digital twins cannot be useful without ML and AI, which is the predictive and analytical technology necessary to handle complex data streams involving multiple sources and predict the behavior of systems in different circumstances (Peladarinos et al., 2023). As an illustration, reservoir computing based on digital twins has been used to observe nonlinear ecological processes, forecast potential failures and detect latent variables. This indicates their potential of real-time monitoring of health and adaptive control, especially phytoremediation scenarios (Kong et al., 2023). Nevertheless, all these challenges remain to be addressed, such as the incorporation of various forms of data, its quality, and the intricacies of large scale digital twins (Mitsanis et al., 2024). Nevertheless, in the future, with the further development of sensor technologies, data fusion algorithms, and AI algorithms, digital twins will become an essential part of hybrid, multiomics-based phytoremediation, and more specific, efficient, and sustainable management of polluted environments will be achieved (Peladarinos et al., 2023).

7. CHALLENGES AND LIMITATIONS

7.1. Data Scarcity and Quality Issues

The lack of data and the quality remains a major challenge in the successful application of AI and ML to deal with heavy metal stress using hydrophytes. The efficiency of the predictive models of metal levels, plant health, and remediation outcomes depends on the availability of large, diverse, and well labelled datasets. Unluckily, such datasets can be scarce because of the expensive nature of field sampling and lab studies, which cost a lot, a lot of labor, and time and result in models that do not generalize well in new environments (Sari et al., 2023). Another significant issue is the quality of data. Environmental data are often characterized by invalid values, irregular formats as well as different measurement standards, which may create noise and bias within the models. As an example, to predict heavy metal changes with time, good quality and complete records on the past are important, however, absence of quality records or incomplete ones can negate the accuracy of predictions. Although the pre-processing process, such as normalization, imputation, and data transformation are important, they cannot fully compensate bad data. In this way, the selection of appropriate models and pre-processing methods is a crucial factor to avoid reduced performance and misinterpretation (Bushra et al., 2023).

In order to address data scarcity, scholars are considering the use of synthetic data generation and augmentation. Hydrodynamic modeling is a conventional environmental simulation, which can provide large, trustworthy neural network datasets to train (Aizhulov et al., 2024). Recent methods such as deep generative models or reinforcement learning directed synthesis have potential, although they require extreme caution to prevent the creation of new biases (Chai et al., 2024). It is also important that the data is shared, and standardized. Standard metadata, quality assurance, and monitoring protocols and the use of common platforms can assist the integration of the various datasets and thus facilitate statistical power and robustness of models (Yuan et al., 2024).

7.2. Model Interpretability and Transferability

Transferability and model interpretability are still giant obstacles with regards to the application of AI and ML in solving heavy metal stress in hydrophytes. The word interpretability concerns the way models come to their predictions and it is extremely significant in the creation of confidence, regulatory acceptance, and the practical application of such models. However, most of the best-performing models, such as deep neural networks and ensemble methods, are black box models, and we do not know much about how environmental factors, plant characteristics and resulting phenomena such as metal uptake or stress responses relate (Xu & Sun, 2025). In order to address this problem, such tools as SHapley Additive exPlanations (SHAP) and Local Interpretable Model explanations (LIME) gain popularity as they can also be used to identify the most important features and provide personalized explanations (Sun et al., 2024). As an example, SHAP has been used to elucidate the effects of the concentration of heavy metal on the health of plants, and feature engineering has enhanced the model interpretability and generalizability (Liu et al., 2024).

Another challenge is transferability or the capacity of a model that is trained in a certain setting to be effective in other settings. Due to the difference in environmental and biological conditions, the models used in particular species or areas of the hydrophytes frequently do not fit in the new area. More diverse datasets, features that reflect crucial mechanisms and an independent validation of datasets should be used to enhance transferability (Shen et al., 2024). There are however instances when the performance may still decline despite the presence of these strategies in place. Symbolic regression and transfer learning are proving effective in developing models that provide a compromise between interpretability and flexibility, but have yet to be applied in phytoremediation (Birky et al., 2023).

7.3. Economic and Infrastructural Constraints

There are major challenges that cost and resource constraints offer to the scaling up of AI driven phytoremediation despite its massive potential in enhancing the clean-up of heavy metal polluted regions with the assistance of plants. In order to accomplish this, we must invest enormously on sophisticated sensing technologies such as multispectral, hyperspectral, LiDAR, and laser induced breakdown spectroscopy (LIBS) and imaging and data collection devices (Singh et al., 2023). Although these technologies are quite indispensable in terms of detection accuracy of heavy metals in the targeted area, they are extremely expensive and can be unattainable in resource poor or remote regions where phytoremediation is the most needed (Wang & Aghajani Delavar, 2023). Although the idea of phyto-remediation is typically considered the solution that is cost-efficient when compared to conventional ones, the integration of AI implies high initial expenses in hardware, software, and infrastructure, not to mention the constant financial cost of data storage, management, and model updating. Where land is a costly factor, the time-intensive phytoremediation process coupled with the extra expense of AI might not be as economically-effective as faster, traditional remediation techniques.

Resource constraints are not only in terms of finances, but also in terms of computational and energy needs. The computation of more advanced AI and ML models such as convolutional neural networks typically requires a high performance computing cluster or cloud services (Lu et al., 2025). To be closer to practice, edge computing and portable LIBS devices are being used as a viable alternative, providing the possibility of localized analysis without depending on a constant connection with a server. In order to make these technologies more affordable, researchers are developing lightweight, energy efficient algorithms, which apply pruning, quantization, and dynamic power management techniques (Hameed, 2025).

7.4. Ethical and Environmental Considerations

The issue of using AI and ML to address heavy metal stress in plants is pursuing more and more importance in terms of ethical and environmental considerations. Although such technologies can enhance phytoremediation work significantly, it is important to make them responsible and sustainable. On the moral dimension, we have to struggle with such issues as transparency, fairness, and equal accessibility. Without the means of the richer part of the population to provide the required tools and skills, the state-of-the-art AI may only increase disparities. Many AI models are also black boxes, which is also something that can cause a lack of trust and acceptance, so it is necessary to emphasize on explainable and inclusive decision making, in particular, when the consequences affect the local communities or ecosystems (R. Sharma & Lenaghan, 2022). Another main concern is data governance, and it is important to consider property rights, informed consent, and integrate local or indigenous knowledge.

In environmental terms, hydrophyte based phytoremediation is a green technology because it helps to utilize the inherent capacity of plants to absorb heavy metals without using chemicals and heavy machinery (Nguyen et al., 2021). Nevertheless, there are challenges associated with the introduction of AI. The computational costs may result in high energy usage which may offset some of the sustainability advantages. Moreover, in case AI contributes to the process of the selection of non-native or genetically modified plants, it may become a threat to the biodiversity unless it is approached cautiously (R. Sharma & Lenaghan, 2022). The problem of the waste of polluted plant biomass also exists, giving rise to the risks of the secondary pollution. Reliability of the models is another issue. In case the datasets they are fed on are small or biased, such models may negatively perform in the changing conditions, leading to ineffective or even harmful results (Coimbra & Borges, 2023). Continuous monitoring, validation and adaptation is a must to avoid ecological destruction.

8. FUTURE PERSPECTIVES

8.1. Role of Explainable AI in Ecological Risk Assessment

The concept of explainable AI (XAI) is stirring up the environment of ecological risk assessment by addressing the pressing demand of transparency, interpretability and trust in AI motivated decisions. Although the traditional AI models are admittedly potent, they can be viewed as black boxes and scientists, policymakers, and communities wonder how it is possible to make predictions regarding such hazards as floods, droughts, landslides, or chemical spills. XAI methods believe that the focus on attention, SHapley Additive Explanations (SHAP) and Generalized Additive Models (GAM) are increasingly being incorporated in climate hazard and ecological risk assessment, increasing predictive effectiveness as well as readability (Reddy, 2025). It is these tools that shed some light on the most impactful environmental factors and prepare more specific and efficient interventions.

Fairness and accountability are other issues that XAI addresses on the ethical front. Opening models will facilitate trusting the stakeholders, provide equitable risk estimations to different groups, and aid in the equitable distribution of resources. In addition, XAI systems offer the regulatory needs of offering transparently documented assumptions, evidence, and decision making processes that are vital in the high stakes environmental management (Hartung, 2023). Nonetheless, it does not mean that the obstacles cannot be overcome. The effective use can be complicated by issues such as data scarcity and quality, reduced transferability to other ecosystems, and the requirement of real time interpretability. To address such issues, we should have elaborate frameworks that incorporate technical rigor with ethical and contextual issues. Future studies will probably concentrate on the development of scaled, flexible models of XAI that are suitable in ecological contexts, and studies on the long-term effects of AI-based decisions on society and the environment (Nannini et al., 2025). Concisely, XAI has the potential of uplifting the ecological risk management through transparency, accountability and inclusivity whereby AI-sourced plans are scientifically considered and also socially accountable.

8.2. AI-Enhanced Breeding and Genetic Engineering of Hydrophytes

There is a promising future of environmental resilience, phytoremediation, and climate change adaptation due to AI-enhanced breeding and genetic engineering of hydrophytes. When paired with AI and ML, breeders of crops can delve into the world of large amounts of multiomics data (genomics, transcriptomics, proteomics, and phenomics) to identify desirable traits and accelerate the development of high-quality plant versions that AI is especially talented at finding out how genetic changes will respond to various environmental conditions (Farooq et al., 2024).

Recent discoveries have demonstrated that AI powered platforms are capable of simulating the behavior of genes within their environment and can simulate the growth of plants under stress and come up with novel combinations of genes to increase tolerance and cleaning performance (Kaushik, 2025). State-of-the-art tools such as graph neural networks (GNNs), generative adversarial networks (GANs), and digital twin simulations allow researchers to test the performance of plants on the computer and save time and resources, which are then taken to the field (Cheng & Wang, 2024). This is particularly important to hydrophytes which gestate well in altering aquatic environments and are imperative in the elimination of heavy metals and the reinstatement of ecosystems. Another use of AI to improve precision gene editing is to predict the effects of genetic variants and guide CRISPR/Cas interventions to produce plants that are more able to absorb pollutants, grow and be resilient (Farooq et al., 2024). Nonetheless, success depends on the presence of high-quality and diverse datasets, powerful and interpretable models and collaboration between plant biologists, data scientists, engineers, and policymakers (Harfouche et al., 2019). Other ethical challenges, such as the environmental harmfulness of AI-modified hydrophytes and the equity of access to AI, also need to be addressed. In the future,

AI-enabled adaptive breeding programs can change phyto-remediation, enhance climate change, and enhance food and water security globally.

8.3. Integration with IoT and Autonomous Aquatic Platforms

AI is collaborating with the IoT and autonomous aqueous systems to revolutionize our aquatic ecosystems management, upkeep, and surveillance. This is an adventurous combination that is called the Artificial Intelligence of Things (AIoT) that makes it possible to make real time and data driven decisions and automate lakes, rivers and the oceans (De Araújo et al., 2024). AIoT-enabled sailboats and underwater vehicles equipped with the latest sensors, edge computing, and powerful communication channels can be left to work on its own in harsh conditions and over a long time (Ananthi et al., 2025).

Modular, application neutral designs such as OpenBoat and its prototype F-Boat with commercial components to support autonomous navigation, AI acceleration and flexible environmental monitoring are recent developments. The field tests have demonstrated that they can be depended upon to work in actual ecosystems in the long run (De Araújo et al., 2024). Similarly, the Adaptive Marine Intelligence and Sensing Architecture (AMISA) of autonomous underwater vehicles integrates AI and IoT to monitor in real time, predictive map, and adaptive sensing based on reinforcement learning and continuous updates to the model to stay abreast with the constantly changing environments (Ananthi et al., 2025). Within the aquaculture industry, AIoT can be used to process IoT sensor data on temperature, pH, dissolved oxygen, and fish behavior to optimize the feeding process, disease detection, and regulation of water quality, and improve sustainability despite challenges such as cost, data privacy, and scalability (Y. P. Huang & Khabusi, 2025). Simultaneously, bio-inspired AUVs will use AI to emulate marine animals to navigate better, become more efficient, and reduce environmental disturbances. In the future, modularity, interoperability, edge computing, and adaptive learning will further improve the independence, stability, and sustainability of AIoT systems, transforming the ecosystem monitoring, aquaculture, and resource management across the globe (Mejia et al., 2025).

8.4. Policy and Regulatory Support for AI-Driven Phytoremediation

The future of AI application in the phytoremediation process depends on the availability of good policies and regulations to promote safety, effectiveness, and fairness. Since AI and ML are increasingly involved in scale, optimizing and monitoring of phytoremediation processes, there is a great need to have policies that can maximize the gains and ensure that the risks involved are contained. Advanced sensing, remote monitoring, and AI analytics should be supported by regulatory measures, and new sensors, imaging, and predictive models that make the process of remediation more efficient and accurate should be introduced (Singh et al., 2023). It is also important to develop data quality standards, interoperability standards, privacy standards, and ethical standards of using environmental and biological data. There is also policy support that we require to stimulate research and development in such areas as genetic engineering, nanotechnology, and AI-aided selection of plants and optimization of treatments (Bas et al., 2025). We can accelerate innovation by promoting partnerships between the population and the private sector, developing financing systems, and providing incentives. Also, specific investments in education and workforce training will contribute to the development of the expertise required to conduct the safe and efficient implementation (Singh et al., 2023). The policies also need to incorporate AI operated phytoremediation into overall environmental and agricultural policies, and these efforts would be connected with food safety, carbon capture, and climate adaptation objectives (Kumar Yadav et al., 2018). Simultaneously, the regulatory control should address the possible hazards, including the unexpected ecological effects, introducing new engineered or invasive species, and the controlled disposal of polluted biomass (Singh et al., 2023). Risk assessment, monitoring and adaptive management should have clear guidelines in place to ensure that they protect the ecosystems and the health of the people. Finally, equitable benefits, community engagement, and international cooperation will be the considerations necessary to ensure the sustainability of these solutions (Kumar Yadav et al., 2018).

Table 5. Challenges and Future Prospects of AI-Integrated Hydrophyte Systems

Challenges	Future Prospects
Data heterogeneity and lack of standardized datasets	Development of unified environmental databases
Limited field validation	Integration of AI-driven models in field trials
Algorithmic transparency and ethical issues	Ethically compliant AI models
Scalability issues	Cost-effective, large-scale phytoremediation projects

9. CONCLUSION

The very high level of heavy metal pollution in our water ecosystems is a grave menace to the biodiversity and the services these ecosystems render as well as the human health. Conventional cleaning processes such as chemical precipitation, ion exchange and filtration may be very costly, energy intensive and attract more garbage hence not sustainable. It is in this regard that hydrophyte based phytoremediation is the more eco-friendly response. *Eichhornia crassipes, Lemna minor* and *Typha spp.* are aquatic plants with an exceptional ability to absorb, stabilize and detoxify Cd, As, Pb and Cr to name but a few. Although they are ecologically and economically useful, their success may be different because of the alterations in the environment, the tolerance of particular plants, the interaction of microorganisms, and co-occurring contaminants, and determining the outcome may be a challenging task. On the positive side, AI and ML promise much in solving these problems. They are useful in predictive modelling, real time monitoring and fine tuning remediation plans. RF, SVM, and Neural Network algorithms are able to simulate the way in which plants react to metals and also predict their ability to absorb them under various conditions. DL, including CNN, can be used to track the growth and stress of plants using images, whereas reinforcement learning can be used to develop adaptive management strategies. In combination with IoT devices and remote sensing technology, AI enables continuous assessment of water quality and plant health, and subsequent adjustments made to enhance success in long term remediation. Besides, AI-based approaches have the potential to contribute to the improved risk assessment by combining information on heavy metal types, plant physiology, and ecosystem processes to predict bioaccumulation risks and ecological impacts. It is a proactive method that minimizes uncertainties and would synchronize the phytoremediation process with sustainable environmental management. The limitations we have, such as the variety of data, non-standardized data, and little real world testing are actually impeding the power of our models. Many AI systems work as black boxes, and it is difficult to comprehend them and believe their results. In order to establish reliability and transparency, we must work on developing explainable AI and continuously enhancing our models based on feedback of field trials.

In the future, AI has great potential in single cell genomics and systems biology, particularly in unravelling the mechanisms of heavy metal tolerances in plants. And by adding it to genome editing technologies such as CRISPR, we can increase the efficiency with which these elements are taken up by plants. Climate sensitive models will play a major role in forecasting the resilience of the aquatic plants as our world transforms and implementing circular economy models could turn biomass harvested into a resource like biochar, biofuels, or industrial materials. Blockchain can also contribute to the safe exchange of data, which will complement cooperation in various spheres. The combination of AI and ML and plant based remediation is quite a breakthrough in the ecological engineering. Through the combination of natural resistance of plants and intelligent computation methods, we can work out adaptive, scalable, cost-effective solutions that will unambiguously contribute to global sustainability objectives, such as clean water, climate action, ecosystem conservation. In comparison to the traditional approach, the innovative approach is more proactive and real-time, as it is based on managing environmental health in a more proactive and timely manner, which leads to long-term and short-term advantages.

REFERENCES

Aakash, D., Jaraline Kirubavathy, K., & Bai, V., T. (2023). Environmental Analysis using Remote Sensing Data. *International Conference on Recent Advances in Science and Engineering Technology, ICRASET 2023*, 1–7. DOI: 10.1109/ICRASET59632.2023.10420225

Aasim, M., Ali, S. A., Aydin, S., Bakhsh, A., Sogukpinar, C., Karatas, M., Khawar, K. M., & Aydin, M. E. (2023). Artificial intelligence–based approaches to evaluate and optimize phytoremediation potential of in vitro regenerated aquatic macrophyte Ceratophyllum demersum L. *Environmental Science and Pollution Research International, 30*(14), 40206–40217. DOI: 10.1007/s11356-022-25081-3 PMID: 36607572

Abubakar, M. Y., Ahmad, K. B., Mathew, T. S., Shamsudden, R., Muhammad, H. M., Haladu, M., & Adam, A. B. (2024). Heavy Metal Pollution in Aquatic Ecosystems: A Review of Toxic Impacts and Remediation Strategies. *Kwaghe International Journal of Sciences and Technology, 1*(1), 416–427. DOI: 10.58578/kijst.v1i1.3621

Aftab, R. A., Zaidi, S., Danish, M., Ansari, K. B., & Danish, M. (2022). Novel Machine Learning (ML) models for predicting the performance of multi-metal binding green adsorbent for the removal of Cd (II), Cu (II), Pb (II) and Zn (II) ions. *Environmental Advances, 9*, 100256. Advance online publication. DOI: 10.1016/j.envadv.2022.100256

Ahmed, Z. (2025). Applying AI/ML for Analyzing Gene Expression Patterns. *Methods in Molecular Biology (Clifton, N.J.), 2880*, 319–330. DOI: 10.1007/978-1-0716-4276-4_16 PMID: 39900767

Aizhulov, D., Tungatarova, M., Kurmanseiit, M., & Shayakhmetov, N. (2024). Artificial Neural Networks for Mineral Production Forecasting in the In Situ Leaching Process: Uranium Case Study. *Processes (Basel, Switzerland), 12*(10), 2285. DOI: 10.3390/pr12102285

Al-Hamdan, M. Z., Crosson, W. L., Economou, S. A., Estes, M. G.Jr, Estes, S. M., Hemmings, S. N., Kent, S. T., Puckett, M., Quattrochi, D. A., Rickman, D. L., Wade, G. M., & McClure, L. A. (2014). Environmental public health applications using remotely sensed data. *Geocarto International, 29*(1), 85–98. DOI: 10.1080/10106049.2012.715209 PMID: 24910505

Ali, M., Yaseen, M., Ali, S., & Kim, H. C. (2025). Deep Learning-Based Approach for Microscopic Algae Classification with Grad-CAM Interpretability. *Electronics (Basel), 14*(3), 442. Advance online publication. DOI: 10.3390/electronics14030442

Ali, M. A. H., Moiduddin, K., Nukman, Y., Razak, B. A., Aboudaif, M. K., & Thangaraj, M. (2024). An automated AI-powered IoT algorithm with data processing and noise elimination for plant monitoring and actuating. *PeerJ. Computer Science, 10*, e2448. Advance online publication. DOI: 10.7717/peerj-cs.2448 PMID: 39650371

Ananthi, M., Lakshmana Kumar, R., Muthu, B., & Punitha, P. (2025). Adaptive marine intelligence and sensing architecture for autonomous underwater ecosystem monitoring using AI and IoT integration. *Intelligent Data Analysis: An International Journal*. DOI: 10.1177/1088467X251339271

Ayaz, T., Khan, S., Khan, A. Z., Lei, M., & Alam, M. (2020). Remediation of industrial wastewater using four hydrophyte species: A comparison of individual (pot experiments) and mix plants (constructed wetland). *Journal of Environmental Management, 255*, 109833. DOI: 10.1016/j.jenvman.2019.109833 PMID: 31747629

Azizi, K., Ayoubi, S., Nabiollahi, K., Garosi, Y., & Gislum, R. (2022). Predicting heavy metal contents by applying machine learning approaches and environmental covariates in west of Iran. *Journal of Geochemical Exploration, 233*, 106921. Advance online publication. DOI: 10.1016/j.gexplo.2021.106921

Bas, S., Aasim, M., Gumus, N. E., Katırcı, R., Ali, S. A., & Karatas, M. (2025). Genetic algorithms assisted machine learning algorithms to optimize nano-phytoremediation of cadmium designed by response surface methodology. *International Journal of Phytoremediation, 27*(8), 1057–1065. DOI: 10.1080/15226514.2025.2473598 PMID: 40055883

Birky, D., Garbrecht, K., Emery, J., Alleman, C., Bomarito, G., & Hochhalter, J. (2023). Generalizing the Gurson model using symbolic regression and transfer learning to relax inherent assumptions. *Modelling and Simulation in Materials Science and Engineering, 31*(8), 085005. Advance online publication. DOI: 10.1088/1361-651X/acfe28

Bushra, B., Bazneh, L., Deka, L., Wood, P. J., McGowan, S., & Das, D. B. (2023). Temporal modelling of long-term heavy metal concentrations in aquatic ecosystems. *Journal of Hydroinformatics, 25*(4), 1188–1209. DOI: 10.2166/hydro.2023.151

Cao, Y., Taghvaie Nakhjiri, A., & Ghadiri, M. (2024). Different applications of machine learning approaches in materials science and engineering: Comprehensive review. *Engineering Applications of Artificial Intelligence, 135*, 108783. DOI: 10.1016/j.engappai.2024.108783

Caudai, C., Galizia, A., Geraci, F., Le Pera, L., Morea, V., Salerno, E., Via, A., & Colombo, T. (2021). AI applications in functional genomics. *Computational and Structural Biotechnology Journal, 19*, 5762–5790. DOI: 10.1016/j.csbj.2021.10.009 PMID: 34765093

Chai, C., Jin, K., Tang, N., Fan, J., Qiao, L., Wang, Y., Luo, Y., Yuan, Y., & Wang, G. (2024). Mitigating Data Scarcity in Supervised Machine Learning Through Reinforcement Learning Guided Data Generation. *Proceedings - International Conference on Data Engineering*, 3613–3626. DOI: 10.1109/ICDE60146.2024.00278

Chang, N.-B., & Bai, K. (2018). *Multisensor Data Fusion and Machine Learning for Environmental Remote Sensing*. Multisensor Data Fusion and Machine Learning for Environmental Remote Sensing., DOI: 10.1201/9781315154602

Cheng, Q., & Wang, X. (2024). Machine Learning for AI Breeding in Plants. *Genomics, Proteomics & Bioinformatics, 22*(4), qzae051. Advance online publication. DOI: 10.1093/gpbjnl/qzae051 PMID: 38954837

Chow, R., Scheidegger, R., Doppler, T., Dietzel, A., Fenicia, F., & Stamm, C. (2020). A review of long-term pesticide monitoring studies to assess surface water quality trends. *Water Research X, 9*, 100064. Advance online publication. DOI: 10.1016/j.wroa.2020.100064 PMID: 32995734

Coimbra, E. C. L., & Borges, A. C. (2023). Removing Mn, Cu and Fe from Real Wastewaters with Macrophytes: Reviewing the Relationship between Environmental Factors and Plants' Uptake Capacity. *Toxics, 11*(2), 158. Advance online publication. DOI: 10.3390/toxics11020158 PMID: 36851032

Connors, K. A., Beasley, A., Barron, M. G., Belanger, S. E., Bonnell, M., Brill, J. L., de Zwart, D., Kienzler, A., Krailler, J., Otter, R., Phillips, J. L., & Embry, M. R. (2019). Creation of a Curated Aquatic Toxicology Database: EnviroTox. *Environmental Toxicology and Chemistry, 38*(5), 1062–1073. DOI: 10.1002/etc.4382 PMID: 30714190

De Araújo, A. P. D., Daniel, D. H. J., Guerra, R., Brandão, D. N., Vasconcellos, E. C., Negreiros, A. P. F., Clua, E. W. G., Gonçalves, L. M. G., & Preux, P. (2024). General System Architecture and COTS Prototyping of an AIoT-Enabled Sailboat for Autonomous Aquatic Ecosystem Monitoring. *IEEE Internet of Things Journal, 11*(3), 3801–3811. DOI: 10.1109/JIOT.2023.3324525

Dhal, S. B., Mahanta, S., Gumero, J., O'Sullivan, N., Soetan, M., Louis, J., Gadepally, K. C., Mahanta, S., Lusher, J., & Kalafatis, S. (2023). An IoT-Based Data-Driven Real-Time Monitoring System for Control of Heavy Metals to Ensure Optimal Lettuce Growth in Hydroponic Set-Ups. *Sensors (Basel), 23*(1), 451. Advance online publication. DOI: 10.3390/s23010451 PMID: 36617048

Durgun, Y. (2024). Real-time water quality monitoring using AI-enabled sensors: Detection of contaminants and UV disinfection analysis in smart urban water systems. *Journal of King Saud University. Science, 36*(9), 103409. Advance online publication. DOI: 10.1016/j.jksus.2024.103409

Ellen, J. S., Graff, C. A., & Ohman, M. D. (2019). Improving plankton image classification using context metadata. *Limnology and Oceanography: Methods, 17*(8), 439–461. DOI: 10.1002/lom3.10324

Farid, N., Ullah, A., Khan, S., Butt, S., Khan, A. Z., Afsheen, Z., El-Serehy, H. A., Yasmin, H., Ayaz, T., & Ali, Q. (2023). Algae and Hydrophytes as Potential Plants for Bioremediation of Heavy Metals from Industrial Wastewater. *Water (Basel), 15*(12), 2142. Advance online publication. DOI: 10.3390/w15122142

Farooq, M. A., Gao, S., Hassan, M. A., Huang, Z., Rasheed, A., Hearne, S., Prasanna, B., Li, X., & Li, H. (2024). Artificial intelligence in plant breeding. *Trends in Genetics, 40*(10), 891–908. DOI: 10.1016/j.tig.2024.07.001 PMID: 39117482

Galanti, R., de Leoni, M., Monaro, M., Navarin, N., Marazzi, A., Di Stasi, B., & Maldera, S. (2023). An explainable decision support system for predictive process analytics. *Engineering Applications of Artificial Intelligence, 120*, 105904. DOI: 10.1016/j.engappai.2023.105904

Ghuge, S. A., Nikalje, G. C., Kadam, U. S., Suprasanna, P., & Hong, J. C. (2023). Comprehensive mechanisms of heavy metal toxicity in plants, detoxification, and remediation. *Journal of Hazardous Materials, 450*, 131039. DOI: 10.1016/j.jhazmat.2023.131039 PMID: 36867909

Górriz, J. M., Álvarez-Illán, I., Álvarez-Marquina, A., Arco, J. E., Atzmueller, M., Ballarini, F., Barakova, E., Bologna, G., Bonomini, P., Castellanos-Dominguez, G., Castillo-Barnes, D., Cho, S. B., Contreras, R., Cuadra, J. M., Domínguez, E., Domínguez-Mateos, F., Duro, R. J., Elizondo, D., Fernández-Caballero, A., & Ferrández-Vicente, J. M. (2023). Computational approaches to Explainable Artificial Intelligence: Advances in theory, applications and trends. *Information Fusion, 100*, 101945. DOI: 10.1016/j.inffus.2023.101945

Guo, H., Liu, H., & Wu, S. (2022). Simulation, prediction and optimization of typical heavy metals immobilization in swine manure composting by using machine learning models and genetic algorithm. *Journal of Environmental Management, 323*, 116266. DOI: 10.1016/j.jenvman.2022.116266 PMID: 36137458

Hama Aziz, K. H., Mustafa, F. S., Omer, K. M., Hama, S., Hamarawf, R. F., & Rahman, K. O. (2023). Heavy metal pollution in the aquatic environment: efficient and low-cost removal approaches to eliminate their toxicity: a review. *RSC Advances, 13*(26), 17595–17610. DOI: 10.1039/D3RA00723E PMID: 37312989

Hameed, H. (2025). Developing Cost-Effective AI Algorithms for Resource-Constrained Devices. *International Journal of Computers and Informatics, 4*(1), 151–168. DOI: 10.59992/IJCI.2025.v4n1p3

Harfouche, A. L., Jacobson, D. A., Kainer, D., Romero, J. C., Harfouche, A. H., Scarascia Mugnozza, G., Moshelion, M., Tuskan, G. A., Keurentjes, J. J. B., & Altman, A. (2019). Accelerating Climate Resilient Plant Breeding by Applying Next-Generation Artificial Intelligence. *Trends in Biotechnology, 37*(11), 1217–1235. DOI: 10.1016/j.tibtech.2019.05.007 PMID: 31235329

Hartung, T. (2023). Artificial intelligence as the new frontier in chemical risk assessment. *Frontiers in Artificial Intelligence, 6*, 1269932. Advance online publication. DOI: 10.3389/frai.2023.1269932 PMID: 37915539

Hu, H., Li, X., Wu, S., & Yang, C. (2020). Sustainable livestock wastewater treatment via phytoremediation: Current status and future perspectives. *Bioresource Technology, 315*, 123809. DOI: 10.1016/j.biortech.2020.123809 PMID: 32682262

Huang, S., Song, Q., Li, Q., Zhang, H., Luo, X., & Zheng, Z. (2020). Damage of heavy metals to Vallisneria natans (V. natans) and characterization of microbial community in biofilm. *Aquatic Toxicology (Amsterdam, Netherlands), 225*, 105515. DOI: 10.1016/j.aquatox.2020.105515 PMID: 32516672

Huang, Y. P., & Khabusi, S. P. (2025). Artificial Intelligence of Things (AIoT) Advances in Aquaculture: A Review. *Processes (Basel, Switzerland), 13*(1), 73. Advance online publication. DOI: 10.3390/pr13010073

Ignatius, A. R., & Flood, M. G. (2021). Integration of remote sensing & environmental science in placed-based learning. *International Archives of the Photogrammetry, Remote Sensing and Spatial Information Sciences - ISPRS Archives, 44*(M–3), 73–78. DOI: 10.5194/isprs-archives-XLIV-M-3-2021-73-2021

Isak, M. A., Bozkurt, T., Tütüncü, M., Dönmez, D., İzgü, T., & Şimşek, Ö. (2024). Leveraging machine learning to unravel the impact of cadmium stress on goji berry micropropagation. *PLoS One*, *19*(6), e0305111. Advance online publication. DOI: 10.1371/journal.pone.0305111 PMID: 38870239

Jeong, H., Byeon, E., Kim, D. H., Maszczyk, P., & Lee, J. S. (2023). Heavy metals and metalloid in aquatic invertebrates: A review of single/mixed forms, combination with other pollutants, and environmental factors. *Marine Pollution Bulletin*, *191*, 114959. DOI: 10.1016/j.marpolbul.2023.114959 PMID: 37146547

Johnson, K., Vermeer, W., Hills, H., Chin-Purcell, L., Barnett, J. T., Burns, T., Dean, M. J., & Hendricks Brown, C. (2024). Model-driven decision support: A community-based meta-implementation strategy to predict population impact. *Annals of Epidemiology*, *95*, 12–18. DOI: 10.1016/j.annepidem.2024.05.002 PMID: 38754571

Kaushik, P. (2025). AI-Driven Breeding Enhances Stress Tolerance in High-Elevation Extremophytes: A Proof-of-Concept Study with Cross-Component Validation. *BioRxiv*. DOI: 10.1101/2025.02.21.639605

Ke, B., Nguyen, H., Bui, X. N., Bui, H. B., Choi, Y., Zhou, J., Moayedi, H., Costache, R., & Nguyen-Trang, T. (2021). Predicting the sorption efficiency of heavy metal based on the biochar characteristics, metal sources, and environmental conditions using various novel hybrid machine learning models. *Chemosphere*, *276*, 130204. DOI: 10.1016/j.chemosphere.2021.130204 PMID: 34088091

Kong, L. W., Weng, Y., Glaz, B., Haile, M., & Lai, Y. C. (2023). Reservoir computing as digital twins for nonlinear dynamical systems. *Chaos (Woodbury, N.Y.)*, *33*(3), 33111. DOI: 10.1063/5.0138661 PMID: 37003826

Kumar Yadav, K., Gupta, N., Kumar, A., Reece, L. M., Singh, N., Rezania, S., & Ahmad Khan, S. (2018). Mechanistic understanding and holistic approach of phytoremediation: A review on application and future prospects. *Ecological Engineering*, *120*, 274–298. DOI: 10.1016/j.ecoleng.2018.05.039

Lakshmi, D., Akhil, D., Kartik, A., Gopinath, K. P., Arun, J., Bhatnagar, A., Rinklebe, J., Kim, W., & Muthusamy, G. (2021). Artificial intelligence (AI) applications in adsorption of heavy metals using modified biochar. *The Science of the Total Environment*, *801*, 149623. DOI: 10.1016/j.scitotenv.2021.149623 PMID: 34425447

Lambin, P., Van Stiphout, R. G. P. M., Starmans, M. H. W., Rios-Velazquez, E., Nalbantov, G., Aerts, H. J. W. L., Roelofs, E., Van Elmpt, W., Boutros, P. C., Granone, P., Valentini, V., Begg, A. C., De Ruysscher, D., & Dekker, A. (2013). Predicting outcomes in radiation oncology-multifactorial decision support systems. *Nature Reviews. Clinical Oncology*, *10*(1), 27–40. DOI: 10.1038/nrclinonc.2012.196 PMID: 23165123

Lee, K. W., Min, H. J., Park, H. W., Manavalan, B., & Jeon, Y.-J. (2025). Abstract 2427: Optimization of preprocessing strategies for developing AI-based disease diagnosis model using whole transcriptomic data. *Cancer Research*, *85*(8, Supplement_1), 2427–2427. DOI: 10.1158/1538-7445.AM2025-2427

Liu, J., Li, X., & Zhu, P. (2024). Effects of Various Heavy Metal Exposures on Insulin Resistance in Non-diabetic Populations: Interpretability Analysis from Machine Learning Modeling Perspective. *Biological Trace Element Research*, *202*(12), 5438–5452. DOI: 10.1007/s12011-024-04126-3 PMID: 38409445

Lu, Y., Tao, Z., Nie, L., Guo, X., Pan, T., Chen, R., Li, T., Kong, W., & Liu, F. (2025). Quantitative elemental mapping of heavy metals translocation and accumulation in hyperaccumulator plant using laser-induced breakdown spectroscopy with interpretable deep learning. *Computers and Electronics in Agriculture*, *230*, 109907. DOI: 10.1016/j.compag.2025.109907

Mejia, C., Nye, R., Contreras, J., Nand, A., & Edgerton, C. (2025). Unmanned Offshore Platforms in the Gulf of Mexico with Ai, IOT ScadaReduces Cost and Increase Productivity. *Proceedings of the Annual Offshore Technology Conference*. DOI: 10.4043/35721-MS

Mian, S. M., Khan, M. S., Shawez, M., & Kaur, A. (2024). Artificial Intelligence (AI), Machine Learning (ML) & Deep Learning (DL): A Comprehensive Overview on Techniques, Applications and Research Directions. *2nd International Conference on Sustainable Computing and Smart Systems, ICSCSS 2024 - Proceedings*, 1404–1409. DOI: 10.1109/ICSCSS60660.2024.10625198

Milosavljević, A., Predić, B., & Milošević, D. (2023). Imagelytics: A Deep Learning-Based Image Classification Tool to Support Bioassessment. *Lecture Notes in Networks and Systems, 742 LNNS*, 42–50. DOI: 10.1007/978-3-031-38616-9_5

Mitsanis, C., Hurst, W., & Tekinerdogan, B. (2024). A 3D functional plant modelling framework for agricultural digital twins. *Computers and Electronics in Agriculture*, *218*, 108733. DOI: 10.1016/j.compag.2024.108733

Nannini, L., Huyskes, D., Panai, E., Pistilli, G., & Tartaro, A. (2025). Nullius in Explanans: An ethical risk assessment for explainable AI. *Ethics and Information Technology*, *27*(1), 5. DOI: 10.1007/s10676-024-09800-7

Nguyen, T. Q., Sesin, V., Kisiala, A., & Emery, R. J. N. (2021). Phytohormonal Roles in Plant Responses to Heavy Metal Stress: Implications for Using Macrophytes in Phytoremediation of Aquatic Ecosystems. *Environmental Toxicology and Chemistry*, *40*(1), 7–22. DOI: 10.1002/etc.4909 PMID: 33074580

Obadimu, C. O., Shaibu, S. E., Ekwere, I. O., & Adelagun, R. O. A. (2024). Machine Learning-Based Forecasting of Bioaccumulation and Histopathological Effects in Aquatic Organisms. *Fudma Journal of Sciences*, *8*(6), 485–496. DOI: 10.33003/fjs-2024-0806-3002

Peladarinos, N., Piromalis, D., Cheimaras, V., Tserepas, E., Munteanu, R. A., & Papageorgas, P. (2023). Enhancing Smart Agriculture by Implementing Digital Twins: A Comprehensive Review. *Sensors (Basel)*, *23*(16), 7128. Advance online publication. DOI: 10.3390/s23167128 PMID: 37631663

Peng, Y., Du, K., Yue, H., Li, H., Li, H., Liu, M., Shangguan, S., He, X., Li, X., & Chang, Y. (2025). Integrated deep eutectic system enrichment and AI-assisted high-throughput visual detection for Hg2+ in environmental samples. *Journal of Advanced Research*. Advance online publication. DOI: 10.1016/j.jare.2025.04.011 PMID: 40220898

Perrin, J. E., Jernigan, S. R., Thayer, J. D., Howell, A. W., Leary, J. K., & Buckner, G. D. (2022). Sensor Fusion with Deep Learning for Autonomous Classification and Management of Aquatic Invasive Plant Species. *Robotics (Basel, Switzerland)*, *11*(4), 68. DOI: 10.3390/robotics11040068

Ponnuru, A., Madhuri, J. V., Saravanan, S., Vijayakumar, T., Manimegalai, V., & Das, A. (2025). Data-Driven Approaches to Water Quality Monitoring: Leveraging AI, Machine Learning, and Management Strategies for Environmental Protection. *Journal of Neonatal Surgery*, *14*(5S), 664–675. DOI: 10.52783/jns.v14.2107

Raghavendra, R., & Kumar, N. (2023). Application of WEKA Machine Learning Tools in Phytoremediation for the Removal of Heavy Metals in Water Treatment Processes. *2023 International Conference on Communication, Security and Artificial Intelligence, ICCSAI 2023*, 219–223. DOI: 10.1109/ICCSAI59793.2023.10421571

Rahman, M. A., & Hasegawa, H. (2011). Aquatic arsenic: Phytoremediation using floating macrophytes. *Chemosphere*, *83*(5), 633–646. DOI: 10.1016/j.chemosphere.2011.02.045 PMID: 21435676

Rajitha, A., Aravinda, K., Nagpal, A., Kalra, R., Maan, P., Kumar, A., & Abdul-Zahra, D. S. (2024). Machine Learning and AI-Driven Water Quality Monitoring and Treatment. *E3S Web of Conferences, 505*. DOI: 10.1051/e3sconf/202450503012

Reddy, C. N.Chalamalla Nikhitha Reddy. (2025). Explainable Artificial Intelligence (XAI) for Climate Hazard Assessment: Enhancing Predictive Accuracy and Transparency in Drought, Flood, and Landslide Modeling. *International Journal on Science and Technology*, *16*(1), 1309. Advance online publication. DOI: 10.71097/IJSAT.v16.i1.1309

Ribeiro, R., Pilastri, A., Moura, C., Morgado, J., & Cortez, P. (2023). A data-driven intelligent decision support system that combines predictive and prescriptive analytics for the design of new textile fabrics. *Neural Computing & Applications*, *35*(23), 17375–17395. DOI: 10.1007/s00521-023-08596-9

Rodrigo, M. A. (2021). Wetland restoration with hydrophytes: A review. *Plants*, *10*(6), 1035. Advance online publication. DOI: 10.3390/plants10061035 PMID: 34063930

Rosati, R., Romeo, L., Cecchini, G., Tonetto, F., Viti, P., Mancini, A., & Frontoni, E. (2023). From knowledge-based to big data analytic model: A novel IoT and machine learning based decision support system for predictive maintenance in Industry 4.0. *Journal of Intelligent Manufacturing*, *34*(1), 107–121. DOI: 10.1007/s10845-022-01960-x

Sahoo, A., Sharma, S., Pattnaik, S., Kumar, M. R., & Shrivastava, V. K. (2024). AI-Driven Crop Health Monitoring and Efficient Irrigation Management System. *2024 IEEE 4th International Conference on Applied Electromagnetics, Signal Processing, and Communication, AESPC 2024*, 1–6. DOI: 10.1109/AESPC63931.2024.10872127

Sari, M., Yalcin, I. E., Taner, M., Cosgun, T., & Ozyigit, I. I. (2023). An investigation on environmental pollution due to essential heavy metals: A prediction model through multilayer perceptrons. *International Journal of Phytoremediation*, *25*(1), 89–97. DOI: 10.1080/15226514.2022.2059056 PMID: 35400247

Shahid, M., Dumat, C., Khalid, S., Schreck, E., Xiong, T., & Niazi, N. K. (2017). Foliar heavy metal uptake, toxicity and detoxification in plants: A comparison of foliar and root metal uptake. *Journal of Hazardous Materials*, *325*, 36–58. DOI: 10.1016/j.jhazmat.2016.11.063 PMID: 27915099

Sharma, M., Kant, R., Sharma, A. K., & Sharma, A. K. (2025). Exploring the impact of heavy metals toxicity in the aquatic ecosystem. *International Journal of Energy and Water Resources*, *9*(1), 267–280. DOI: 10.1007/s42108-024-00284-1

Sharma, R., & Lenaghan, S. C. (2022). Duckweed: A potential phytosensor for heavy metals. *Plant Cell Reports*, *41*(12), 2231–2243. DOI: 10.1007/s00299-022-02913-7 PMID: 35980444

Shen, T., Peng, H., Yuan, X., Liang, Y., Liu, S., Wu, Z., Leng, L., & Qin, P. (2024). Feature engineering for improved machine-learning-aided studying heavy metal adsorption on biochar. *Journal of Hazardous Materials*, *466*, 133442. DOI: 10.1016/j.jhazmat.2024.133442 PMID: 38244458

Singh, P., Pani, A., Mujumdar, A. S., & Shirkole, S. S. (2023). New strategies on the application of artificial intelligence in the field of phytoremediation. *International Journal of Phytoremediation*, *25*(4), 505–523. DOI: 10.1080/15226514.2022.2090500 PMID: 35802802

Soliman, M. I., Ibrahim, A. A., Rizk, R. M., & Naser, N. S. (2019). Phytoremediation, Biochemical and Molecular Studies of Some Selected Hydrophytes in Egypt. *Journal of Applied Sciences (Faisalabad)*, *19*(7), 708–717. DOI: 10.3923/jas.2019.708.717

Song, Y., Zhan, D., He, Z., Li, W., Duan, W., Yang, Z., & Lu, M. (2023). HPO-empowered machine learning with multiple environment variables enables spatial prediction of soil heavy metals in coastal delta farmland of China. *Computers and Electronics in Agriculture*, *213*, 108254. DOI: 10.1016/j.compag.2023.108254

Sun, Y., Chen, S., Jiang, H., Qin, B., Li, D., Jia, K., & Wang, C. (2024). Towards interpretable machine learning for observational quantification of soil heavy metal concentrations under environmental constraints. *The Science of the Total Environment*, *926*, 171931. DOI: 10.1016/j.scitotenv.2024.171931 PMID: 38531447

Thakur, T. K., Barya, M. P., Dutta, J., Mukherjee, P., Thakur, A., Swamy, S. L., & Anderson, J. T. (2023). Integrated Phytobial Remediation of Dissolved Pollutants from Domestic Wastewater through Constructed Wetlands: An Interactive Macrophyte-Microbe-Based Green and Low-Cost Decontamination Technology with Prospective Resource Recovery. *Water (Basel)*, *15*(22), 3877. Advance online publication. DOI: 10.3390/w15223877

van Oosterhout, C. (2024). AI-informed conservation genomics. *Heredity*, *132*(1), 1–4. DOI: 10.1038/s41437-023-00666-x PMID: 38151537

Verdouw, C., Tekinerdogan, B., Beulens, A., & Wolfert, S. (2021). Digital twins in smart farming. *Agricultural Systems*, *189*, 103046. Advance online publication. DOI: 10.1016/j.agsy.2020.103046

Wagh, M. S., Sivarajan, S., & Osborne, J. W. (2024). Deciphering the enhanced translocation of Pb, Ni and Cd from artificially polluted soil to Chrysopogon zizanioides augmented with Bacillus xiamenensis VITMSJ3. *3 Biotech, 14*(7), 180. DOI: 10.1007/s13205-024-04001-x

Wang, J., & Aghajani Delavar, M. (2023). Techno-economic analysis of phytoremediation: A strategic rethinking. *The Science of the Total Environment*, *902*, 165949. DOI: 10.1016/j.scitotenv.2023.165949 PMID: 37536595

Wani, K. A., Manzoor, J., Dar, A. A., Shuab, R., & Lone, R. (2020). Heavy Metal Intrusion and Accumulation in Aquatic Ecosystems. *Fresh Water Pollution Dynamics and Remediation*, 83–104. DOI: 10.1007/978-981-13-8277-2_6

Wei, X., Liu, Y., Shen, L., Lu, Z., Ai, Y., & Wang, X. (2024). Machine learning insights in predicting heavy metals interaction with biochar. *Biochar*, *6*(1), 1–11. DOI: 10.1007/s42773-024-00304-7

Wissgott, P. (2025). Genetic AI: Evolutionary Simulation for Data Analysis. *ArXiv, abs/2501.1*. https://doi.org//arXiv.2501.19113DOI: 10.48550

Xie, P., Zahoor, F., Iqbal, S. S., Zahoor, , Ullah, S., Noman, M., Din, Z. U., & Yang, W. (2022). Elimination of toxic heavy metals from industrial polluted water by using hydrophytes. *Journal of Cleaner Production*, *352*, 131358. Advance online publication. DOI: 10.1016/j.jclepro.2022.131358

Xu, S., & Sun, M. (2025). The interpretable machine learning model for depression associated with heavy metals via EMR mining method. *Scientific Reports*, *15*(1), 10811. Advance online publication. DOI: 10.1038/s41598-025-95938-3 PMID: 40155474

Yang, Z., Yang, F., Liu, J. L., Wu, H. T., Yang, H., Shi, Y., Liu, J., Zhang, Y. F., Luo, Y. R., & Chen, K. M. (2022). Heavy metal transporters: Functional mechanisms, regulation, and application in phytoremediation. *The Science of the Total Environment*, *809*, 151099. DOI: 10.1016/j.scitotenv.2021.151099 PMID: 34688763

Yaseen, Z. M., & Alhalimi, F. L. (2025). Heavy metal adsorption efficiency prediction using biochar properties: A comparative analysis for ensemble machine learning models. *Scientific Reports*, *15*(1), 13434. Advance online publication. DOI: 10.1038/s41598-025-96271-5 PMID: 40251173

Yu, K., Fang, S., & Zhao, Y. (2021). Heavy metal Hg stress detection in tobacco plant using hyperspectral sensing and data-driven machine learning methods. *Spectrochimica Acta. Part A: Molecular and Biomolecular Spectroscopy*, *245*, 118917. DOI: 10.1016/j.saa.2020.118917 PMID: 32949945

Yuan, X., Li, J., Lim, J. Y., Zolfaghari, A., Alessi, D. S., Wang, Y., Wang, X., & Ok, Y. S. (2024). Machine Learning for Heavy Metal Removal from Water: Recent Advances and Challenges. *ACS ES&T Water*, *4*(3), 820–836. DOI: 10.1021/acsestwater.3c00215

Zhai, Y., Zhou, L., Qi, H., Gao, P., & Zhang, C. (2023). Application of Visible/Near-Infrared Spectroscopy and Hyperspectral Imaging with Machine Learning for High-Throughput Plant Heavy Metal Stress Phenotyping: A Review. *Plant Phenomics (Washington, D.C.)*, *5*, 0124. Advance online publication. DOI: 10.34133/plantphenomics.0124 PMID: 38239738

Compilation of References

Aakash, D., Jaraline Kirubavathy, K., & Bai, V., T. (2023). Environmental Analysis using Remote Sensing Data. *International Conference on Recent Advances in Science and Engineering Technology, ICRASET 2023*, 1–7. DOI: 10.1109/ICRASET59632.2023.10420225

Aasim, M., Ali, S. A., Aydin, S., Bakhsh, A., Sogukpinar, C., Karatas, M., Khawar, K. M., & Aydin, M. E. (2023). Artificial intelligence–based approaches to evaluate and optimize phytoremediation potential of in vitro regenerated aquatic macrophyte Ceratophyllum demersum L. *Environmental Science and Pollution Research International, 30*(14), 40206–40217. DOI: 10.1007/s11356-022-25081-3 PMID: 36607572

Abbas, S. H., Ismail, I. M., Mostafa, T. M., & Sulaymon, A. H. (2014). Biosorption of heavy metals: A review. *J Chem Sci Technol, 3*, 74–102.

Abbaszadeh-Dahaji, P., Omidvari, M., & Ghorbanpour, M. (2017). Increasing phytoremediation efficiency of heavy metal-contaminated soil using PGPR for sustainable agriculture. In *Plant-microbe interaction: An approach to sustainable agriculture* (pp. 187–204). Springer Singapore.

Abdelaal, M., Mashaly, I. A., Srour, D. S., Dakhil, M. A., El-Liethy, M. A., El-Keblawy, A., El-Barougy, R. F., Halmy, M. W. A., & El-Sherbeny, G. A. (2021). Phytoremediation perspectives of seven aquatic macrophytes for removal of heavy metals from polluted drains in the Nile Delta of Egypt. *Biology (Basel), 10*(6), 560. DOI: 10.3390/biology10060560 PMID: 34203088

Abubakar, M. Y., Ahmad, K. B., Mathew, T. S., Shamsudden, R., Muhammad, H. M., Haladu, M., & Adam, A. B. (2024). Heavy Metal Pollution in Aquatic Ecosystems: A Review of Toxic Impacts and Remediation Strategies. *Kwaghe International Journal of Sciences and Technology, 1*(1), 416–427. DOI: 10.58578/kijst.v1i1.3621

Aftab, R. A., Zaidi, S., Danish, M., Ansari, K. B., & Danish, M. (2022). Novel Machine Learning (ML) models for predicting the performance of multi-metal binding green adsorbent for the removal of Cd (II), Cu (II), Pb (II) and Zn (II) ions. *Environmental Advances*, *9*, 100256. Advance online publication. DOI: 10.1016/j.envadv.2022.100256

Agunbiade, F. O., Olu-Owolabi, B. I., & Adebowale, K. O. (2009). Phytoremediation potential of Eichhorniacrassipes in metal-contaminated coastal water. *Bioresource Technology*, *100*(19), 4521–4526. DOI: 10.1016/j.biortech.2009.04.011 PMID: 19414252

Ahmad, M. (2024). Mapping Green Infrastructure Harnessing OSM Data for Sustainable Development. In *Marketing and Resource Management for Green Transitions in Economies* (pp. 105–125). IGI Global. DOI: 10.4018/979-8-3693-3439-3.ch005

Ahmed, M., Mavukkandy, M. O., Giwa, A., Elektorowicz, M., Katsou, E., Khelifi, O., Naddeo, V., & Hasan, S. W. (2022). Recent developments in hazardous pollutants removal from wastewater and water reuse within a circular economy. *NPJ Clean Water*, *5*(1), 12. Advance online publication. DOI: 10.1038/s41545-022-00154-5

Ahmed, S. F., Mofijur, M., Nuzhat, S., Chowdhury, A. T., Rafa, N., Uddin, M., Inayat, A., Mahlia, T. M. I., Ong, H. C., Chia, W. Y., & Show, P. L. (2021). Recent developments in physical, biological, chemical, and hybrid treatment techniques for removing emerging contaminants from wastewater. *Journal of Hazardous Materials*, *416*, 125912. DOI: 10.1016/j.jhazmat.2021.125912 PMID: 34492846

Ahmed, Z. (2025). Applying AI/ML for Analyzing Gene Expression Patterns. *Methods in Molecular Biology (Clifton, N.J.)*, *2880*, 319–330. DOI: 10.1007/978-1-0716-4276-4_16 PMID: 39900767

Aizhulov, D., Tungatarova, M., Kurmanseiit, M., & Shayakhmetov, N. (2024). Artificial Neural Networks for Mineral Production Forecasting in the In Situ Leaching Process: Uranium Case Study. *Processes (Basel, Switzerland)*, *12*(10), 2285. DOI: 10.3390/pr12102285

Aken, B. V., Correa, P. A., & Schnoor, J. L. (2009). Phytoremediation of polychlorinated biphenyls: New trends and promises. *Environmental Science & Technology*, *43*(20), 7166–7176. PMID: 20384372

Akhtar, F. Z., Archana, K. M., Krishnaswamy, V. G., & Rajagopal, R. (2020). Remediation of heavy metals (Cr, Zn) using physical, chemical and biological methods: A novel approach. *SN Applied Sciences*, *2020*(2), 1–14. DOI: 10.1007/s42452-019-1918-x

Akratos, C. S., & Tsihrintzis, V. A. (2007). Effect of temperature, HRT, vegetation and porous media on removal efficiency of pilot-scale horizontal subsurface flow constructed wetlands. *Ecological Engineering*, *29*(2), 173–191. DOI: 10.1016/j.ecoleng.2006.06.013

Alam, M. Z., Ahmad, S., Malik, A., & Khan, Z. (2021). Phytoremediation potential of *Typha* and *Phragmites* for heavy metal-contaminated wetlands. *Environmental Science and Pollution Research International*, *28*(12), 14527–14539. DOI: 10.1007/s11356-020-11463-1

Al-Baldawi, I. A., Yasin, S. R., Jasim, S. S., Abdullah, S. R. S., Almansoory, A. F., & Ismail, N. I. (2022). Removal of copper by Azolla filiculoides and Lemna minor: Phytoremediation potential, adsorption kinetics and isotherms. *Heliyon*, *8*(11), e11456. DOI: 10.1016/j.heliyon.2022.e11456 PMID: 36406685

Al-Emran, M., & Griffy-Brown, C. (2023). The role of technology adoption in sustainable development: Overview, opportunities, challenges, and future research agendas. *Technology in Society*, *73*, 102240. DOI: 10.1016/j.techsoc.2023.102240

Al-Hamdan, M. Z., Crosson, W. L., Economou, S. A., Estes, M. G.Jr, Estes, S. M., Hemmings, S. N., Kent, S. T., Puckett, M., Quattrochi, D. A., Rickman, D. L., Wade, G. M., & McClure, L. A. (2014). Environmental public health applications using remotely sensed data. *Geocarto International*, *29*(1), 85–98. DOI: 10.1080/10106049.2012.715209 PMID: 24910505

Alharthi, F., Althagafi, H. A., Jafri, I., Oyouni, A. A. A., Althaqafi, M. M., Al-Hijab, L. Y. A., Al-Hazmi, N. E., Elagib, S. M., & Naguib, D. M. (2024). Phytochemical composition and bioactivities of aquatic plants: Antioxidant, antiparasitic, antibacterial, and anticancer properties and mechanisms. *Plants*, *13*(15), 2148. DOI: 10.3390/plants13152148 PMID: 39124266

Ali, H., & Khan, E. (2019). Trophic transfer, bioaccumulation, and biomagnification of non-essential hazardous heavy metals and metalloids in food chains/webs—Concepts and implications for wildlife and human health. *Human and Ecological Risk Assessment*, *25*(6), 1353–1376. DOI: 10.1080/10807039.2018.1469398

Ali, H., Khan, E., & Ilahi, I. (2019). Environmental chemistry and ecotoxicology of hazardous heavy metals: Environmental persistence, toxicity, and bioaccumulation. *Journal of Chemistry*, *2019*, 1–14. DOI: 10.1155/2019/6730305

Ali, H., Khan, E., & Ilahi, I. (2021). Green technologies for sustainable environmental remediation. *Environmental Chemistry Letters*, *19*, 1403–1421.

Ali, H., Khan, E., & Sajad, M. A. (2013). Phytoremediation of heavy metals—Concepts and applications. *Chemosphere, 91*(7), 869–881. DOI: 10.1016/j.chemosphere.2013.01.075 PMID: 23466085

Ali, H., Khan, E., & Sajad, M. A. (2018). Nanotechnology and plant sciences: Nanoparticles and their impact on plant growth. In *Nanotechnology and plant sciences* (pp. 19–35). Springer., DOI: 10.1007/978-3-319-99651-4_2

Alikhani, S., Nummi, P., & Ojala, A. (2021). Urban wetlands: A review on ecological and cultural values. *Water (Basel), 13*(22), 3301. DOI: 10.3390/w13223301

Ali, M. A. H., Moiduddin, K., Nukman, Y., Razak, B. A., Aboudaif, M. K., & Thangaraj, M. (2024). An automated AI-powered IoT algorithm with data processing and noise elimination for plant monitoring and actuating. *PeerJ. Computer Science, 10*, e2448. Advance online publication. DOI: 10.7717/peerj-cs.2448 PMID: 39650371

Ali, M. M., Ali, M. L., Islam, M. S., & Rahman, M. Z. (2020a). Preliminary assessment of heavy metals in water and sediment of Karnaphuli River, Bangladesh. *Environmental Nanotechnology, Monitoring & Management, 14*, 100333. DOI: 10.1016/j.enmm.2020.100333

Ali, M., & Nas, F. S. (2018). The effect of lead on plants in terms of growing and biochemical parameters: A review. *MOJ Ecology & Environmental Sciences, 3*(4). Advance online publication. DOI: 10.15406/mojes.2018.03.00098

Ali, M., Yaseen, M., Ali, S., & Kim, H. C. (2025). Deep Learning-Based Approach for Microscopic Algae Classification with Grad-CAM Interpretability. *Electronics (Basel), 14*(3), 442. Advance online publication. DOI: 10.3390/electronics14030442

Ali, S., Rizwan, M., Qayyum, M. F., Ok, Y. S., Ibrahim, M., Riaz, M., & Arif, M. S. (2020b). Application of floating aquatic plants in phytoremediation of heavy metals polluted water: A review. *Sustainability (Basel), 12*(5), 1927. DOI: 10.3390/su12051927

Alloway, B. J. (2013). *Heavy metals in soils: Trace metals and metalloids in soils and their bioavailability* (3rd ed.). Springer., DOI: 10.1007/978-94-007-4470-7

Alsafran, M., Usman, K., Ahmed, B., Rizwan, M., Saleem, M. H., & Al Jabri, H. (2022). Understanding the Phytoremediation Mechanisms of Potentially Toxic Elements: A Proteomic Overview of Recent Advances. *Frontiers in Plant Science, 13*, 881242. DOI: 10.3389/fpls.2022.881242 PMID: 35646026

Alvarez, L. M., Ruberto, L. A. M., Gurevich, J., & Mac Cormack, W. P. (2020). Environmental factors affecting reproducibility of bioremediation field assays in Antarctica. *Cold Regions Science and Technology, 169*, 102915. DOI: 10.1016/j.coldregions.2019.102915

Alves, A. R. A., Yin, Q., Oliveira, R. S., Silva, E. F., & Novo, L. A. B. (2022). Plant growth-promoting bacteria in phytoremediation of metal-polluted soils: Current knowledge and future directions. *The Science of the Total Environment, 838*, 156435. DOI: 10.1016/j.scitotenv.2022.156435 PMID: 35660615

Alzate Zuluaga, M. Y., Fattorini, R., Cesco, S., & Pii, Y. (2024). Plant-microbe interactions in the rhizosphere for smarter and more sustainable crop fertilization: The case of PGPR-based biofertilizers. *Frontiers in Microbiology, 15*, 1440978. Advance online publication. DOI: 10.3389/fmicb.2024.1440978 PMID: 39176279

Al-Zurfi, S. K. L., Al-Tabatabai, H. H. M., & Abojassim, A. A. (2021). Cadmium metal effects physiological and anatomical variations for aquatic plant Hydrilla verticillata (L. f.) Royle. *Plant Physiology Reports, 26*(3), 457–465. DOI: 10.1007/s40502-021-00600-5

Amlon Group. (2024). Maximize sustainability and ROI through industrial metals recovery. *Amlon Group.* https://amlongroup.com/news/the-circular-economy-maximizing-sustainability-through-valued-metals-recovery/

Ananthi, M., Lakshmana Kumar, R., Muthu, B., & Punitha, P. (2025). Adaptive marine intelligence and sensing architecture for autonomous underwater ecosystem monitoring using AI and IoT integration. *Intelligent Data Analysis: An International Journal.* DOI: 10.1177/1088467X251339271

Angelova, V., Ivanova, R., & Todorov, T. (2004). Use of *Gossypium hirsutum* for phytoextraction of heavy metals from contaminated soils. *Journal of Environmental Protection and Ecology, 5*(4), 886–891.

Anjum, N. A., Hasanuzzaman, M., Hossain, M. A., Thangavel, P., Roychoudhury, A., Gill, S. S., Rodrigo, M. A. M., Adam, V., Fujita, M., Kizek, R., Duarte, A. C., Pereira, E., & Ahmad, I. (2015). Jacks of metal/metalloid chelation trade in plantsâ€"an overview. *Frontiers in Plant Science, 6.* Advance online publication. DOI: 10.3389/fpls.2015.00192 PMID: 25883598

Anna, MK., Vimala, KS., Raiby, PP., Priyalatha, B., & Priya, S. (2023). Phytoremediation potential of a few hydrophytic medicinal plants: A review. International Journal of Scholarly Research in Biology and Pharmacy, 2, 005-009.

Apel, K., & Hirt, H. (2004). Reactive oxygen species: Metabolism, oxidative stress, and signal transduction. *Annual Review of Plant Biology, 55*(1), 373–399. DOI: 10.1146/annurev.arplant.55.031903.141701 PMID: 15377225

Aravind, P., Prasad, M. N. V., Malec, P., Waloszek, A., & Strzałka, K. (2009). Zinc protects Ceratophyllum demersum L.(free-floating hydrophyte) against reactive oxygen species induced by cadmium. *Journal of Trace Elements in Medicine and Biology, 23*(1), 50–60. DOI: 10.1016/j.jtemb.2008.10.002 PMID: 19203717

Arguello, J. M., Raimunda, D., & Gonzalez-Guerrero, M. (2012). Metal transport across biomembranes: Emerging models for a distinct chemistry. *The Journal of Biological Chemistry, 287*(17), 13510–13517. DOI: 10.1074/jbc.R111.319343 PMID: 22389499

Arif, N., Yadav, V., Singh, S., Singh, S., Ahmad, P., Mishra, R. K., Sharma, S., Tripathi, D. K., Dubey, N. K., & Chauhan, D. K. (2016). Influence of High and Low Levels of Plant-Beneficial Heavy Metal Ions on Plant Growth and Development. *Frontiers in Environmental Science, 4*. Advance online publication. DOI: 10.3389/fenvs.2016.00069

Arsalan, M. H., Mubin, O., Al Mahmud, A., Khan, I. A., & Hassan, A. J. (2025, April). Mapping Data-Driven Research Impact Science: The Role of Machine Learning and Artificial Intelligence. In *Metrics* (Vol. 2, No. 2, p. 5). MDPI.

Aryal, M. (2024). Phytoremediation strategies for mitigating environmental toxicants. *Heliyon, 10*(19), e38683. DOI: 10.1016/j.heliyon.2024.e38683 PMID: 39430524

Asiminicesei, D. M., Fertu, D. I., & Gavrilescu, M. (2024). Impact of heavy metal pollution in the environment on the metabolic profile of medicinal plants and their therapeutic potential. *Plants, 13*(6), 913. DOI: 10.3390/plants13060913 PMID: 38592933

Aslanzadeh, M., Saboora, A., & Moradlou, O. (2024). Phytoremediation potential of duckweed (Lemna minor L.) for hexavalent chromium removal in synthetic wastewater: Unveiling physiological response and defense mechanisms against excessive heavy metal uptake. *International Journal of Environmental Science and Technology, 21*(16), 10155–10174. DOI: 10.1007/s13762-024-05721-6

Ayaz, T., Khan, S., Khan, A. Z., Lei, M., & Alam, M. (2020). Remediation of industrial wastewater using four hydrophyte species: A comparison of individual (pot experiments) and mix plants (constructed wetland). *Journal of Environmental Management, 255*, 109833. DOI: 10.1016/j.jenvman.2019.109833 PMID: 31747629

Ayilara, M. S., & Babalola, O. O. (2023). Bioremediation of environmental wastes: The role of microorganisms. *Frontiers in Agronomy*, *5*, 1183691. DOI: 10.3389/fagro.2023.1183691

Azizi, K., Ayoubi, S., Nabiollahi, K., Garosi, Y., & Gislum, R. (2022). Predicting heavy metal contents by applying machine learning approaches and environmental covariates in west of Iran. *Journal of Geochemical Exploration*, *233*, 106921. Advance online publication. DOI: 10.1016/j.gexplo.2021.106921

Aziz, M. (2023). Heavy metal pollution in the aquatic environment: efficient and low-cost removal approaches to eliminate their toxicity: a review. *RSC Advances*, *13*(26), 17595–17610. DOI: 10.1039/D3RA00723E PMID: 37312989

Baba, K.. (2016). Vacuolar sequestration mechanisms of heavy metals in plants. *Plant Science Review*, *5*(2), 101–118.

Bakhshoodeh, R., Alavi, N., Oldham, C., Santos, R. M., Babaei, A. A., Vymazal, J., & Paydary, P. (2020). Constructed wetlands for landfill leachate treatment: A review. *Ecological Engineering*, *146*, 105725. DOI: 10.1016/j.ecoleng.2020.105725

Bakrim, W. B., Ezzariai, A., Karouach, F., Sobeh, M., Kibret, M., Hafidi, M., Kouisni, L., & Yasri, A. (2022). *Eichhornia crassipes* (Mart.) Solms: A comprehensive review of its chemical composition, traditional use, and value-added products. *Frontiers in Pharmacology*, *13*, 842511. DOI: 10.3389/fphar.2022.842511 PMID: 35370709

Bala, S., Garg, D., Thirumalesh, B. V., Sharma, M., Sridhar, K., Inbaraj, B. S., & Tripathi, M. (2022). Recent strategies for bioremediation of emerging pollutants: A review for a green and sustainable environment. *Toxics*, *10*(8), 484. DOI: 10.3390/toxics10080484 PMID: 36006163

Bali, A. S., & Sidhu, G. P. S. (2021). Arsenic acquisition, toxicity and tolerance in plants - From physiology to remediation: A review. *Chemosphere*, *283*, 131050. DOI: 10.1016/j.chemosphere.2021.131050 PMID: 34147983

Balzano, S., Sardo, A., Blasio, M., Chahine, T. B., Dell'Anno, F., Sansone, C., & Brunet, C. (2020). Microalgal Metallothioneins and Phytochelatins and Their Potential Use in Bioremediation. *Frontiers in Microbiology*, *11*, 517. Advance online publication. DOI: 10.3389/fmicb.2020.00517 PMID: 32431671

Banerjee, P., Hazra, A., Ghosh, P., Ganguly, A., Murmu, N. C., & Chatterjee, P. K. (2019). Solid waste management in India: a brief review. In Ghosh, S. (Ed.), *Waste management and resource efficiency*. Springer. DOI: 10.1007/978-981-10-7290-1_86

Banerjee, S., Chatterjee, S., & Gupta, S. K. (2022). IoT-based smart constructed wetlands: A review on remote sensing and automation in wastewater treatment. *Journal of Environmental Management, 317*, 115423.

Barakat, M. A. (2011). New trends in removing heavy metals from industrial wastewater. *Arabian Journal of Chemistry, 4*(4), 361–377. DOI: 10.1016/j.arabjc.2010.07.019

Barathi, S., Lee, J., Venkatesan, R., & Vetcher, A. A. (2023). Current status of biotechnological approaches to enhance the phytoremediation of heavy metals in India—A review. *Plants, 12*(22), 3816. DOI: 10.3390/plants12223816 PMID: 38005713

Barbierato, E., & Gatti, A. (2024). The challenges of machine learning: A critical review. *Electronics (Basel), 13*(2), 416. DOI: 10.3390/electronics13020416

Bar-Peled, L., & Kory, N. (2022). Principles and functions of metabolic compartmentalization. *Nature Metabolism, 4*(10), 1232–1244. DOI: 10.1038/s42255-022-00645-2 PMID: 36266543

Bas, S., Aasim, M., Gumus, N. E., Katırcı, R., Ali, S. A., & Karatas, M. (2025). Genetic algorithms assisted machine learning algorithms to optimize nano-phytoremediation of cadmium designed by response surface methodology. *International Journal of Phytoremediation, 27*(8), 1057–1065. DOI: 10.1080/15226514.2025.2473598 PMID: 40055883

Batty, L. C., & Younger, P. L. (2004). Seasonal variability of iron and manganese in surface water receiving mine drainage from abandoned coal mines. *Environmental Geochemistry and Health, 26*(1), 33–42.

Bauer, LH. (2020). Nature-based solutions: the use of a floating wetland as an alternative to wastewater treatment.

Bayuo, J., Karthik, C., & Bharagava, R. N. (2024). Recent and sustainable advances in phytoremediation of heavy metals from wastewater using aquatic plant species. *The Science of the Total Environment, 925*, 171657. DOI: 10.1016/j.scitotenv.2024.171657 PMID: 39305882

Bayuo, J., Rwiza, M. J., Choi, J. W., Njau, K. N., & Mtei, K. M. (2024). Recent and sustainable advances in phytoremediation of heavy metals from wastewater using aquatic plant species: Green approach. *Journal of Environmental Management, 370*, 122523. DOI: 10.1016/j.jenvman.2024.122523 PMID: 39305882

Beckett, D. C., Aartila, T. P., & Miller, A. C. (1992). Contrasts in Density of Benthic Invertebrates Between Macrophyte Beds and Open Littoral Patches in Eau-Galle Lake, Wisconsin. *American Midland Naturalist, 127*(1), 77–90. DOI: 10.2307/2426324

Beesley, L., Moreno-Jiménez, E., & Gomez-Eyles, J. L. (2011). Effects of biochar and greenwaste compost amendments on mobility, bioavailability, and toxicity of inorganic and organic contaminants in a multi-element polluted soil. *Environmental Pollution, 158*(6), 2282–2287. DOI: 10.1016/j.envpol.2010.02.003 PMID: 20219274

Bharagava, R. N., & Mishra, S. (2018). Hexavalent chromium reduction potential of Cellulosimicrobium sp. isolated from common effluent treatment plant of tannery industries. *Ecotoxicology and Environmental Safety, 147*, 102–109. DOI: 10.1016/j.ecoenv.2017.08.040 PMID: 28841524

Bhat, S. A., Bashir, O., Ul Haq, S. A., Amin, T., Rafiq, A., Ali, M., Américo-Pinheiro, J. H. P., & Sher, F. (2022). Phytoremediation of heavy metals in soil and water: An eco-friendly, sustainable and multidisciplinary approach. *Chemosphere, 303*, 134788. DOI: 10.1016/j.chemosphere.2022.134788 PMID: 35504464

Bhatt, P., Rene, E. R., Kumar, A. J., Kumar, A. J., Zhang, W., & Chen, S. (2020). Binding interaction of allethrin with esterase: Bioremediation potential and mechanism. *Bioresource Technology, 315*, 13845. DOI: 10.1016/j.biortech.2020.123845 PMID: 32707504

Bienert, G. P., Thorsen, M., Schüssler, M. D., Nilsson, H. R., Wagner, A., Tamás, M. J., & Jahn, T. P. (2008). A subgroup of plant aquaporins facilitates the bidirectional diffusion of $As(OH)_3$ and $Sb(OH)_3$ across membranes. *BMC Biology, 6*(1), 26. DOI: 10.1186/1741-7007-6-26 PMID: 18544156

Bini, C., & Bech, J. (2014). PHEs, environment, and human health. In Sanchez, F., & Alvaro, R. (Eds.), *Environmental geochemistry: Site characterization, data analysis and case histories* (pp. 59–88). Elsevier., DOI: 10.1016/B978-0-08-099364-7.00004-3

Birky, D., Garbrecht, K., Emery, J., Alleman, C., Bomarito, G., & Hochhalter, J. (2023). Generalizing the Gurson model using symbolic regression and transfer learning to relax inherent assumptions. *Modelling and Simulation in Materials Science and Engineering, 31*(8), 085005. Advance online publication. DOI: 10.1088/1361-651X/acfe28

Biswal, T. (2025). Rhizofiltration: A Sustainable Green Technology for Remediation of Heavy Metals from Aquatic Systems. In *Recent Advances in Bioremediation and Phytoremediation* (pp. 1–21). Springer Nature Switzerland. DOI: 10.1007/978-3-031-77884-1_1

Blampey, Q., Mulder, K., Gardet, M., Christodoulidis, S., Dutertre, C. A., André, F., Ginhoux, F., & Cournède, P. H. (2024). Sopa: A technology-invariant pipeline for analyses of image-based spatial omics. *Nature Communications, 15*(1), 4981. DOI: 10.1038/s41467-024-48981-z PMID: 38862483

Bolan, N. S., Adriano, D. C., & Naidu, R. (2011). Role of phosphorus in (im)mobilization and bioavailability of heavy metals in the soil-plant system. *Reviews of Environmental Contamination and Toxicology*, *177*, 1–44. PMID: 12666817

Bolan, N., Kunhikrishnan, A., Thangarajan, R., Kumpiene, J., Park, J., Makino, T., Kirkham, M. B., & Scheckel, K. (2014). Remediation of heavy metal(loid)s contaminated soils—To mobilize or to immobilize? *Journal of Hazardous Materials*, *266*, 141–166. DOI: 10.1016/j.jhazmat.2013.12.018 PMID: 24394669

Bonanno, G., & Lo Giudice, R. (2010). Heavy metal bioaccumulation by the organs of *Phragmites australis* (common reed) and their potential use as contamination indicators. *Ecological Indicators*, *10*(3), 639–645. DOI: 10.1016/j.ecolind.2009.11.002

Borisova, G. G., Chukina, N. V., Maleva, M. G., & Levchenko, U. A. (2017). Accumulation of heavy metals in leaves of submerged hydrophytes (Elodea canadensis Michx. and Potamogeton perfoliatus L.) and their responses to the effect of the wastewater of a metallurgical plant. *Inland Water Biology*, *10*(2), 176–181. DOI: 10.1134/S1995082917020031

Bornette, G., & Puijalon, S. (2011). Response of aquatic plants to abiotic factors. *Aquatic Sciences*, *73*(1), 1–14. DOI: 10.1007/s00027-010-0162-7

Bragato, C., Brix, H., & Malagoli, M. (2006). Accumulation of nutrients and heavy metals in *Phragmites australis* and *Bolboschoenus maritimus* in a constructed wetland of the Venice Lagoon watershed. *Ecological Engineering*, *26*(3), 268–285. PMID: 16574288

Brix, H. (1997). Do macrophytes play a role in constructed treatment wetlands? *Water Science and Technology*, *35*(5), 11–17. DOI: 10.2166/wst.1997.0154

Brix, H., Arias, C. A., & Del Bubba, M. (2001). Media selection for sustainable phosphorus removal in subsurface flow constructed wetlands. *Water Science and Technology*, *44*(11–12), 47–54. DOI: 10.2166/wst.2001.0808 PMID: 11804137

Brix, H., Arias, C. A., & Johansen, N. H. (2007). Treatment of agricultural drainage water in a two-stage constructed wetland system. *Water Science and Technology*, *56*(3), 79–85. DOI: 10.2166/wst.2007.512 PMID: 17410843

Brix, H., & Schierup, H. H. (1989). The use of aquatic macrophytes in water pollution control. *Ambio*, *18*(2), 100–107.

Burton, E. D., & Johnston, S. G. (2010). Impact of sediment geochemistry on trace metal bioavailability. *Environmental Chemistry*, *7*(6), 451–464. DOI: 10.1071/EN10075

Bushra, B., Bazneh, L., Deka, L., Wood, P. J., McGowan, S., & Das, D. B. (2023). Temporal modelling of long-term heavy metal concentrations in aquatic ecosystems. *Journal of Hydroinformatics*, *25*(4), 1188–1209. DOI: 10.2166/hydro.2023.151

Buta, E., Torok, A., Zongo, B., Cantor, M., & Majdik, C. (2014). Comparative Studies of the Phytoextraction Capacity of Five Aquatic Plants in Heavy Metal Contaminated Water. *Notulae Botanicae Horti Agrobotanici Cluj-Napoca*, *42*(1), 173–179. DOI: 10.15835/nbha4219341

Calheiros, C. S. C., Rangel, A. O. S. S., & Castro, P. M. L. (2007). Constructed wetland systems vegetated with different plants applied to the treatment of tannery wastewater. *Water Research*, *41*(8), 1790–1798. DOI: 10.1016/j.watres.2007.01.012 PMID: 17320926

Calheiros, C. S. C., Rangel, A. O. S. S., & Castro, P. M. L. (2009). Treatment of industrial wastewater with two-stage constructed wetlands planted with *Typha latifolia* and *Phragmites australis*. *Bioresource Technology*, *100*(13), 3205–3213. DOI: 10.1016/j.biortech.2009.02.017 PMID: 19289277

Cao, Y., Taghvaie Nakhjiri, A., & Ghadiri, M. (2024). Different applications of machine learning approaches in materials science and engineering: Comprehensive review. *Engineering Applications of Artificial Intelligence*, *135*, 108783. DOI: 10.1016/j.engappai.2024.108783

Carpenter, S. R., & Lodge, D. M. (1986). Effects of submerged macrophytes on ecosystem processes. *Aquatic Botany*, *26*, 341 370. DOI: 10.1016/0304-3770(86)90031-8

Caudai, C., Galizia, A., Geraci, F., Le Pera, L., Morea, V., Salerno, E., Via, A., & Colombo, T. (2021). AI applications in functional genomics. *Computational and Structural Biotechnology Journal*, *19*, 5762–5790. DOI: 10.1016/j.csbj.2021.10.009 PMID: 34765093

Cervantes-Avilés, P., Huang, X., & Keller, A. A. (2021). Dissolution and aggregation of metal oxide nanoparticles in root exudates and soil leachate: Implications for nanoagrochemical application. *Environmental Science & Technology*, *55*(20), 13443–13451. DOI: 10.1021/acs.est.1c00767 PMID: 34029070

Chai, C., Jin, K., Tang, N., Fan, J., Qiao, L., Wang, Y., Luo, Y., Yuan, Y., & Wang, G. (2024). Mitigating Data Scarcity in Supervised Machine Learning Through Reinforcement Learning Guided Data Generation. *Proceedings - International Conference on Data Engineering*, 3613–3626. DOI: 10.1109/ICDE60146.2024.00278

Chai, Y., Chen, A., Bai, M., Peng, L., Shao, J., Yuan, J., Shang, C., Zhang, J., Huang, H., & Peng, C. (2022). Valorization of heavy metal contaminated biomass: Recycling and expanding to functional materials. *Journal of Cleaner Production, 366*, 132771. DOI: 10.1016/j.jclepro.2022.132771

Chakraborty, A., Bhakat, K., Islam, E., & Murmu, R. (2025). Arthrobacter sp. mediated chromium remediation and copper accumulation from leached liquor for E-waste management. *The Microbe, 6*, 100277. DOI: 10.1016/j.microb.2025.100277

Chambers, P. A., Lacoul, P., Murphy, K. J., & Thomaz, S. M. (2008). Global diversity of aquatic macrophytes in freshwater. *Hydrobiologia, 595*(1), 9–26. DOI: 10.1007/s10750-007-9154-6

Chandanshive, N. E., Kadam, S. K., Patil, S. M., & Jadhav, J. P. (2017). Integrated phytoremediation of heavy metals and decolorization of textile effluent using *Typha latifolia* in constructed wetland. *Ecotoxicology and Environmental Safety, 139*, 47–53.

Chandel, S., Dar, R. A., Singh, D., Thakur, S., Kaur, R., & Singh, K. (2023). Plant assisted bioremediation of heavy metal polluted soils. In *Bio-Inspired Land Remediation* (pp. 85-114). DOI: 10.1007/978-3-031-04931-6_4

Chang, N.-B., & Bai, K. (2018). *Multisensor Data Fusion and Machine Learning for Environmental Remote Sensing*. Multisensor Data Fusion and Machine Learning for Environmental Remote Sensing., DOI: 10.1201/9781315154602

Chan, M. Y., Tee, C. S., Chai, T. T., Sim, Y. L., & Beh, W. L. (2022). Evaluation of electro-assisted phytoremediation (EAPR) system for heavy metal removal from synthetic leachate using Pistia stratiotes. *International Journal of Phytoremediation, 24*(13), 1376–1384. DOI: 10.1080/15226514.2022.2031863 PMID: 35191343

Chan, Y. C., Brown, P., & Chan, M. (2004). Phytoremediation of metal-contaminated soils. *Environmental Pollution, 133*(3), 393–401. DOI: 10.1016/S0269-7491(01)00286-X PMID: 12009137

Chaudhary, E., & Sharma, P. (2019). Chromium and cadmium removal from wastewater using duckweed-Lemna gibba L. and ultrastructural deformation due to metal toxicity. *International Journal of Phytoremediation, 21*(3), 279–286. DOI: 10.1080/15226514.2018.1522614 PMID: 30698020

Chaudhary, P., Shukla, P., & Sharma, A. K. (2020). Nanotechnology in wastewater management: A new avenue for environmental cleanup. *Environmental Technology & Innovation, 19*, 100830.

Che, J., Kumar, A., Gupta, S., & Khatri, N. (2024). Comparative performance of emergent macrophytes in treating heavy metal-contaminated water. *Environmental Science and Pollution Research International, 31*(2), 2841–2853. DOI: 10.1007/s11356-023-29235-1

Cheng, J. J., & Stomp, A. M. (2009). Growing duckweed to recover nutrients from wastewaters and for production of fuel ethanol and animal feed. *Clean (Weinheim), 37*(1), 17–26. DOI: 10.1002/clen.200800210

Cheng, Q., & Wang, X. (2024). Machine Learning for AI Breeding in Plants. *Genomics, Proteomics & Bioinformatics, 22*(4), qzae051. Advance online publication. DOI: 10.1093/gpbjnl/qzae051 PMID: 38954837

Cheng, S., Grosse, W., Karrenbrock, F., & Thoennessen, M. (2002). Efficiency of constructed wetlands in decontamination of water polluted by heavy metals. *Ecological Engineering, 18*(3), 317–325. DOI: 10.1016/S0925-8574(01)00091-X

Cheng, Z., Wang, C., Tang, F., Zhou, Y., Zhu, C., & Ding, Y. (2025). The cell wall functions in plant heavy metal response. *Ecotoxicology and Environmental Safety, 299*, 118326. DOI: 10.1016/j.ecoenv.2025.118326 PMID: 40403694

Chen, Y., Zhou, Q., Chen, Z., & Xu, J. (2021). Transcriptional responses and detoxification mechanisms of *Nelumbo nucifera* under arsenic stress. *Environmental and Experimental Botany, 186*, 104450.

Cherian, S., & Oliveira, M. M. (2005). Transgenic plants in phytoremediation: Recent advances and new possibilities. *Environmental Science & Technology, 39*(24), 9377–9390. DOI: 10.1021/es051134l PMID: 16475312

Chibuike, G., & Obiora, S. (2014). Heavy metal polluted soils: Effect on plants and bioremediation methods. *Applied and Environmental Soil Science, 2014*, 2014. DOI: 10.1155/2014/752708

Choudhary, D. K., & Varma, A. (Eds.). (2016). *Microbial-mediated induced systemic resistance in plants* (pp. 147–162). Springer. DOI: 10.1007/978-981-10-0388-2

Chow, R., Scheidegger, R., Doppler, T., Dietzel, A., Fenicia, F., & Stamm, C. (2020). A review of long-term pesticide monitoring studies to assess surface water quality trends. *Water Research X, 9*, 100064. Advance online publication. DOI: 10.1016/j.wroa.2020.100064 PMID: 32995734

Cilliers, C. J., Zeller, D. A., & Strydom, G. (2003). Short-term response of aquatic macrophyte communities following the removal of *Eichhornia crassipes* using herbicide and biological control. *African Journal of Aquatic Science, 28*(1), 73–76.

Clemens, S. (2006). Toxic metal accumulation, responses to exposure and mechanisms of tolerance in plants. *Biochimie, 88*(11), 1707–1719. DOI: 10.1016/j.biochi.2006.07.003 PMID: 16914250

Clemens, S., Aarts, M. G. M., Thomine, S., & Verbruggen, N. (2013). Plant science: The key to preventing slow cadmium poisoning. *Trends in Plant Science, 18*(2), 92–99. DOI: 10.1016/j.tplants.2012.08.003 PMID: 22981394

Clijsters, H., & Van Assche, F. (1985). Inhibition of photosynthesis by heavy metals. *Photosynthesis Research, 7*(1), 31–40. DOI: 10.1007/BF00032920 PMID: 24443011

Cobbett, C., & Goldsbrough, P. (2002). Phytochelatins and metallothioneins: Roles in heavy metal detoxification and homeostasis. *Annual Review of Plant Biology, 53*(1), 159–182. DOI: 10.1146/annurev.arplant.53.100301.135154 PMID: 12221971

Coetzee, J. A., Hill, M. P., Julien, M. H., Center, T. D., & Cordo, H. (2009). *Eichhornia crassipes* (Mart.) Solms–Laub. (Pontederiaceae). In Muniappan, R., Reddy, G. V. P., & Raman, A. (Eds.), *Biological control of tropical weeds using arthropods* (pp. 183–210). Cambridge University Press., DOI: 10.1017/CBO9780511576348.011

Cohen-Shacham, E., Walters, G., Janzen, C., & Maginnis, S. (2016). Nature-based Solutions to address global societal challenges. Gland, Switzerland: International Union for Conservation of Nature (IUCN). DOI: 10.2305/IUCN.CH.2016.13.en

Coimbra, E. C. L., & Borges, A. C. (2023). Removing Mn, Cu and Fe from Real Wastewaters with Macrophytes: Reviewing the Relationship between Environmental Factors and Plants' Uptake Capacity. *Toxics, 11*(2), 158. Advance online publication. DOI: 10.3390/toxics11020158 PMID: 36851032

Collin, A.. (2022). Heavy metal interference with seed germination, water uptake, and early seedling growth. *Frontiers in Ecology and Evolution, 11*, 130894. DOI: 10.3389/fevo.2023.130894

Collin, M. S., Venkatraman, S. K., Vijayakumar, N., Kanimozhi, V., Arbaaz, S. M., Stacey, R. S., & Swamiappan, S. (2022). Bioaccumulation of lead (Pb) and its effects on human: A review. *Journal of Hazardous Materials Advances, 7*, 100094. DOI: 10.1016/j.hazadv.2022.100094

Colmer, T. D. (2003). Long-distance transport of gases in plants: A perspective on internal aeration and radial oxygen loss from roots. *Plant, Cell & Environment, 26*(1), 17–36. DOI: 10.1046/j.1365-3040.2003.00846.x

Colmer, T. D., & Flowers, T. J. (2008). Flooding tolerance in halophytes. *The New Phytologist, 179*(4), 964–974. DOI: 10.1111/j.1469-8137.2008.02483.x PMID: 18482227

Connors, K. A., Beasley, A., Barron, M. G., Belanger, S. E., Bonnell, M., Brill, J. L., de Zwart, D., Kienzler, A., Krailler, J., Otter, R., Phillips, J. L., & Embry, M. R. (2019). Creation of a Curated Aquatic Toxicology Database: EnviroTox. *Environmental Toxicology and Chemistry, 38*(5), 1062–1073. DOI: 10.1002/etc.4382 PMID: 30714190

Cooney, C. M. (1996). Sunflowers remove radionuclides from water in ongoing phytoremediation field tests. *Environmental Science & Technology, 30*(5), 194A. DOI: 10.1021/es962219c

Cronk, J. K., & Fennessy, M. S. (2016). *Wetland plants: Biology and ecology* (2nd ed.). CRC Press. DOI: 10.1201/9781420032925

Cubas, P. (2020). Plant Seasonal Growth: How Perennial Plants Sense That Winter Is Coming. *Current Biology, 30*(1), R21–R23. DOI: 10.1016/j.cub.2019.11.044 PMID: 31910371

Cunningham, S. D., & Ow, D. W. (1996). Promises and challenges in the development of phytoremediation. *Plant Physiology, 110*(3), 715–719. DOI: 10.1104/pp.110.3.715 PMID: 12226213

da Silva, D. F., Dias, O. B. S., de Carli Souza, A., Parnian, A., Borgo, L., Melo, L. C. A., & Rabêlo, F. H. S. (2025). Transforming contaminated biomass from phytoremediation into biochar. In *Biochar Ecotechnology for Sustainable Agriculture and Environment* (pp. 131–155). Elsevier., DOI: 10.1016/B978-0-443-29855-4.00006-0

Dalahmeh, S. S., Lalander, C. H., Pell, M., Vinnerås, B., & Jönsson, H. (2014). Floating treatment wetlands for domestic wastewater treatment: Effects of different plants and aeration strategies. *Water Science and Technology, 69*(4), 754–762.

DalCorso, G., Farinati, S., Maistri, S., & Furini, A. (2019). How plants cope with cadmium: Staking all on metabolism and gene expression. *Journal of Integrative Plant Biology, 61*(6), 653–676. DOI: 10.1111/jipb.12738 PMID: 19017114

DalCorso, G., Fasani, E., Manara, A., Visioli, G., & Furini, A. (2019). Heavy metal pollutions: State of the art and innovation in phytoremediation. *International Journal of Molecular Sciences, 20*(14), 3412. DOI: 10.3390/ijms20143412 PMID: 31336773

Davidson, N. C. (2014). How much wetland has the world lost? *Marine and Freshwater Research, 65*(10), 934–941. DOI: 10.1071/MF14173

De Araújo, A. P. D., Daniel, D. H. J., Guerra, R., Brandão, D. N., Vasconcellos, E. C., Negreiros, A. P. F., Clua, E. W. G., Gonçalves, L. M. G., & Preux, P. (2024). General System Architecture and COTS Prototyping of an AIoT-Enabled Sailboat for Autonomous Aquatic Ecosystem Monitoring. *IEEE Internet of Things Journal, 11*(3), 3801–3811. DOI: 10.1109/JIOT.2023.3324525

de Lima, J. P. M., Amaral, M. C. S., & de Lima, S. C. R. B. (2025). Sustainable water management in the mining industry: Paving the way for the future. *Journal of Water Process Engineering, 71*, 107239. DOI: 10.1016/j.jwpe.2025.107239

Degryse, F., Smolders, E., & Parker, D. R. (2009). Partitioning of metals (Cd, Co, Cu, Ni, Pb, Zn) in soils: Concepts, methodologies, prediction and applications. *European Journal of Soil Science, 60*(4), 590–612. DOI: 10.1111/j.1365-2389.2009.01142.x

De, M., Roy, C., Medda, S., Roy, S., & Dey, S. R. (2019). Diverse role of Macrophytes in aquatic ecosystems: A brief review. *International Journal of Experimental Research and Review, 19*, 40–48. DOI: 10.52756/ijerr.2019.v19.005

Deng, G., Li, M., Li, H., Yin, L., & Li, W. (2014). Exposure to cadmium causes declines in growth and photosynthesis in the endangered aquatic fern (Ceratopteris pteridoides). *Aquatic Botany, 112*, 23–32. DOI: 10.1016/j.aquabot.2013.07.003

Deng, H., Ye, Z. H., & Wong, M. H. (2004). Accumulation of lead, zinc, copper and cadmium by 12 wetland plant species thriving in metal-contaminated sites in China. *Environmental Pollution, 132*(1), 29–40. DOI: 10.1016/j.envpol.2004.03.030 PMID: 15276271

Dhal, S. B., Mahanta, S., Gumero, J., O'Sullivan, N., Soetan, M., Louis, J., Gadepally, K. C., Mahanta, S., Lusher, J., & Kalafatis, S. (2023). An IoT-Based Data-Driven Real-Time Monitoring System for Control of Heavy Metals to Ensure Optimal Lettuce Growth in Hydroponic Set-Ups. *Sensors (Basel), 23*(1), 451. Advance online publication. DOI: 10.3390/s23010451 PMID: 36617048

Dhir, B. (2013). *Phytoremediation: Role of aquatic plants in environmental clean-up*. Springer. DOI: 10.1007/978-81-322-1307-9

Dhir, B., Sharmila, P., & Saradhi, P. P. (2004). Hydrophytes lack potential to exhibit cadmium stress induced enhancement in lipid peroxidation and accumulation of proline. *Aquatic Toxicology (Amsterdam, Netherlands), 66*(2), 141–147. DOI: 10.1016/j.aquatox.2003.08.005 PMID: 15036869

Di Stasio, L., Gentile, A., Tangredi, D. N., Piccolo, P., Oliva, G., Vigliotta, G., Cicatelli, A., Guarino, F., Guidi Nissim, W., Labra, M., & Castiglione, S. (2025). Urban Phytoremediation: A Nature-Based Solution for Environmental Reclamation and Sustainability. *Plants, 14*(13), 2057. DOI: 10.3390/plants14132057 PMID: 40648066

Diaconu, M., Pavel, L. V., Hlihor, R. M., Rosca, M., Fertu, D. I., Lenz, M., Corvini, P. X., & Gavrilescu, M. (2020). Characterization of heavy metal toxicity in some plants and microorganisms-a preliminary approach for environmental bioremediation. *New Biotechnology, 56*, 130–139. DOI: 10.1016/j.nbt.2020.01.003 PMID: 31945501

Diaz, L. F., Golueke, C. G., Savage, G. M., & Eggerth, L. L. (2020). *Composting and recycling municipal solid waste*. CRC Press. DOI: 10.4324/9781315150444

Díaz-Torres, R. D. C., Alonso-Castro, A. J., Carrillo-Inungaray, M. L., & Carranza-Alvarez, C. (2021). Bioactive compounds obtained from plants, their pharmacological applications and encapsulation. In *Phytomedicine* (pp. 181–205). Elsevier., DOI: 10.1016/B978-0-12-824109-7.00017-0

Dos Santos, C. M., Vromman, V., & Bastien, C. (2016). Life cycle assessment of constructed wetlands for wastewater treatment: A review. *Environmental Progress & Sustainable Energy, 35*(6), 1680–1692.

Dotro, G., Langergraber, G., Molle, P., Nivala, J., Puigagut, J., Stein, O., & Von Sperling, M. (2017). *Treatment Wetlands. Biological Wastewater Treatment Series*. IWA Publishing., DOI: 10.2166/9781780408774

Doyle, R. D. (2001). Effects of *Vallisneria americana* (L.) on community structure and ecosystem function in a shallow aquatic system. *Hydrobiologia, 448*(1), 139–146.

Du Laing, G., Rinklebe, J., Vandecasteele, B., Meers, E., & Tack, F. M. G. (2009). Trace metal behaviour in estuarine and riverine floodplain soils and sediments: A review. *The Science of the Total Environment, 407*(13), 3972–3985. DOI: 10.1016/j.scitotenv.2008.07.025 PMID: 18786698

Dumanović, J., Nepovimova, E., Natić, M., Kuča, K., & Jaćević, V. (2021). The Significance of Reactive Oxygen Species and Antioxidant Defense System in Plants: A Concise Overview. *Frontiers in Plant Science, 11*, 552969. Advance online publication. DOI: 10.3389/fpls.2020.552969 PMID: 33488637

Durgun, Y. (2024). Real-time water quality monitoring using AI-enabled sensors: Detection of contaminants and UV disinfection analysis in smart urban water systems. *Journal of King Saud University. Science, 36*(9), 103409. Advance online publication. DOI: 10.1016/j.jksus.2024.103409

Dushenkov, V., Kumar, P. B. A. N., Motto, H., & Raskin, I. (1995). Rhizofiltration: The use of plants to remove heavy metals from aqueous streams. *Environmental Science & Technology, 29*(5), 1239–1245. DOI: 10.1021/es00005a015 PMID: 22192017

Ebrahimbabaie, P., Smith, A., Zahran, E. M., & Pichtel, J. (2023). Phytoremediation of Engineered Nanoparticles Using *Typha latifolia* and *Carex rostrata*. Journal of Nanotechnology. *Applied and Environmental Soil Science, 3417525*, 1–16. Advance online publication. DOI: 10.1155/2023/3417525

Eggleton, J., & Thomas, K. V. (2004). A review of factors affecting the release and bioavailability of contaminants during sediment disturbance events. *Environment International, 30*(7), 973–980. DOI: 10.1016/j.envint.2004.03.001 PMID: 15196845

Eid, E. M., Galal, T. M., Sewelam, N. A., Talha, N. I., & Abdallah, S. M. (2020). Phytoremediation of heavy metals by four aquatic macrophytes and their potential use as contamination indicators: A comparative assessment. *Environmental Science and Pollution Research International, 27*(11), 12138–12151. DOI: 10.1007/s11356-020-07839-9 PMID: 31984462

Ekoko, W. A., Yao, Y. L., Shan, Y. Q., Liu, B., & Shabani, I. E. (2021). Vascular plants from the Sanjiang Plain, Northeast China. *Applied Ecology and Environmental Research, 19*(4), 2689–2699. DOI: 10.15666/aeer/1904_26892699

El Awady, F. R., Abbas, M. A., Abdelghany, A. M., El-Amier, Y. A., Darwish, H., & Fakhry, H. (2021). Silver modified hydrophytes for heavy metal removal from different water resources. *Biointerface Research in Applied Chemistry, 11*(2), 14555–14563. DOI: 10.33263/BRIAC116.1455514563

Ellen, J. S., Graff, C. A., & Ohman, M. D. (2019). Improving plankton image classification using context metadata. *Limnology and Oceanography: Methods, 17*(8), 439–461. DOI: 10.1002/lom3.10324

El-Liethy, M. A., Dakhil, M. A., El-Keblawy, A., Abdelaal, M., Halmy, M. W. A., Elgarhy, A. H., Kamika, I., El-Sherbeny, G. A., & Mwaheb, M. A. (2022). Temporal phytoremediation potential for heavy metals and bacterial abundance in drainage water. *Scientific Reports, 12*(1), 8223. DOI: 10.1038/s41598-022-11951-w PMID: 35581245

El-Sappah, A. H., Zhu, Y., Huang, Q., Chen, B., Soaud, S. A., Abd Elhamid, M. A., Yan, K., Li, J., & El-Tarabily, K. A. (2024). Plants' molecular behavior to heavy metals: From criticality to remediation. *Frontiers in Plant Science, 15*, 1423625. DOI: 10.3389/fpls.2024.1423625 PMID: 39280950

Emenike, E. C., Iwuozor, K. O., & Anidiobi, S. U. (2022). Heavy Metal Pollution in Aquaculture: Sources, Impacts and Mitigation Techniques. *Biological Trace Element Research, 200*(10), 4476–4492. DOI: 10.1007/s12011-021-03037-x PMID: 34813030

Erwin, K. L. (2009). Wetlands and global climate change: The role of wetland restoration in a changing world. *Wetlands Ecology and Management, 17*(1), 71–84. DOI: 10.1007/s11273-008-9119-1

European Commission. (2000). *Directive 2000/60/EC of the European Parliament and of the Council establishing a framework for the Community action in the field of water policy.*

FAO. (2020). *State of the World's Land and Water Resources for Food and Agriculture.* Food and Agriculture Organization of the United Nations.

Farchi, A., Laloyaux, P., Bonavita, M., & Bocquet, M. (2021). Using machine learning to correct model error in data assimilation and forecast applications. *Quarterly Journal of the Royal Meteorological Society, 147*(739), 3067–3084. DOI: 10.1002/qj.4116

Farid, N., Ullah, A., Khan, S., Butt, S., Khan, A. Z., Afsheen, Z., El-Serehy, H. A., Yasmin, H., Ayaz, T., & Ali, Q. (2023). Algae and Hydrophytes as Potential Plants for Bioremediation of Heavy Metals from Industrial Wastewater. *Water (Basel), 15*(12), 2142. DOI: 10.3390/w15122142

Faroon, O., Ashizawa, A., Wright, S., ... (2012). *Toxicological Profile for Cadmium.* Atlanta (GA): Agency for Toxic Substances and Disease Registry (US); Chapter: POTENTIAL FOR HUMAN EXPOSURE.

Farooq, M. A., Gao, S., Hassan, M. A., Huang, Z., Rasheed, A., Hearne, S., Prasanna, B., Li, X., & Li, H. (2024). Artificial intelligence in plant breeding. *Trends in Genetics, 40*(10), 891–908. DOI: 10.1016/j.tig.2024.07.001 PMID: 39117482

Favas, P. J. C., Pratas, J., Paul, M. S., Sarkar, S. K., & Prasad, M. N. V. (2016). Phytofiltration of metal (loid)-contaminated water: The potential of native aquatic plants. *Phytoremediation: Management of Environmental Contaminants, 3*, 305–343. DOI: 10.1007/978-3-319-40148-5_11

Finlayson, C. M., Arthington, A. H., & Pittock, J. (2017). Freshwater ecosystems in protected areas: From 'ripples' to 'rivers' of conservation responsibility. *Aquatic Conservation, 27*, 96–113. DOI: 10.1002/aqc.2798

Fisher, J., & Acreman, M. C. (2004). Wetland nutrient removal: A review of the evidence. *Hydrology and Earth System Sciences, 8*(4), 673–685. DOI: 10.5194/hess-8-673-2004

Fletcher, J., Willby, N., Oliver, D. M., & Quilliam, R. S. (2022). Resource recovery and freshwater ecosystem restoration—Prospecting for phytoremediation potential in wild macrophyte stands. *Resources. Environmental Sustainability*, *7*, 100050. Advance online publication. DOI: 10.1016/j.resenv.2022.100050

Fonder, N., & Headley, T. (2013). Systematic classification, structure, and size relationships in constructed wetlands for pollutant removal. *Ecological Engineering*, *53*, 144–158.

Food and Agriculture Organization of the United Nations (FAO). (2017). *The future of food and agriculture – Trends and challenges*. FAO.

Food and Agriculture Organization of the United Nations (FAO). (2020). *Financing sustainable agriculture*. FAO.

Förstner, U., & Salomons, W. (2010). Sediment research, management, and policy. *Journal of Soils and Sediments*, *10*(8), 1440–1452. DOI: 10.1007/s11368-010-0310-7

Förstner, U., & Wittmann, G. T. W. (2012). *Metal pollution in the aquatic environment* (3rd ed.). Springer., DOI: 10.1007/978-3-642-69385-5

Foyer, C. H., & Noctor, G. (2011). Ascorbate and glutathione: The heart of the redox hub. *Plant Physiology*, *155*(1), 2–18. DOI: 10.1104/pp.110.167569 PMID: 21205630

Fu, F., & Wang, Q. (2011). Removal of heavy metal ions from wastewaters: A review. *Journal of Environmental Management*, *92*(3), 407–418. DOI: 10.1016/j.jenvman.2010.11.011 PMID: 21138785

Gadd, G. M. (2000). Bioremedial potential of microbial mechanisms of metal mobilization and immobilization. *Current Opinion in Biotechnology*, *11*(3), 271–279. DOI: 10.1016/S0958-1669(00)00095-1 PMID: 10851150

Gadd, G. M. (2004). Microbial influence on metal mobility and application for bioremediation. *Geoderma*, *122*(2–4), 109–119. DOI: 10.1016/j.geoderma.2004.01.002

Gaiero, J. R., McCall, C. A., Thompson, K. A., Day, N. J., Best, A. S., & Dunfield, K. E. (2013). Inside the root microbiome: Bacterial root endophytes and plant growth promotion. *American Journal of Botany*, *100*(9), 1738–1750. DOI: 10.3732/ajb.1200572 PMID: 23935113

Galanti, R., de Leoni, M., Monaro, M., Navarin, N., Marazzi, A., Di Stasi, B., & Maldera, S. (2023). An explainable decision support system for predictive process analytics. *Engineering Applications of Artificial Intelligence*, *120*, 105904. DOI: 10.1016/j.engappai.2023.105904

Garbisu, C., & Alkorta, I. (2001). Phytoextraction: A cost-effective plant-based technology for the removal of metals from the environment. *Bioresource Technology, 77*(3), 229–236. DOI: 10.1016/S0960-8524(00)00108-5 PMID: 11272009

Gasic, K., & Korban, S. S. (2007). Expression of *Arabidopsis* phytochelatin synthase in Indian mustard (*Brassica juncea*) plants enhances tolerance for and accumulation of cadmium and arsenic. *Plant Molecular Biology, 65*, 519–529. DOI: 10.1007/s11103-007-9213-2

Genchi, G., Sinicropi, M. S., Lauria, G., Carocci, A., & Catalano, A. (2020, May 26). The Effects of Cadmium Toxicity. *International Journal of Environmental Research and Public Health, 17*(11), 3782. DOI: 10.3390/ijerph17113782 PMID: 32466586

Gershenzon, J., & Dudareva, N. (2007). The function of terpene natural products in the natural world. *Nature Chemical Biology, 3*(7), 408–414. DOI: 10.1038/nchembio.2007.5 PMID: 17576428

Gharineiat, Z., Tarsha Kurdi, F., & Campbell, G. (2022). Review of automatic processing of topography and surface feature identification LiDAR data using machine learning techniques. *Remote Sensing (Basel), 14*(19), 4685. DOI: 10.3390/rs14194685

Ghosh, M., & Singh, S. P. (2005). A review on phytoremediation of heavy metals and utilization of it's by products. *Asian Journal of Energy and Environment, 6*(4), 18.

Ghosh, M., & Singh, S. P. (2005). A review on phytoremediation of heavy metals and utilization of its byproducts. *Applied Ecology and Environmental Research, 3*(1), 1–18. DOI: 10.15666/aeer/0301_001018

Ghuge, S. A., Nikalje, G. C., Kadam, U. S., Suprasanna, P., & Hong, J. C. (2023). Comprehensive mechanisms of heavy metal toxicity in plants, detoxification, and remediation. *Journal of Hazardous Materials, 450*, 131039. DOI: 10.1016/j.jhazmat.2023.131039 PMID: 36867909

Gilmour, C. C., Henry, E. A., & Mitchell, R. (1992). Sulfate stimulation of mercury methylation in freshwater sediments. *Environmental Science & Technology, 26*(11), 2281–2287. DOI: 10.1021/es00035a029

Glick, B. R. (2010). Using soil bacteria to facilitate phytoremediation. *Biotechnology Advances, 28*(3), 367–374. DOI: 10.1016/j.biotechadv.2010.02.001 PMID: 20149857

Goddijn-Murphy, L., Martínez-Vicente, V., Dierssen, H. M., Raimondi, V., Gandini, E., Foster, R., & Chirayath, V. (2024). Emerging technologies for remote sensing of floating and submerged plastic litter. *Remote Sensing (Basel), 16*(10), 1770. DOI: 10.3390/rs16101770

Goncharuk, E. A., & Zagoskina, N. V. (2023). Heavy metals, their phytotoxicity, and the role of phenolic antioxidants in plant stress responses with focus on cadmium. *Molecules (Basel, Switzerland)*, *28*(9), 3921. DOI: 10.3390/molecules28093921 PMID: 37175331

Gong, Y., Zhao, D., & Wang, Q. (2018). An overview of field-scale studies on remediation of soil contaminated with heavy metals and metalloids: Technical progress over the last decade. *Water Research*, *147*, 440–460. DOI: 10.1016/j.watres.2018.10.024 PMID: 30343201

Gonzalez Mendoza, D., Moreno Sanchez, R., & Santamaria, J. M. (2007). Metallothionein induction in aquatic plants under Cd/Cu stress. *Environmental Botany*, *43*(4), 415–422. DOI: 10.1016/j.envexpbot.2006.12.005

Gopal, B. (1987). *Water hyacinth*. Elsevier.

Górriz, J. M., Álvarez-Illán, I., Álvarez-Marquina, A., Arco, J. E., Atzmueller, M., Ballarini, F., Barakova, E., Bologna, G., Bonomini, P., Castellanos-Dominguez, G., Castillo-Barnes, D., Cho, S. B., Contreras, R., Cuadra, J. M., Domínguez, E., Domínguez-Mateos, F., Duro, R. J., Elizondo, D., Fernández-Caballero, A., & Ferrández-Vicente, J. M. (2023). Computational approaches to Explainable Artificial Intelligence: Advances in theory, applications and trends. *Information Fusion*, *100*, 101945. DOI: 10.1016/j.inffus.2023.101945

Grifoni, M., Franchi, E., Fusini, D., Vocciante, M., Barbafieri, M., Pedron, F., Rosellini, I., & Petruzzelli, G. (2022). Soil remediation: Towards a resilient and adaptive approach to deal with the ever-changing environmental challenges. *Environments (Basel, Switzerland)*, *9*(2), 18. DOI: 10.3390/environments9020018

Grill, E., Winnacker, E. L., & Zenk, M. H. (1985). Phytochelatins: The principal heavy-metal complexing peptides of higher plants. *Science*, *230*(4726), 674–676. DOI: 10.1126/science.230.4726.674 PMID: 17797291

Guerinot, M. L. (2000). The ZIP family of metal transporters. *Biochimica et Biophysica Acta*, *1465*(1–2), 190–198. DOI: 10.1016/S0005-2736(00)00138-3 PMID: 10748254

Gunatilake, S. K. (2015). Methods of removing heavy metals from industrial wastewater. *Methods (San Diego, Calif.)*, *1*, 14.

Guo, H., Liu, H., & Wu, S. (2022). Simulation, prediction and optimization of typical heavy metals immobilization in swine manure composting by using machine learning models and genetic algorithm. *Journal of Environmental Management*, *323*, 116266. DOI: 10.1016/j.jenvman.2022.116266 PMID: 36137458

Gustin, J. L., Loureiro, M. E., Kim, D., Na, G., Tikhonova, M., & Salt, D. E. (2009). MTP1-dependent Zn sequestration into shoot vacuoles suggests dual roles in Zn tolerance and accumulation in *Arabidopsis thaliana*. *The Plant Journal*, *57*(6), 1116–1127. DOI: 10.1111/j.1365-313X.2008.03754.x PMID: 19054361

Haferburg, G., & Kothe, E. (2007). Microbes and metals: Interactions in the environment. *Journal of Basic Microbiology*, *47*(6), 453–467. DOI: 10.1002/jobm.200700275 PMID: 18072246

Haghshenas, S. M. (2020). An exploratory view of water quality and sediment characteristics in Musa Estuary, the Persian Gulf. *arXiv*. https://doi.org//arXiv.2004.12548DOI: 10.48550

Haghshenas, S. S. (2020). Atmospheric deposition of heavy metals in aquatic ecosystems: Sources, processes, and consequences. *Environmental Monitoring and Assessment*, *192*(8), 517. DOI: 10.1007/s10661-020-08525-w PMID: 32666386

Hall, J. L. (2002). Cellular mechanisms for heavy metal detoxification and tolerance. *Journal of Experimental Botany*, *53*(366), 1–11. DOI: 10.1093/jexbot/53.366.1 PMID: 11741035

Hameed, H. (2025). Developing Cost-Effective AI Algorithms for Resource-Constrained Devices. *International Journal of Computers and Informatics*, *4*(1), 151–168. DOI: 10.59992/IJCI.2025.v4n1p3

Ha, N. T. T.. (2019). Heavy metal accumulation in *Nelumbo nucifera* from contaminated water bodies. *Environmental Monitoring and Assessment*, *191*(2), 1–10. DOI: 10.1007/s10661-019-7226-4

Hanafiah, M. M., Zainuddin, M. F., Mohd Nizam, N. U., Halim, A. A., & Rasool, A. (2020). Phytoremediation of Aluminum and Iron from Industrial Wastewater Using *Ipomoea aquatica* and *Centella asiatica. Applied Sciences (Basel, Switzerland)*, *10*(9), 3064. DOI: 10.3390/app10093064

Hanafy, R. R., Eweda, W. E., Zayed, M. S., & Khalil, H. (2018). Potentiality of using a. pinnata to bioremediate different heavy metals from polluted draining water. *Magallat Itihad al-Gami'at al-'Arabiyyat Lil-Dirasat Wa-al-Buhut al-Zira'iyyat*, *26*(1), 359–372. DOI: 10.21608/ajs.2018.14022

Han, H., Wu, X., Yao, L., & Chen, Z. (2020). Heavy metal-immobilizing bacteria combined with calcium polypeptides reduced the uptake of Cd in wheat and shifted the rhizosphere bacterial communities. *Environmental Pollution*, *267*, 115432. DOI: 10.1016/j.envpol.2020.115432 PMID: 32841909

Haque, S. E., & Gazi-Khan, L. (2025). Limitations and challenges of bioremediation approach: alternative solutions. In *Biotechnologies for Wastewater Treatment and Resource Recovery* (pp. 147–157). Elsevier., DOI: 10.1016/B978-0-443-27376-6.00023-2

Harfouche, A. L., Jacobson, D. A., Kainer, D., Romero, J. C., Harfouche, A. H., Scarascia Mugnozza, G., Moshelion, M., Tuskan, G. A., Keurentjes, J. J. B., & Altman, A. (2019). Accelerating Climate Resilient Plant Breeding by Applying Next-Generation Artificial Intelligence. *Trends in Biotechnology, 37*(11), 1217–1235. DOI: 10.1016/j.tibtech.2019.05.007 PMID: 31235329

Hartung, T. (2023). Artificial intelligence as the new frontier in chemical risk assessment. *Frontiers in Artificial Intelligence, 6*, 1269932. Advance online publication. DOI: 10.3389/frai.2023.1269932 PMID: 37915539

Hasan, Md. M., Uddin, Md. N., Ara-Sharmeen, I., F. Alharby, H., Alzahrani, Y., Hakeem, K. R., & Zhang, L. (. (2019). Assisting Phytoremediation of Heavy Metals Using Chemical Amendments. *Plants, 8*(9), 295. DOI: 10.3390/plants8090295 PMID: 31438507

Hasan, S. A., Fariduddin, Q., Ali, B., Hayat, S., & Ahmad, A. (2009). Cadmium: Toxicity and tolerance in plants. *Journal of Environmental Biology, 30*(2), 165–174. PMID: 20121012

Haslam, E. (1996). Natural polyphenols (vegetable tannins) as drugs: Possible modes of action. *Journal of Natural Products, 59*(2), 205–215. DOI: 10.1021/np960040+ PMID: 8991956

Hassinen, V. H., Tervahauta, A. I., Schat, H., & Kärenlampi, S. O. (2009). Plant metallothioneins – Metal chelators with ROS scavenging activity? *Plant Biology, 11*(4), 418–426. DOI: 10.1111/j.1438-8677.2008.00158.x

Heisi, H. D., Awosusi, A. A., Nkuna, R., & Matambo, T. S. (2023). Phytoextraction of anthropogenic heavy metal contamination of the Blesbokspruit wetland: Potential of wetland macrophytes. *Journal of Contaminant Hydrology, 253*, 104101. Advance online publication. DOI: 10.1016/j.jconhyd.2022.104101 PMID: 36379730

Heliyon. (2024). Phytoremediation potential of *Pistia stratiotes, Eichhornia crassipes*, and *Typha latifolia* for chromium removal from tannery wastewater. *Heliyon, 10*(2), e13578. https://doi.org/DOI: 10.1016/j.heliyon.2024.e13578

He, M., Yin, H., Liu, Z., Luo, F., Zhao, X., Li, H., & Song, X. (2024). Root exudate–assisted phytoremediation of copper and lead contamination using Rumex acetosa L. and Rumex K-1. *Ecotoxicology and Environmental Safety*, *284*, 117036. DOI: 10.1016/j.ecoenv.2024.117036 PMID: 39270415

Hisam, N. I. B., Zakaria, M. Z., Azid, A., Bakar, M. F. A., & Samsudin, M. S. (2022). Phytoremediation process of water spinach (Ipomoea aquatica) in absorbing heavy metal concentration in wastewater. *Journal Of Agrobiotechnology*, *13*(1S), 131–144. DOI: 10.37231/jab.2022.13.1S.322

Horrigan, L., Lawrence, R. S., & Walker, P. (2002). How sustainable agriculture can address the environmental and human health harms of industrial agriculture. *Environmental Health Perspectives*, *110*(5), 445–456. DOI: 10.1289/ehp.02110445 PMID: 12003747

Hou, D., Lin, Z., Wang, R., Ge, J., Wei, S., Xie, R., Wang, H., Wang, K., Hu, Y., Yang, X., Lu, L., & Tian, S. (2018). Cadmium Exposure-Sedum alfredii Planting Interactions Shape the Bacterial Community in the Hyperaccumulator Plant Rhizosphere. *Applied and Environmental Microbiology*, *84*(12), e02797-17. Advance online publication. DOI: 10.1128/AEM.02797-17 PMID: 29654182

Houwing, E. J., Tánczos, I. C., Kroon, A., & de Vries, M. B. (2002). Interaction of Submerged Vegetation, Hydrodynamics and Turbidity: Analysis of Field and Laboratory Studies. In Winterwerp, J. C., & Kranenburg, C. (Eds.), *Proceedings in Marine Science* (Vol. 5, pp. 441–453). Elsevier.

Huang, Q., Zheng, X., Pan, Z., He, J., Ma, J., Wang, J., Chen, B., & Chu, C. (2025). A synthetic hydrophyte for sunlight-driven water purification. *Environmental Science & Technology*, *59*(28), 14597–14604. Advance online publication. DOI: 10.1021/acs.est.5c00765 PMID: 40631453

Huang, S., Song, Q., Li, Q., Zhang, H., Luo, X., & Zheng, Z. (2020). Damage of heavy metals to Vallisneria natans (V. natans) and characterization of microbial community in biofilm. *Aquatic Toxicology (Amsterdam, Netherlands)*, *225*, 105515. DOI: 10.1016/j.aquatox.2020.105515 PMID: 32516672

Huang, Y. P., & Khabusi, S. P. (2025). Artificial Intelligence of Things (AIoT) Advances in Aquaculture: A Review. *Processes (Basel, Switzerland)*, *13*(1), 73. Advance online publication. DOI: 10.3390/pr13010073

Huang, Y., Pan, H., Wang, Q., Ge, Y., Liu, W., & Christie, P. (2019). Enrichment of the soil microbial community in the bioremediation of a petroleum-contaminated soil amended with rice straw or sawdust. *Chemosphere*, *224*, 265–271. DOI: 10.1016/j.chemosphere.2019.02.148 PMID: 30825852

Hudson-Edwards, K. A. (2016). Tackling mine wastes. *Science, 352*(6283), 288–290. DOI: 10.1126/science.aaf3354 PMID: 27081053

Hu, H., Li, X., Wu, S., & Yang, C. (2020). Sustainable livestock wastewater treatment via phytoremediation: Current status and future perspectives. *Bioresource Technology, 315*, 123809. DOI: 10.1016/j.biortech.2020.123809 PMID: 32682262

Hu, H., Zhou, Q., Li, X., Lou, W., Du, C., Teng, Q., Zhang, D. M., Liu, H. Y., Zhong, Y. Y., & Yang, C. P. (2019). Phytoremediation of anaerobically digested swine wastewater contaminated by oxytetracycline via *Lemna aequinoctialis*: Nutrient removal, growth characteristics and degradation pathways. *Bioresource Technology, 291*, 121853. DOI: 10.1016/j.biortech.2019.121853 PMID: 31377510

Huser, B. J., Futter, M. N., Bärlund, I., & Ecke, F. (2021). Managing sediment and nutrient interactions with metals in aquatic ecosystems. *The Science of the Total Environment, 757*, 143860. DOI: 10.1016/j.scitotenv.2020.143860

Hussain, D., Haydon, M. J., Wang, Y., Wong, E., Sherson, S. M., Young, J., Camakaris, J., Harper, J. F., & Cobbett, C. S. (2004). P-type ATPase heavy metal transporters with roles in essential zinc homeostasis in *Arabidopsis. The Plant Cell, 16*(5), 1327–1339. DOI: 10.1105/tpc.020487 PMID: 15100400

Hussain, S., Hussain, E., Saxena, P., Sharma, A., Thathola, P., & Sonwani, S. (2024). Navigating the impact of climate change in India: A perspective on climate action (SDG13) and sustainable cities and communities (SDG11). *Frontiers in Sustainable Cities, 5*, 1308684. DOI: 10.3389/frsc.2023.1308684

Iasur-Kruh, L., Hadar, Y., & Minz, D. (2010). Seasonal dynamics of bacterial community composition in sediments of a constructed wetland. *Microbial Ecology, 60*(3), 492–503. DOI: 10.1007/s00248-010-9721-0

Igiri, B. E., Okoduwa, S. I., Idoko, G. O., Akabuogu, E. P., Adeyi, A. O., & Ejiogu, I. K. (2018). Toxicity and bioremediation of heavy metals contaminated ecosystem from tannery wastewater: A review. *Journal of Toxicology, 1*, 16. DOI: 10.1155/2018/2568038 PMID: 30363677

Ignatius, A. R., & Flood, M. G. (2021). Integration of remote sensing & environmental science in placed-based learning. *International Archives of the Photogrammetry, Remote Sensing and Spatial Information Sciences - ISPRS Archives, 44*(M–3), 73–78. DOI: 10.5194/isprs-archives-XLIV-M-3-2021-73-2021

Ikkonen, E., & Kaznina, N. (2022). Physiological responses of lettuce (Lactuca sativa L.) to soil contamination with Pb. *Horticulturae, 8*(10), 951. DOI: 10.3390/horticulturae8100951

Imron, M. F., Kurniawan, S. B., Ismail, N. I., & Abdullah, S. R. S. (2020). Future challenges in diesel biodegradation by bacteria isolates: A review. *Journal of Cleaner Production, 251*, 119716. DOI: 10.1016/j.jclepro.2019.119716

Inelova, Z., Zayadan, B., Zaparina, Y., Aitzhan, M., & Boros, E. (2023). Perspectives for the application of aquatic and semi-aquatic plants in biomonitoring of freshwater, saline and soda aquatic ecosystems. *Pakistan Journal of Botany, 55*(3), 1099–1115. DOI: 10.30848/PJB2023-3(33)

Interstate Technology & Regulatory Council. (2008). *Phytotechnologies technical and regulatory guidance and decision trees* (Revised). Phytotechnologies Team.

Interstate Technology & Regulatory Council. (2009). *Phytotechnologies: A technical and regulatory guidance document.*

IPCC. (2022). *Climate Change: impacts, adaptation and vulnerability, contribution of working group ii to the sixth assessment report of the intergovernmental panel on climate change.* Cambridge University Press., DOI: 10.1017/9781009325844

Isak, M. A., Bozkurt, T., Tütüncü, M., Dönmez, D., İzgü, T., & Şimşek, Ö. (2024). Leveraging machine learning to unravel the impact of cadmium stress on goji berry micropropagation. *PLoS One, 19*(6), e0305111. Advance online publication. DOI: 10.1371/journal.pone.0305111 PMID: 38870239

Islam, M. S., Ueno, Y., Sikder, M. T., & Kurasaki, M. (2013). Phytofiltration of arsenic and cadmium from the water environment using Micranthemum umbrosum (JF Gmel) SF Blake as a hyperaccumulator. *International Journal of Phytoremediation, 15*(10), 1010–1021. DOI: 10.1080/15226514.2012.751356 PMID: 23819292

Iyyappan, J., Baskar, G., Deepanraj, B., Vivek Anand, A., Saravanan, R., & Awasthi, M. K. (2023). Promising strategies of circular bioeconomy using heavy metal phytoremediated plants: A critical review. *Chemosphere, 313*, 137097. Advance online publication. DOI: 10.1016/j.chemosphere.2022.137097 PMID: 36334740

Jachym, S., Uhlik, O., Viktorova, J., & Macek, T. (2018). Phytoextraction of Heavy Metals: A Promising Tool for Clean-Up of Polluted Environment? *Frontiers in Plant Science*, •••, 9. PMID: 30459775

Jacob, J. M., Karthik, C., Saratale, R. G., Kumar, S. S., Prabakar, D., Kadirvelu, K., & Pugazhendhi, A. (2018). Biological approaches to tackle heavy metal pollution: A survey of literature. *Journal of Environmental Management, 217*, 56–70. DOI: 10.1016/j.jenvman.2018.03.077 PMID: 29597108

Jafari, S. A., Cheraghi, S., Mirbakhsh, M., Mirza, R., & Maryamabadi, A. (2015). Employing response surface methodology for optimization of mercury bioremediation by Vibrio parahaemolyticus PG02 in coastal sediments of Bushehr, Iran. *Clean (Weinheim)*, *43*(1), 118–126. DOI: 10.1002/clen.201300616

Jain, S., & Arnepalli, D. (2016). Biominerlisation as a remediation technique: A critical review; *Proceedings of the Indian Geotechnical Conference (IGC2016)*; Chennai, India. 15:17.

Jaishankar, M., Tseten, T., Anbalagan, N., Mathew, B. B., & Beeregowda, K. N. (2014). Toxicity, mechanism, and health effects of some heavy metals. *Interdisciplinary Toxicology*, *7*(2), 60–72. DOI: 10.2478/intox-2014-0009 PMID: 26109881

Jamil Emon, F., Rohani, M. F., Sumaiya, N., Tuj Jannat, M. F., Akter, Y., Shahjahan, M., Abdul Kari, Z., Tahiluddin, A. B., & Goh, K. W. (2023, June 6). Bioaccumulation and Bioremediation of Heavy Metals in Fishes — A Review. *Toxics*, *11*(6), 510. DOI: 10.3390/toxics11060510 PMID: 37368610

Javed, M. T., Tanwir, K., Akram, M. S., Shahid, M., Niazi, N. K., & Lindberg, S. (2019). Phytoremediation of cadmium-polluted water/sediment by aquatic macrophytes: role of plant-induced pH changes. *Cadmium Toxicity and Tolerance in Plants*, 495–529.

Jeong, H., Byeon, E., Kim, D. H., Maszczyk, P., & Lee, J. S. (2023). Heavy metals and metalloid in aquatic invertebrates: A review of single/mixed forms, combination with other pollutants, and environmental factors. *Marine Pollution Bulletin*, *191*, 114959. DOI: 10.1016/j.marpolbul.2023.114959 PMID: 37146547

Jeppesen, E.. (2012). Impacts of climate warming on lake fish community structure and potential effects on ecosystem function. *Hydrobiologia*, *694*(1), 1–39. DOI: 10.1007/s10750-012-1182-1

Jibril, S. A., Hassan, S. A., Ishak, C. F., & Megat Wahab, P. E. (2017). Cadmium toxicity affects phytochemicals and nutrient elements composition of lettuce (Lactuca sativa L.). *Advances in Agriculture*, *2017*(1), 1236830. DOI: 10.1155/2017/1236830

Johnson, K., Vermeer, W., Hills, H., Chin-Purcell, L., Barnett, J. T., Burns, T., Dean, M. J., & Hendricks Brown, C. (2024). Model-driven decision support: A community-based meta-implementation strategy to predict population impact. *Annals of Epidemiology*, *95*, 12–18. DOI: 10.1016/j.annepidem.2024.05.002 PMID: 38754571

Jonak, C., Nakagami, H., & Hirt, H. (2004). Heavy metal stress: Activation of distinct mitogen-activated protein kinase pathways by copper and cadmium. *Plant Physiology*, *136*(2), 3276–3283. DOI: 10.1104/pp.104.045724 PMID: 15448198

Jones, S. F., Arias-Ortiz, A., Baldocchi, D., Eagle, M., Friess, D. A., Gore, C., Noe, G., Nolte, S., Oikawa, P., Paytan, A., Raw, J. L., Roberts, B. J., Rogers, K., Schutte, C., Stagg, C. L., Thorne, K. M., Ward, E. J., Windham-Myers, L., & Yando, E. S. (2024). When and where can coastal wetland restoration increase carbon sequestration as a natural climate solution? *Cambridge Prisms: Coastal Futures*, *2*, e13. Advance online publication. DOI: 10.1017/cft.2024.14 PMID: 40881330

Julien, M. H., Hill, M. P., Center, T. D., & Ding, J. (2001). *Biological and integrated control of water hyacinth (Eichhornia crassipes)*. ACIAR Proceedings, 102.

Justin, L. D., Olukanni, D. O., & Babaremu, K. O. (2022). Performance assessment of local aquatic macrophytes for domestic wastewater treatment in Nigerian communities: A review. *Heliyon*, *8*(8), e10093. DOI: 10.1016/j.heliyon.2022.e10093 PMID: 36042728

Kabata-Pendias, A. (2011). *Trace elements in soils and plants* (4th ed.). CRC Press.

Kabeer, R., VP, S., CS, P. K., AP, T., V, S., EK, R., & KR, B. (. (2022). Role of heavy metal tolerant rhizosphere bacteria in the phytoremediation of Cu and Pb using Eichhornia crassipes (Mart.) Solms. *International Journal of Phytoremediation*, *24*(11), 1120–1132. DOI: 10.1080/15226514.2021.2007215 PMID: 34846266

Kadlec, R. H., & Wallace, S. D. (2009). *Treatment wetlands* (2nd ed.). CRC Press.

Kadlec, R. H., & Wallace, S. D. (2009). *Treatment Wetlands* (2nd ed.). CRC Press.

Kafle, A., Timilsina, A., Gautam, A., Adhikari, K., Bhattarai, A., & Aryal, N. (2022). Phytoremediation: Mechanisms, plant selection and enhancement by natural and synthetic agents. *Environmental Advances*, *8*, 100203. DOI: 10.1016/j.envadv.2022.100203

Kah, M., Kookana, R. S., Gogos, A., & Bucheli, T. D. (2018). A critical evaluation of nanopesticides and nanofertilizers against their conventional analogues. *Nature Nanotechnology*, *13*(8), 677–684. DOI: 10.1038/s41565-018-0131-1 PMID: 29736032

Kamal, M., Ghaly, A. E., Mahmoud, N., & Côté, R. (2004). Phytoaccumulation of heavy metals by aquatic plants. *Environment International*, *29*(8), 1029–1039. DOI: 10.1016/S0160-4120(03)00091-6 PMID: 14680885

Karnwal, A., Kumar, G., El Din Mahmoud, A., Dutta, J., Singh, R., Mohammad Said Al-Tawaha, A. R., & Malik, T. (2025). Eco-engineered remediation: Microbial and rhizosphere-based strategies for heavy metal detoxification. *Current Research in Biotechnology*, *9*, 100297. DOI: 10.1016/j.crbiot.2025.100297

Karnwal, A., Martolia, S., Dohroo, A., Al-Tawaha, A. R. M. S., & Malik, T. (2024). Exploring bioremediation strategies for heavy metals and POPs pollution: The role of microbes, plants, and nanotechnology. *Frontiers in Environmental Science, 12,* 1397850. DOI: 10.3389/fenvs.2024.1397850

Karwal, T., Patel, Y., Bartlett, A., Poulin, V., Smith, T. L., & Pfeffer, D. N. (2024). Procoli: Profiles of cosmological likelihoods. arXiv preprint arXiv:2401.14225.

Kaur, R., Mahajan, A., Kaur, R., & Sinha, D. (2012). Phytoremediation potential of water lettuce (*Pistia stratiotes*) for the removal of Pb and Cd. *African Journal of Environmental Science and Technology*, *6*(6), 207–213.

Kaushik, P. (2025). AI-Driven Breeding Enhances Stress Tolerance in High-Elevation Extremophytes: A Proof-of-Concept Study with Cross-Component Validation. *BioRxiv*. DOI: 10.1101/2025.02.21.639605

Ke, B., Nguyen, H., Bui, X. N., Bui, H. B., Choi, Y., Zhou, J., Moayedi, H., Costache, R., & Nguyen-Trang, T. (2021). Predicting the sorption efficiency of heavy metal based on the biochar characteristics, metal sources, and environmental conditions using various novel hybrid machine learning models. *Chemosphere*, *276*, 130204. DOI: 10.1016/j.chemosphere.2021.130204 PMID: 34088091

Keunen, E.. (2011). Role of ROS and antioxidant systems in heavy metal stress. *Plant Physiology and Biochemistry*, *49*(6), 649–658. DOI: 10.1016/j.plaphy.2011.03.017

Khan, A. H. A., Kiyani, A., Santiago-Herrera, M., Ibáñez, J., Yousaf, S., Iqbal, M., Martel-Martín, S., & Barros, R. (2023). Sustainability of phytoremediation: Post-harvest stratagems and economic opportunities for the produced metals contaminated biomass. *Journal of Environmental Management, 326*(Pt. B). *Journal of Environmental Management, 116700*, 116700. Advance online publication. DOI: 10.1016/j.jenvman.2022.116700 PMID: 36423411

Khan, H. N., & Faisal, M. (2024). Bioremediation of heavy metals using the symbiosis of hydrophytes and metal-resistant bacteria. In *Membrane technologies for heavy metal removal from water* (pp. 126–142). CRC Press.

Khan, M. I. R., Asgher, M., & Khan, N. A. (2020). Phytoremediation of heavy metals and role of plant growth promoting microbes in the rhizosphere. In *Plant-microbe interactions in agro-ecological perspectives* (pp. 309–337). Springer., DOI: 10.1007/978-981-15-2696-6_13

Khan, M. I. R., Nazir, F., Asgher, M., Per, T. S., & Khan, N. A. (2009). Selenium and sulfur influence ethylene formation and alleviate cadmium-induced oxidative stress by improving proline and glutathione production in wheat. *Journal of Plant Physiology*, *170*(2), 167–174. DOI: 10.1016/j.jplph.2012.08.018 PMID: 25462073

Khan, M. I., Ahmad, A., Iqbal, M., & Hussain, M. (2020). Synergistic effects of microbes and nanoparticles in phytoremediation: Recent advances and challenges. *Environmental Research*, *188*, 109765. DOI: 10.1016/j.envres.2020.109765

Khan, S., Ahmad, I., Shah, M. T., Rehman, S., & Khaliq, A. (2009). Use of constructed wetland for the removal of heavy metals from industrial wastewater. *Journal of Environmental Management*, *90*(11), 3451–3457. DOI: 10.1016/j.jenvman.2009.05.026 PMID: 19535201

Khan, S., Kamal, M., Noor, S., & Afzal, S. M. (2024). Assessment of heavy metals and its treatment through phytoremediation in groundwater along River Kabul in district Charsadda. *Frontiers in Environmental Science*, *12*, 1392892. Advance online publication. DOI: 10.3389/fenvs.2024.1392892

Khan, T., Shah, L., Khan, S., Wani, O. A., Sheikh, Z. N., Afroza, B., Rashid, R., & Baloch, F. S. (2025). Comprehensive review of multiomics applications and remediation of plant heavy metal toxicity. *Frontiers in Plant Science*, *16*, 12440847. DOI: 10.3389/fpls.2025.12440847 PMID: 40956384

Kharangate-Lad A., D'Souza N. C. (2021). Current approaches in bioremediation of toxic contaminants by application of microbial cells; biosurfactants and bioemulsifiers of microbial origin. Rhizobiont Bioremediat. Hazard. Waste, 217–263

Khatoon, H., Rai, J. P. N., & Jillani, A. (2021). Role of fungi in bioremediation of contaminated soil. In: Fungi bio-prospects in sustainable agriculture, environment and nano-technology, pp 121-156 DOI: 10.1016/B978-0-12-821925-6.00007-1

Khatoon, Z., Orozco-Mosqueda, M., & Santoyo, G., del C., & Santoyo, G. (2024). Microbial Contributions to Heavy Metal Phytoremediation in Agricultural Soils: A Review. *Microorganisms*, *12*(10), 1945. Advance online publication. DOI: 10.3390/microorganisms12101945 PMID: 39458255

Kim, D. Y., Bovet, L., Maeshima, M., Martinoia, E., & Lee, Y. (2006). Plant ABC transporters sequester PC-metal complexes into vacuoles. *The Plant Journal*, *46*(3), 685–697. DOI: 10.1111/j.1365-313X.2006.02731.x

Kim, J. Y., Hwang, J.-H., Lee, T., Kim, Y., Bae, S., Kim, D., Lee, J., Lee, J., Tsang, Y. F., & Kwon, E. E. (2025). A review of sustainable metal recovery from e-waste via bioleaching: Current status and strategies for improvement. *Chemical Engineering Journal, 516*, 164149. DOI: 10.1016/j.cej.2025.164149

Kim, Y., Park, K., Bak, J., & Choi, S. (2024). *Iris pseudacorus* and *Lythrumanceps* as Plants Supporting the Process of Removing Microplastics from Aquatic Environments—Preliminary Research. *Horticulturae, 10*(6), 631. DOI: 10.3390/horticulturae10060631

Kivaisi, A. K. (2001). The potential for constructed wetlands for wastewater treatment and reuse in developing countries: A review. *Ecological Engineering, 16*(4), 545–560. DOI: 10.1016/S0925-8574(00)00113-0

Kong, L. W., Weng, Y., Glaz, B., Haile, M., & Lai, Y. C. (2023). Reservoir computing as digital twins for nonlinear dynamical systems. *Chaos (Woodbury, N.Y.), 33*(3), 33111. DOI: 10.1063/5.0138661 PMID: 37003826

Konnerup, D., Koottatep, T., & Brix, H. (2009). Treatment of domestic wastewater in tropical, subsurface flow constructed wetlands planted with Canna and Heliconia. *Ecological Engineering, 35*(2), 248–257. DOI: 10.1016/j.ecoleng.2008.04.018

Korzeniowska, J., & Stanislawska-Glubiak, E. (2019). Phytoremediation potential of Phalaris arundinacea, Salix viminalis and Zea mays for nickel-contaminated soils. *International Journal of Environmental Science and Technology, 16*(4), 1999–2008. DOI: 10.1007/s13762-018-1823-7

Kosakivska, I.. (2020). Molecular mechanisms of heavy metal tolerance in aquatic plants. *International Journal of Aquatic Plant Biology, 12*(4), 183–195.

Kosal, E. (2023). *Plant Kingdom. Introductory Biology: Ecology*. Evolution, and Biodiversity.

Kristanti, R. A., & Hadibarata, T. (2023). Phytoremediation of contaminated water using aquatic plants, its mechanism and enhancement. *Current Opinion in Environmental Science & Health, 32*, 100451. DOI: 10.1016/j.coesh.2023.100451

Kristensen, E., Penha-Lopes, G., Delefosse, M., Valdemarsen, T., Quintana, C. O., & Banta, G. T. (2012). What is bioturbation? The a need for a precise definition of fauna in aquatic sciences. *Marine Ecology Progress Series, 446*, 285–302. DOI: 10.3354/meps09506

Krzesłowska, M. (2011). The cell wall in plant cell response to trace metals: Polysaccharide remodeling and its role in defense strategy. *Acta Physiologiae Plantarum, 33*(1), 35–51. DOI: 10.1007/s11738-010-0581-z

Kumar Yadav, K., Gupta, N., Kumar, A., Reece, L. M., Singh, N., Rezania, S., & Ahmad Khan, S. (2018). Mechanistic understanding and holistic approach of phytoremediation: A review on application and future prospects. *Ecological Engineering*, *120*, 274–298. DOI: 10.1016/j.ecoleng.2018.05.039

Kumar, A., Dadhwal, M., Mukherjee, G., Srivastava, A., Gupta, S., &Ahuja, V. (2024). Phytoremediation: Sustainable Approach for Heavy Metal Pollution. *Scientifica (Cairo).*

Kumar, A., Dadhwal, M., Mukherjee, G., Srivastava, A., Gupta, S., & Ahuja, V. (2024). Phytoremediation: Sustainable approach for heavy metal pollution. *Scientifica*, *2024*(1), 3909400. Advance online publication. DOI: 10.1155/2024/3909400 PMID: 39430119

Kumar, A., Kumar, V., Chawla, M., Thakur, M., Bhardwaj, R., Wang, J., & Rinklebe, J. (2024). Bioremediation of mercury contaminated soil and water: A review. *Land Degradation & Development*, *35*(4), 1261–1283. DOI: 10.1002/ldr.4989

Kumara, M., Singhab, M., & Mukherjeec, P. (2020). Economic importance of Hydrophytes. *RE:view*.

Kumaran, N. S., Sundaramanicam, A., & Bragadeeswaran, S. (2011). Adsorption studies on heavy metals by isolated cyano bacterial strain (Nostoc sp.) from Uppanar estuarine water, southeast coast of India. *Journal of Applied Sciences Research*, *7*(11), 1609–1615.

Kumari, M., & Tripathi, B. D. (2015). Efficiency of Phragmites australis and Typha latifolia for heavy metal removal from wastewater. *Ecotoxicology and Environmental Safety*, *112*, 80–86. DOI: 10.1016/j.ecoenv.2014.10.034 PMID: 25463857

Kumar, K. S., Dahms, H. U., Lee, J. S., Kim, H. C., Lee, W. C., & Shin, K. H. (2014). Algal photosynthetic responses to toxic metals and herbicides assessed by chlorophyll *a* fluorescence. *Ecotoxicology and Environmental Safety*, *104*, 51–71. DOI: 10.1016/j.ecoenv.2014.01.042 PMID: 24632123

Kumar, M., Ghosh, M., & Pandey, R. (2022). Microbial synthesized silver nanoparticles and their impact on phytoremediation of chromium by *Hydrilla verticillata*. *Environmental Science and Pollution Research International*, *28*, 41635–41646. DOI: 10.1007/s11356-021-14292-5

Kumar, U., Werners, S. E., Paparrizos, S., Datta, D. K., & Ludwig, F. (2021). Co-producing climate information services with smallholder farmers in the Lower Bengal Delta: How forecast visualization and communication support farmers' decision-making. *Climate Risk Management*, *33*, 100346. DOI: 10.1016/j.crm.2021.100346

Kumar, V., Shahi, S. K., & Singh, S. (2018). Bioremediation: An eco-sustainable approach for restoration of contaminated sites. In Singh, J., Sharma, D., Kumar, G., & Sharma, N. (Eds.), *Microbial bioprospecting for sustainable development* (pp. 123–146). Springer., DOI: 10.1007/978-981-13-0053-0_6

Kwok, Hong-Ching, & Chan, Wan (2022). Wheatgrass (Triticum aestivum) as an Efficient Phytoremediation Plant for Aristolochic Acid-Contaminated Water and Arable Soil. *ACS Publications.* doi:. 2c00057.s001.DOI: 10.1021/acsagscitech

Lakshmi, D., Akhil, D., Kartik, A., Gopinath, K. P., Arun, J., Bhatnagar, A., Rinklebe, J., Kim, W., & Muthusamy, G. (2021). Artificial intelligence (AI) applications in adsorption of heavy metals using modified biochar. *The Science of the Total Environment, 801*, 149623. DOI: 10.1016/j.scitotenv.2021.149623 PMID: 34425447

Lal, R. (2008). Carbon sequestration. *Philosophical Transactions of the Royal Society of London. Series B, Biological Sciences, 363*(1492), 815–830. DOI: 10.1098/rstb.2007.2185 PMID: 17761468

Lal, R. (2015). Restoring soil quality to mitigate soil degradation. *Sustainability (Basel), 7*(5), 5875–5895. DOI: 10.3390/su7055875

Lambin, P., Van Stiphout, R. G. P. M., Starmans, M. H. W., Rios-Velazquez, E., Nalbantov, G., Aerts, H. J. W. L., Roelofs, E., Van Elmpt, W., Boutros, P. C., Granone, P., Valentini, V., Begg, A. C., De Ruysscher, D., & Dekker, A. (2013). Predicting outcomes in radiation oncology-multifactorial decision support systems. *Nature Reviews. Clinical Oncology, 10*(1), 27–40. DOI: 10.1038/nrclinonc.2012.196 PMID: 23165123

Lasat, M. M. (2002). Phytoremediation of metals. *Brazilian Journal of Plant Physiology, 14*(3), 109–119.

Leaphart, J. C., Abercrombie, S. A., Borchert, E. J., Bryan, A. L.Jr, & Beasley, J. C. (2022). Bioaccumulation of Mercury and Radiocesium in Waterfowl Introduced to a Site with Legacy Contamination. *Environmental Toxicology and Chemistry, 41*(10), 2479–2487. DOI: 10.1002/etc.5444 PMID: 35866467

Lee, K. W., Min, H. J., Park, H. W., Manavalan, B., & Jeon, Y.-J. (2025). Abstract 2427: Optimization of preprocessing strategies for developing AI-based disease diagnosis model using whole transcriptomic data. *Cancer Research, 85*(8, Supplement_1), 2427–2427. DOI: 10.1158/1538-7445.AM2025-2427

Lemming, G., Hauschild, M. Z., & Bjerg, P. L. (2010). Life cycle assessment of soil and groundwater remediation technologies: Literature review. *The International Journal of Life Cycle Assessment, 15*(1), 115–127. DOI: 10.1007/s11367-009-0129-x

Leszczyszyn, O. I., Imam, H. T., & Blindauer, C. A. (2007). Diversity and distribution of plant metallothioneins: A review of structure, properties and functions. *Metallomics*, *1*(5), 416–430. PMID: 23694960

Le, T. T., Vodden, K., Wu, J., Bullock, R., & Sabau, G. (2023). Payments for ecosystem services programs: A global review of contributions towards sustainability. *Heliyon*, *10*(1), e22361. DOI: 10.1016/j.heliyon.2023.e22361 PMID: 38173537

Li, C., Wang, H., Liao, X., Xiao, R., Liu, K., Bai, J., Li, B., He, Q. (2022). Heavy metal pollution in coastal wetlands: A systematic review of studies globally over the past three decades. *Journal of Hazardous Materials*, 424(Pt A), 127312. .DOI: 10.1016/j.jhazmat.2021.127312

Li, H., Zhang, Q., Li, Y., & Shen, Z. (2021). Arbuscular mycorrhizal fungi enhanced phytoremediation potential of *Typha angustifolia* under metal stress. *The Science of the Total Environment*, *764*, 142870. DOI: 10.1016/j.scitotenv.2020.142870

Lima, H. de P., & Asencios, Y. J. O. (2021). Eichhornia crassipes (Mart.) Solms (natural or carbonized) as biosorbent to remove pollutants in water. *SN Applied Sciences*, *3*(8), 750. DOI: 10.1007/s42452-021-04736-9 PMID: 34337324

Lin, Y.. (2022). Adaptations and functions of aquatic macrophytes in wetland ecosystems. *Wetlands Ecology and Management*, *30*(4), 1–16. DOI: 10.1007/s11273-022-09887-3

Lin, Y., Zhang, T., Liu, S., & Wang, M. (2022). Phytoremediation of aquatic ecosystems: Mechanisms and future perspectives. *Environmental Science and Pollution Research International*, *29*(12), 17301–17319. DOI: 10.1007/s11356-021-16924-y

Liu, C., Gu, W., Dai, Z., Li, J., Jiang, H., & Zhang, Q. (2018). Boron accumulation by Lemna minor L. under salt stress. *Scientific Reports*, *8*(1), 8954. DOI: 10.1038/s41598-018-27343-y PMID: 29895991

Liu, D., Wang, W., & Liu, Y. (2014). Enhanced mercury and cadmium tolerance and accumulation in transgenic *Lemna minor* by overexpression of a glutathione synthetase gene. *Plant Cell, Tissue and Organ Culture*, *119*(1), 127–137. DOI: 10.1007/s11240-014-0514-7

Liu, H., Zhang, J., Zhang, J., & Chen, J. (2016). Constructed wetlands in China: Recent developments and future challenges. *Frontiers of Environmental Science & Engineering*, *10*, 478–490. DOI: 10.1007/s11783-016-0831-0

Liu, J., Li, X., & Zhu, P. (2024). Effects of Various Heavy Metal Exposures on Insulin Resistance in Non-diabetic Populations: Interpretability Analysis from Machine Learning Modeling Perspective. *Biological Trace Element Research, 202*(12), 5438–5452. DOI: 10.1007/s12011-024-04126-3 PMID: 38409445

Liu, J., Zhu, Q., Zhang, Z., Xu, J., & Yang, J. (2014). Enhancement of phytoremediation potential of *Lemna minor* by genetic modification with glutathione synthetase gene. *Journal of Environmental Sciences (China), 26*(5), 962–970. DOI: 10.1016/S1001-0742(13)60488-8 PMID: 25079626

Liu, L., Bilal, M., Duan, X., & Iqbal, H. M. (2019). Mitigation of environmental pollution by genetically engineered bacteria—Current challenges and future perspectives. *The Science of the Total Environment, 667*, 444–454. DOI: 10.1016/j.scitotenv.2019.02.390 PMID: 30833243

Liu, L., Li, W., Song, W., & Guo, M. (2018). Remediation techniques for heavy metal-contaminated soils: Principles and applicability. *The Science of the Total Environment, 633*, 206–219. DOI: 10.1016/j.scitotenv.2018.03.161 PMID: 29573687

Liu, W., Wang, B., Wang, Q., Hou, J., Wu, L., Wood, J. L., Luo, Y., & Franks, A. E. (2016). Characteristics of metal-tolerant plant growth-promoting yeast (Cryptococcus sp. NSE1) and its influence on Cd hyperaccumulator Sedum plumbizincicola. *Environmental Science and Pollution Research International, 23*(18), 18621–18629. DOI: 10.1007/s11356-016-7041-2 PMID: 27306207

Liu, X., Sathishkumar, K., Zhang, H., Saxena, K. K., Zhang, F., Naraginiti, S., & Guo, X. (2024). Frontiers in environmental cleanup: Recent advances in remediation of emerging pollutants from soil and water. *Journal of Hazardous Materials Advances, 16*, 100461. DOI: 10.1016/j.hazadv.2024.100461

Li, W., Wang, Z., & Li, J. (2019). Microbial reduction of hexavalent chromium: Mechanisms and applications. *Environmental Science & Technology, 53*(1), 14–34.

Li, X., Peng, W., Jia, Y., Lu, L., & Fan, W. (2016). Removal of cadmium and zinc from contaminated wastewater using *Rhodobacter sphaeroides. Water Science and Technology, 75*(11), 2489–2498. DOI: 10.2166/wst.2016.608 PMID: 28617267

Li, X., Shen, Z., Wai, O. W. H., & Li, Y. S. (2013). Chemical forms of Pb, Zn, and Cu in the sediment profiles of the Pearl River Estuary. *Marine Pollution Bulletin, 47*(1–6), 107–116. DOI: 10.1016/S0025-326X(02)00418-4 PMID: 11381876

Li, X., Yang, C. P., Zeng, G. M., Wu, S. H., Lin, Y., Zhou, Q., Lou, W., Du, C., Nie, L. J., & Zhong, Y. Y. (2020). Nutrient removal from swine wastewater with growing microalgae at various zinc concentrations. *Algal Research, 46*, 101804. DOI: 10.1016/j.algal.2020.101804

Li, X., Yang, W. L., He, H. J., Wu, S. H., Zhou, Q., Yang, C. P., Zeng, G. M., Luo, L., & Lou, W. (2018). Responses of microalgae *Coelastrella* sp. to stress of cupric ions in treatment of anaerobically digested swine wastewater. *Bioresource Technology, 251*, 274–279. DOI: 10.1016/j.biortech.2017.12.058 PMID: 29288955

Li, Y., Shao, Y., Xu, J., Wei, Y., & Yu, Z. (2021). Biochar–microbe–plant interactions in heavy metal phytoremediation: A review. *Ecotoxicology and Environmental Safety, 208*, 111699.

Li, Y., Wang, C., & Zhang, W. (2011). Biofilms and their role in the removal of pollutants in constructed wetlands. *Environmental Science & Technology, 45*(3), 1158–1164. PMID: 21166456

Li, Z., Huang, Y., Zhu, Z., Yu, M., Cheng, H., Shi, H., Zuo, W., Zhou, H., & Wang, S. (2023). Co-pyrolysis of industrial sludge with phytoremediation residue: Improving immobilization of heavy metals at high temperature. *Fuel, 351*, 128942. DOI: 10.1016/j.fuel.2023.128942

Lobell, D. B., Schlenker, W., & Costa-Roberts, J. (2011). Climate trends and global crop production since 1980. *Science, 333*(6042), 616–620. DOI: 10.1126/science.1204531 PMID: 21551030

Lone, M. I., He, Z. L., Stoffella, P. J., Yang, X. E. (2008 Mar). Phytoremediation of heavy metal polluted soils and water: Progresses and perspectives. *Journal of Zhejiang University Science B*, 9(3), 210-220. doi:. B0710633.DOI: 10.1631/jzus

Lone, M. I., He, Z. L., Stoffella, P. J., & Yang, X. E. (2008). Phytoremediation of heavy metal polluted soils and water: Progresses and perspectives. *Journal of Zhejiang University. Science. B., 9*(3), 210–220. DOI: 10.1631/jzus.B0710633 PMID: 18357623

López-Chuken, U. J. (2012). Hydroponics and environmental clean-up. *Hydroponics–a standard methodology for plant biological researches*, 181.

Lottermoser, B. G. (2010). *Mine wastes: Characterization, treatment and environmental impacts* (3rd ed.). Springer., DOI: 10.1007/978-3-642-12419-8

Lowe, S., Browne, M., Boudjelas, S., & De Poorter, M. (2000). *100 of the world's worst invasive alien species*. IUCN/SSC Invasive Species Specialist Group.

Lu, B., Xu, Z., Li, J., & Chai, X. (2018). Removal of water nutrients by different aquatic plant species: An alternative way to remediate polluted rural rivers. *Ecological Engineering*, *110*, 18–26. DOI: 10.1016/j.ecoleng.2017.09.016

Luo, L., He, H. J., Yang, C. P., Wen, S., Zeng, G. M., Wu, M. J., Zhou, Z. L., & Lou, W. (2016). Nutrient removal and lipid production by *Coelastrella* sp. in anaerobically and aerobically treated swine wastewater. *Bioresource Technology*, *216*, 135–141. DOI: 10.1016/j.biortech.2016.05.059 PMID: 27236400

Luo, L., Ma, Y., Zhang, S., Wei, D., & Zhu, Y. G. (2012). An inventory of trace element inputs to agricultural soils in China. *Journal of Environmental Management*, *100*, 124–129. DOI: 10.1016/j.jenvman.2011.11.013 PMID: 19246150

Luoma, S. N., & Rainbow, P. S. (2008). *Metal contamination in aquatic environments: Science and lateral management*. Cambridge University Press., DOI: 10.1017/CBO9780511805360

Luongo, T., & Cutright, T. J. (2011). Characterization of plant root exudates and their impact on microbial communities. *International Journal of Phytoremediation*, *13*(3), 211–224.

Lu, Q., He, Z. L., & Stoffella, P. J. (2017). Land application of biosolids in the USA: A review. *Applied and Environmental Soil Science*, *2012*, 1–11. DOI: 10.1155/2012/201462

Lu, X., Kruatrachue, M., Pokethitiyook, P., & Homyok, K. (2010). Removal of cadmium and zinc by water hyacinth, *Eichhornia crassipes*. *ScienceAsia*, *36*(3), 205–209. DOI: 10.2306/scienceasia1513-1874.2010.36.210

Lu, Y., Tao, Z., Nie, L., Guo, X., Pan, T., Chen, R., Li, T., Kong, W., & Liu, F. (2025). Quantitative elemental mapping of heavy metals translocation and accumulation in hyperaccumulator plant using laser-induced breakdown spectroscopy with interpretable deep learning. *Computers and Electronics in Agriculture*, *230*, 109907. DOI: 10.1016/j.compag.2025.109907

Lv, J. P., Liu, Y., Feng, J., Liu, Q., Nan, F. R., & Xie, S. L. (2018). Nutrient removal from undiluted cattle farm wastewater by the two-stage process of microalgae-based wastewater treatment. *Bioresource Technology*, *264*, 311–318. DOI: 10.1016/j.biortech.2018.05.085 PMID: 29857286

Ma, T., Zhang, W., Fan, H., Huang, L., Xu, Q., Ma, G., Zhou, G., Meng, Y., Sun, J., Qiu, J., Gu, J., & Kang, X. (2020). Study on phytoremediation for heavy-metal-contaminated sediments by hydrophytes. *E3S Web of Conferences*, *143*, 02020. https://doi.org/DOI: 10.1051/e3sconf/202014302020

Maberly, S. C., & Madsen, T. V. (2002). Freshwater angiosperm carbon concentrating mechanisms: Processes and patterns. *Functional Plant Biology*, *29*(3), 393–405. DOI: 10.1071/PP01181 PMID: 32689484

Macek, T., Macková, M., & Kas, J. (2008). Exploitation of plants for the removal of organics in environmental remediation. *Biotechnology Advances*, *26*(5), 436–448. DOI: 10.1016/j.biotechadv.2008.05.007 PMID: 18565714

Machineni, L. (2019). Review on biological wastewater treatment and resources recovery: Attached and suspended growth systems. *Water Science and Technology*, *80*(11), 2013–2026. DOI: 10.2166/wst.2020.034 PMID: 32198320

Madeła, M., & Skuza, M. (2021). Towards a circular economy: Analysis of the use of biowaste as biosorbent for the removal of heavy metals. *Energies*, *14*(17), 5427. DOI: 10.3390/en14175427

Madsen, J. D., Chambers, P. A., James, W. F., Koch, E. W., & Westlake, D. F. (2001). The interaction between water movement, sediment dynamics and submersed macrophytes. *Hydrobiologia*, *444*(1–3), 71–84. DOI: 10.1023/A:1017520800568

Mahar, A., Wang, P., Ali, A., Awasthi, M. K., Lahori, A. H., Wang, Q., Li, R., & Zhang, Z. (2016). Challenges and opportunities in the phytoremediation of heavy metals contaminated soils: A review. *Ecotoxicology and Environmental Safety*, *131*, 218–228. DOI: 10.1016/j.ecoenv.2015.12.023 PMID: 26741880

Mahmoud, M., Ramadan, M., Naher, S., Pullen, K., & Olabi, A. G. (2021). The impacts of different heating systems on the environment: A review. *The Science of the Total Environment*, *766*, 142625. DOI: 10.1016/j.scitotenv.2020.142625 PMID: 33077224

Maine, M. A., Hadad, H. R., Camaño Silvestrini, N. E., Nocetti, E., Sanchez, G. C., & Campagnoli, M. A. (2022). Cr, Ni, and Zn removal from landfill leachate using vertical flow wetlands planted with Typha domingensis and Canna indica. *International Journal of Phytoremediation*, *24*(1), 66–75. DOI: 10.1080/15226514.2021.1926909 PMID: 34077330

Ma, L. Q., Komar, K. M., Tu, U. C., Zhang, W., Cai, Y., & Kennelley, E. D. (2001). A fern that hyperaccumulates arsenic. *Nature*, *409*(6820), 579. DOI: 10.1038/35054664 PMID: 11214308

Malik, A. (2007). Environmental challenge vis-à-vis opportunity: The case of water hyacinth. *Environment International*, *33*(1), 122–138. DOI: 10.1016/j.envint.2006.08.004 PMID: 17010439

Ma, M., Du, H., & Wang, D. (2019). Mercury methylation by anaerobic microorganisms: A review. *Critical Reviews in Environmental Science and Technology, 49*(20), 1893–1936. DOI: 10.1080/10643389.2019.1594517

Mani S, Chowdhary P, Zainith S (2020). Microbes mediated approaches for environmental waste management. Microorg Sustain Environ Health 17

Mansoor, S., Ali, A., Kour, N., Bornhorst, J., AlHarbi, K., Rinklebe, J., Abd El Moneim, D., Ahmad, P., & Chung, Y. S. (2023). Heavy Metal Induced Oxidative Stress Mitigation and ROS Scavenging in Plants. *Plants, 12*(16), 3003. DOI: 10.3390/plants12163003 PMID: 37631213

Mao, J., Zhang, X., Chen, Y., Liu, D., & Wang, L. (2024). Industrial wastewater treatment using floating wetlands: Performance and influencing factors. *Journal of Environmental Management, 352*, 120875.

Mao, L., Long, L., Qin, Q., Huang, Q., Yu, Y., Ding, L., Liu, G., Zhang, J., Cui, B., & Tan, X. (2025). *Nymphaea "Eldorado"* flower extract targets SERPINE1 to attenuate inflammatory and antioxidant crosstalk in zebrafish. *Frontiers in Pharmacology, 16*, 1612233. DOI: 10.3389/fphar.2025.1612233 PMID: 40717968

Marchand, L., Mench, M., Jacob, D. L., & Otte, M. L. (2010). Metal and metalloid removal in constructed wetlands, with emphasis on the importance of plants and standardized measurements: A review. *Environmental Pollution, 158*(12), 3447–3461. DOI: 10.1016/j.envpol.2010.08.018 PMID: 20869144

Marisa, L., de Reyniès, A., Duval, A., Selves, J., Gaub, M. P., Vescovo, L., Etienne-Grimaldi, M.-C., Schiappa, R., Guenot, D., Ayadi, M., Kirzin, S., Chazal, M., Fléjou, J.-F., Benchimol, D., Berger, A., Lagarde, A., Pencreach, E., Piard, F., Elias, D., & Boige, V. (2013). Gene expression classification of colon cancer into molecular subtypes: Characterization, validation, and prognostic value. *PLoS Medicine, 10*(5), e1001453. DOI: 10.1371/journal.pmed.1001453 PMID: 23700391

Markou, G., Wang, L., Ye, J. F., & Unc, A. (2018). Using agro-industrial wastes for the cultivation of microalgae and duckweeds: Contamination risks and biomass safety concerns. *Biotechnology Advances, 36*(4), 1238–1254. DOI: 10.1016/j.biotechadv.2018.04.003 PMID: 29673973

Martinoia, E., Maeshima, M., & Neuhaus, H. E. (2007). Vacuolar transporters and their essential role in plant metabolism. *Journal of Experimental Botany, 58*(1), 83–102. DOI: 10.1093/jxb/erl183 PMID: 17110589

Masamba, W. R. L., & Mazvimavi, D. (2008). The use of natural wetlands for wastewater treatment in developing countries: A case study of the Okavango Delta. *Physics and Chemistry of the Earth Parts A/B/C*, *33*(8–13), 684–688.

Masi, F., Rizzo, A., & Regelsberger, M. (2017a). The role of constructed wetlands in a new circular economy, resource oriented, and ecosystem services paradigm. *Journal of Environmental Management*, *216*, 275–284. DOI: 10.1016/j.jenvman.2017.11.086 PMID: 29224716

Masi, F., Rizzo, A., Regelsberger, M., & Bresciani, R. (2017b). Constructed wetlands for combined sewer overflow treatment: Ecosystem services at Gorla Maggiore, Italy. *Ecological Engineering*, *98*, 427–438. DOI: 10.1016/j.ecoleng.2016.03.043

Maurya, N. K., Singh, R. K., & Yadav, P. (2025). Phytoremediation of metal contamination using *Pistia stratiotes* and *Eichhornia crassipes*: A comparative analysis. *International Journal of Environment and Climate Change*, *15*(4), 205–221. DOI: 10.9734/ijecc/2025/v15i44804

Ma, Y., Prasad, M. N. V., Rajkumar, M., & Freitas, H. J. B. A. (2011). Plant growth promoting rhizobacteria and endophytes accelerate phytoremediation of metalliferous soils. *Biotechnology Advances*, *29*(2), 248–258. DOI: 10.1016/j.biotechadv.2010.12.001 PMID: 21147211

Ma, Y., Rajkumar, M., & Freitas, H. (2011). Inoculation of plant growth promoting bacterium *Achromobacter xylosoxidans* strain Ax10 for the improvement of phytoremediation of metal-contaminated soils. *Chemosphere*, *82*(10), 1383–1390. PMID: 21474162

Ma, Y., Rajkumar, M., Zhang, C., & Freitas, H. (2016). Beneficial role of bacterial endophytes in heavy metal phytoremediation. *Journal of Environmental Management*, *174*, 14–25. DOI: 10.1016/j.jenvman.2016.02.047 PMID: 26989941

Mays, P. A., & Edwards, G. S. (2001). Comparison of heavy metal accumulation in a natural wetland and constructed wetlands receiving acid mine drainage. *Ecological Engineering*, *16*(4), 487–500. DOI: 10.1016/S0925-8574(00)00112-9

McInnes, R. (2020). *Global Wetland Outlook* (Special Edition 2021). Ramsar Convention Secretariat.

Meagher, R. B. (2000). Phytoremediation of toxic elemental and organic pollutants. *Current Opinion in Plant Biology*, *3*(2), 153–162. DOI: 10.1016/S1369-5266(99)00054-0 PMID: 10712958

Meagher, R. B. (2000). Phytoremediation of toxic metals. *Bioscience*, *50*(6), 469–478.

Meena, R. K., Jatav, G. K., & Meena, H. (2020). Emerging role of nanotechnology in bioremediation: An innovative and eco-friendly approach. *Journal of Environmental Management, 260*, 110023.

Mejia, C., Nye, R., Contreras, J., Nand, A., & Edgerton, C. (2025). Unmanned Offshore Platforms in the Gulf of Mexico with Ai, IOT ScadaReduces Cost and Increase Productivity. *Proceedings of the Annual Offshore Technology Conference.* DOI: 10.4043/35721-MS

Mendez, M. O., & Maier, R. M. (2008). Phytostabilization of Mine Tailings in Arid and Semiarid Environments—An Emerging Remediation Technology. *Environmental Health Perspectives, 116*(3), 278–283. DOI: 10.1289/ehp.10608 PMID: 18335091

Mian, S. M., Khan, M. S., Shawez, M., & Kaur, A. (2024). Artificial Intelligence (AI), Machine Learning (ML) & Deep Learning (DL): A Comprehensive Overview on Techniques, Applications and Research Directions. *2nd International Conference on Sustainable Computing and Smart Systems, ICSCSS 2024 - Proceedings*, 1404–1409. DOI: 10.1109/ICSCSS60660.2024.10625198

Michalak, A. (2006). Phenolic compounds and their antioxidant activity in plants growing under heavy metal stress. *Polish Journal of Environmental Studies, 15*(4), 523–530.

Milosavljević, A., Predić, B., & Milošević, D. (2023). Imagelytics: A Deep Learning-Based Image Classification Tool to Support Bioassessment. *Lecture Notes in Networks and Systems, 742 LNNS*, 42–50. DOI: 10.1007/978-3-031-38616-9_5

Ministry of Environment. Forest and Climate Change (MoEFCC). (2017). *National plan for conservation of aquatic ecosystems (NPCA).* Government of India.

Mishra, A., & Malik, A. (2013). Recent advances in microbial metal bioaccumulation. *Critical Reviews in Environmental Science and Technology, 43*(11), 1162–1222. DOI: 10.1080/10934529.2011.627044

Mishra, R. K., Kumar, J., Srivastava, P. K., Bashri, G., & Prasad, S. M. (2017). PSII photochemistry, oxidative damage and anti-oxidative enzymes in arsenate-stressed Oryza sativa L. seedlings. *Chemistry and Ecology, 33*(1), 34–50. DOI: 10.1080/02757540.2016.1265516

Mishra, S.. (2023). Phytoremediation of heavy metals by wetland plants: A review. *Environmental Technology & Innovation, 30*, 103152. DOI: 10.1016/j.eti.2023.103152

Mishra, S., Srivastava, S., Tripathi, R. D., Kumar, R., Seth, C. S., & Gupta, D. K. (2006). Lead detoxification by coontail (Ceratophyllum demersum L.) involves the induction of phytochelatins and the antioxidant system in response to its accumulation. *Chemosphere, 65*(6), 1027–1039. DOI: 10.1016/j.chemosphere.2006.03.033 PMID: 16682069

Mishra, V. K., & Tripathi, B. D. (2008). Heavy metal pollution load in water and sediments of the Ganga River at Varanasi, India. *Environmental Monitoring and Assessment, 142*(1–3), 395–402. DOI: 10.1007/s10661-007-9930-4

Mishra, V. K., Tripathi, B. D., & Srivastava, S. (2008). Efficiency of constructed wetlands for heavy metal removal. *Ecological Engineering, 34*(2), 123–130. DOI: 10.1016/j.ecoleng.2008.07.004

Mishra, V. K., Upadhyay, A. R., & Pandey, S. K. (2008). Heavy metal pollution induced due to coal mining effluent on surrounding aquatic ecosystem and its management through naturally occurring aquatic macrophytes. *Bioresource Technology, 99*(5), 930–936. DOI: 10.1016/j.biortech.2007.03.010 PMID: 17475484

Mishra, V. K., Upadhyay, A. R., Pandey, S. K., & Tripathi, B. D. (2008). Heavy metal pollution in a tropical river, *Ganga*: A case study in a small stretch at Varanasi. *Environmental Monitoring and Assessment, 142*(1-3), 419–425. DOI: 10.1007/s10661-007-9938-1

Mitra, I., De, K., & Pal, D. (2025). Biochar: Insinuations in Agriculture and Bioremediation. *Food and Scientific Reports, 6*(1), 62–66.

Mitsanis, C., Hurst, W., & Tekinerdogan, B. (2024). A 3D functional plant modelling framework for agricultural digital twins. *Computers and Electronics in Agriculture, 218*, 108733. DOI: 10.1016/j.compag.2024.108733

Mitsch, W. J., Bernal, B., & Hernandez, M. E. (2010). Ecosystem services of wetlands. *The International Journal of Biodiversity Science, Ecosystem Services & Management, 6*(1–2), 1–4.

Mitsch, W. J., & Gosselink, J. G. (2015). *Wetlands* (5th ed.). Wiley.

Mittler, R., Vanderauwera, S., Gollery, M., & Van Breusegem, F. (2004). Reactive oxygen species signaling in plants. *Trends in Plant Science, 9*(10), 490–498. DOI: 10.1016/j.tplants.2004.08.009 PMID: 15465684

Mkandawire, M., & Dudel, E. G. (2005). Accumulation of arsenic in *Lemna gibba* L. (duckweed) in tailing waters of two abandoned uranium mining sites in Saxony, Germany. *The Science of the Total Environment, 336*(1–3), 81–89. DOI: 10.1016/j.scitotenv.2004.06.002 PMID: 15589251

Mocek-Płóciniak, A., Mencel, J., Zakrzewski, W., & Roszkowski, S. (2023). Phytoremediation as an Effective Remedy for Removing Trace Elements from Ecosystems. *Plants, 12*(8), 1653. DOI: 10.3390/plants12081653 PMID: 37111876

Mohamed, M. (2023). Agricultural sustainability in the age of deep learning: Current trends, challenges, and future trajectories. *Sustainable Machine Intelligence Journal, 4*, 2–1. DOI: 10.61185/SMIJ.2023.44102

Mohan, D., Sarswat, A., Ok, Y. S., & Pittman, C. U.Jr. (2014). Organic and inorganic contaminants removal from water with biochar, a renewable, low-cost and sustainable adsorbent – A critical review. *Bioresource Technology, 160*, 191–202. DOI: 10.1016/j.biortech.2014.01.120 PMID: 24636918

Mohanty, S., Pandey, P. C., Pandey, M., Srivastava, P. K., & Dwivedi, C. S. (2024). Wetlands contribution and linkage to support SDGs, its indicators and targets-A critical review. *Sustainable Development (Bradford), 32*(5), 5348–5392. DOI: 10.1002/sd.2948

Mohd, S., Kushwaha, A. S., Shukla, J., Mandrah, K., Shankar, J., Arjaria, N., Saxena, P. N., Khare, P., Narayan, R., Dixit, S., Siddiqui, M. H., Tuteja, N., Das, M., Roy, S. K., & Kumar, M. (2019). Fungal mediated biotransformation reduces toxicity of arsenic to soil dwelling microorganism and plant. *Ecotoxicology and Environmental Safety, 176*, 108–118. DOI: 10.1016/j.ecoenv.2019.03.053 PMID: 30925326

Molle, P., Liénard, A., Grasmick, A., Iwema, A., & Boucherie, H. (2008). A pilot-scale study of constructed wetlands for secondary and tertiary treatment of wastewater: Design and performance. *Ecological Engineering, 34*(2), 145–152.

Mondal, K., Jain, A., & Chakraborty, A. (2025). *Phytoremediation Meets Nanotechnology: A Cutting-Edge Solution for Environmental Detoxification.*

Montenegro, R. C., Araújo, A. J., Molina, M. T., Marinho Filho, J. D. B., Rocha, D. D., Lopéz-Montero, E., Goulart, M. O. F., Bento, E. S., Alves, A. P. N. N., Pessoa, C., de Moraes, M. O., & Costa-Lotufo, L. V. (2010). Cytotoxic activity of naphthoquinones with special emphasis on juglone and its 5-O-methyl derivative. *Chemico-Biological Interactions, 184*(3), 439–448. DOI: 10.1016/j.cbi.2010.01.041 PMID: 20138029

Moreira, H., Pereira, S. I., Mench, M., Garbisu, C., Kidd, P., & Castro, P. M. (2021). Phytomanagement of metal (loid)-contaminated soils: Options, efficiency and value. *Frontiers in Environmental Science, 9*, 661423. DOI: 10.3389/fenvs.2021.661423

Morel, F. M. M., Kraepiel, A. M. L., & Amyot, M. (1998). The chemical cycle and bioaccumulation of mercury. *Annual Review of Ecology and Systematics*, *29*(1), 543–566. DOI: 10.1146/annurev.ecolsys.29.1.543

Muramoto, S., & Oki, Y. (1983). Removal of some heavy metals from polluted water by water hyacinth (*Eichhorniacrassipes*). *Bulletin of Environmental Contamination and Toxicology*, *30*(1), 170–177. DOI: 10.1007/BF01610117 PMID: 6839042

Mustafa, H. M., & Hayder, G. (2021). Recent studies on applications of aquatic weed plants in phytoremediation of wastewater: A review article. *Ain Shams Engineering Journal*, *12*(1), 355–365. DOI: 10.1016/j.asej.2020.05.009

Muszyńska, E., Labudda, M., & Kral, A. (2020). Ecotype-Specific Pathways of Reactive Oxygen Species Deactivation in Facultative Metallophyte Silene vulgaris (Moench) Garcke Treated with Heavy Metals. *Antioxidants*, *9*(2), 102. DOI: 10.3390/antiox9020102 PMID: 31991666

Nagajyoti, P. C., Lee, K. D., & Sreekanth, T. V. M. (2010). Heavy metals, occurrence and toxicity for plants: A review. *Environmental Chemistry Letters*, *8*(3), 199–216. DOI: 10.1007/s10311-010-0297-8

Naidoo, S., & Olaniran, A. O. (2014). Treated wastewater effluent as a source of microbial pollution of surface water resources. *International Journal of Environmental Research and Public Health*, *11*(1), 249–270. DOI: 10.3390/ijerph110100249 PMID: 24366046

Nancharaiah, Y. V., Venkata Mohan, S., & Lens, P. N. L. (2016). Metals removal and recovery in bioreactors. *Environmental Science. Water Research & Technology*, *2*(5), 800–817.

Nandgude, N., Singh, T. P., Nandgude, S., & Tiwari, M. (2023). Drought prediction: A comprehensive review of different drought prediction models and adopted technologies. *Sustainability (Basel)*, *15*(15), 11684. DOI: 10.3390/su151511684

Nannini, L., Huyskes, D., Panai, E., Pistilli, G., & Tartaro, A. (2025). Nullius in Explanans: An ethical risk assessment for explainable AI. *Ethics and Information Technology*, *27*(1), 5. DOI: 10.1007/s10676-024-09800-7

National Material. (2025). *The circular economy of metal processing sustainability.* https://www.nationalmaterial.com/the-circular-economy-of-metal-processing-sustainability/

Ndeddy Aka, R. J., & Babalola, O. O. (2017). Identification and characterization of Cr-, Cd-, and Ni-tolerant bacteria isolated from mine tailings. *Bioremediation Journal*, *21*(1), 1–19. DOI: 10.1080/10889868.2017.1282933

Ndimele, P. E., Kumolu-Johnson, C. A., & Anetekhai, M. A. (2011). The invasive aquatic macrophyte *Eichhornia crassipes* (Mart.) Solms: Problems and prospects. *Research Journal of Environmental Sciences*, *5*(6), 509–520. DOI: 10.3923/rjes.2011.509.520

Nedjimi, B. (2021). Phytoremediation: A sustainable environmental technology for heavy metals decontamination. *SN Applied Sciences*, *3*(3), 286. DOI: 10.1007/s42452-021-04301-4

Neubauer, S. C., Franklin, R. B., & Berrier, D. J. (2008). Saltwater intrusion into tidal freshwater marshes alters the biogeochemical processing of organic carbon. *Biogeosciences*, *5*(2), 215–221. DOI: 10.5194/bg-5-215-2008

Nevo, Y., & Nelson, N. (2006). The NRAMP family of metal-ion transporters. *Biochimica et Biophysica Acta*, *1763*(7), 609–620. DOI: 10.1016/j.bbamcr.2006.05.007 PMID: 16908340

New Hampshire Department of Environmental Services. (2019). *Aquatic Plants and Their Role in Lake Ecology*.

Nguyen, T. Q., Sesin, V., Kisiala, A., & Emery, R. J. N. (2021). Phytohormonal Roles in Plant Responses to Heavy Metal Stress: Implications for Using Macrophytes in Phytoremediation of Aquatic Ecosystems. *Environmental Toxicology and Chemistry*, *40*(1), 7–22. DOI: 10.1002/etc.4909 PMID: 33074580

Nnaji, N. D., Onyeaka, H., Miri, T., & Ugwa, C. (2023). Bioaccumulation for heavy metal removal: A review. *SN Applied Sciences*, *5*(5), 125. DOI: 10.1007/s42452-023-05351-6

Nordstrom, D. K. (2011). Hydrogeochemical processes governing the origin, transport, and fate of major and trace elements from mine wastes and mineralized rock. *Applied Geochemistry*, *26*(11), 1777–1791. DOI: 10.1016/j.apgeochem.2011.06.002

Noyma, N. P., Cotovicz, L. C., & Brandini, N. (2020). Climate change effects on aquatic ecosystems: Impacts of extreme events on metal mobility. *Environmental Research Letters*, *15*(11), 113006. DOI: 10.1088/1748-9326/abb4cb

Nriagu, J. O. (1996). A history of global metal pollution. *Science*, *272*(5259), 223–224. DOI: 10.1126/science.272.5259.223

Ntagia, E.. (2023). Pilot and full-scale applications of floating treatment wetlands for urban and industrial effluents. *The Science of the Total Environment*, *872*, 162053.

Nwagbara, V. U. (2023). *Comparative Analysis of Water Hyacinth Efficiency as Biosorbent and Phyto remediating Plant for Removal of Lead (Pb)*. Water Contaminants.

Nyamunda, B. C., Chivhanga, T., Guyo, U., & Chigondo, F. (2019). Removal of Zn (II) and Cu (II) ions from industrial wastewaters using magnetic biochar derived from water hyacinth. *Journal of Engineering*, *2019*, 1–11. Advance online publication. DOI: 10.1155/2019/5656983

Nyquist, J., & Greger, M. (2007). Uptake of Zn, Cu, and Cd in metal loaded Elodea canadensis. *Environmental and Experimental Botany*, *60*(2), 219–226. DOI: 10.1016/j.envexpbot.2006.10.009

Nzengung, V. A., & Jeffers, P. (2001). Sequestration, Phytoreduction, and Phytooxidation of Halogenated Organic Chemicals by Aquatic and Terrestrial Plants. *International Journal of Phytoremediation*, *3*(1), 13–40. DOI: 10.1080/15226510108500048

Obadimu, C. O., Shaibu, S. E., Ekwere, I. O., & Adelagun, R. O. A. (2024). Machine Learning-Based Forecasting of Bioaccumulation and Histopathological Effects in Aquatic Organisms. *Fudma Journal of Sciences*, *8*(6), 485–496. DOI: 10.33003/fjs-2024-0806-3002

Ogunmayowa, O. T. (2015). *Coupling bio/phytoremediation with switchgrass to biofuel feedstock production in mixed-contaminant soils* (Master's thesis, Tennessee State University).

Ojuederie, O. B., & Babalola, O. O. (2017). Microbial and plant-assisted bioremediation of heavy metal polluted environments: A review. *International Journal of Environmental Research and Public Health*, *14*(12), 1504. DOI: 10.3390/ijerph14121504 PMID: 29207531

Osman, M. E., & Fadhlallah, N. (2010). Impact of heavy metals on seed germination and stress physiology. *Plant Stress Physiology*, *34*, 45–56.

Oursel, B., Garnier, C., Durrieu, G., Mounier, S., Omanović, D., & Lucas, Y. (2014). Dynamics and fates of trace metals in an urbanized coastal zone: The case study of Marseille Bay (France). *Marine Pollution Bulletin*, *84*(1–2), 29–38. DOI: 10.1016/j.marpolbul.2014.05.053

Pacyna, J. M., & Pacyna, E. G. (2001). An assessment of global and regional emissions of trace metals to the atmosphere from anthropogenic sources worldwide. *Environmental Reviews*, *9*(4), 269–298. DOI: 10.1139/a01-012

Pajević, S., Borišev, M., Rončević, S., Vukov, D., & Igić, R. (2008). Heavy metal accumulation of Danube River aquatic plants — Indication of chemical contamination. *Open Life Sciences*, *3*(3), 285–294. DOI: 10.2478/s11535-008-0017-6

Pal, D., Bandyopadhyay, P., Bhowmick, S., & Maji, P. (2025). Heavy Metal Detection and Antibacterial Activity of Phytosynthesized Copper oxide (CuO) Nanoparticles using Leaf Extract of Colocasia esculenta. *International Journal of Ecology and Environmental Sciences*, *51*(1), 161–166. DOI: 10.55863/ijees.2025.0524

Pandey, A., Tripathi, N., & Tripathi, R. D. (2023). Engineered nano-bio interactions for enhanced phytoremediation: Current insights and future prospects. *The Science of the Total Environment*, *882*, 163600.

Pang, Y. L., Quek, Y. Y., Lim, S., & Shuit, S. H. (2023). Review on phytoremediation potential of floating aquatic plants for heavy metals: A promising approach. *Sustainability (Basel)*, *15*(2), 1290. DOI: 10.3390/su15021290

Pan, J., Zou, X., Wang, Q., Song, Y., & Liu, Y. (2011b). Carbon sequestration through constructed wetlands: A meta-analysis. *Ecological Engineering*, *37*, 1083–1091. DOI: 10.1016/j.ecoleng.2011.02.010

Pantelopoulos, A., & Bourbakis, N. G. (2010). A survey on wearable sensor-based systems for health monitoring and prognosis. *IEEE Transactions on Systems, Man, and Cybernetics. Part C, Applications and Reviews*, *40*(1), 1–12. DOI: 10.1109/TSMCC.2009.2032660

Pan, Y., Birdsey, R. A., Fang, J., Houghton, R., Kauppi, P. E., Kurz, W. A., Phillips, O. L., Shvidenko, A., Lewis, S. L., Canadell, J. G., Ciais, P., Jackson, R. B., Pacala, S. W., McGuire, A. D., Piao, S., Rautiainen, A., Sitch, S., & Hayes, D. (2011a). A Large and Persistent Carbon Sink in the World's Forests. *Science*, *333*(6045), 988–993. DOI: 10.1126/science.1201609 PMID: 21764754

Pan, Y., & Zhang, L. (2023). Integrating BIM and AI for smart construction management: Current status and future directions. *Archives of Computational Methods in Engineering*, *30*(2), 1081–1110. DOI: 10.1007/s11831-022-09830-8

Parida, T., Riyazuddin, S., Kolli, S. K., Chakraborty, A., Srinivas, N., Kundu, P., Bhattacharya, S., Seth, C. S., & Biswas, J. K. (2024). Phytoremediation of heavy metal(loid)s with integral involvement of the endogenous metal-chelators: Present state-of-the-art and future prospect. *Discover Environment*, *2*(1), 139. DOI: 10.1007/s44274-024-00170-x

Pasricha, S., Mathur, V., Garg, A., Lenka, S., Verma, K., & Agarwal, S. (2021). Molecular mechanisms underlying heavy metal uptake, translocation and tolerance in hyperaccumulators-an analysis. *Environmental Challenges*, *4*, 100197. DOI: 10.1016/j.envc.2021.100197

Patel A. K., Singhania R. R., Albarico F. P. J. B., Pandey A., Chen C.-W., Dong C.-D. (2022). Organic wastes bioremediation and its changing prospects. Sci. Total Environ. 824, 153889.

Patel, S. (2012). Threats, management and envisaged utilizations of aquatic weed *Eichhornia crassipes*: An overview. *Reviews in Environmental Science and Biotechnology, 11*(3), 249–259. DOI: 10.1007/s11157-012-9289-4

Paudel, K. R., & Panth, N. (2015). Phytochemical profile and biological activity of *Nelumbo nucifera. Evidence-Based Complementary and Alternative Medicine : eCAM, 2015*, 789124. Advance online publication. DOI: 10.1155/2015/789124 PMID: 27057194

Paz-Alberto, A. M., & Sigua, G. C. (2013). Phytoremediation: A Green Technology to Remove Environmental Pollutants. *American Journal of Climate Change, 02*(01), 71–86. DOI: 10.4236/ajcc.2013.21008

Peladarinos, N., Piromalis, D., Cheimaras, V., Tserepas, E., Munteanu, R. A., & Papageorgas, P. (2023). Enhancing Smart Agriculture by Implementing Digital Twins: A Comprehensive Review. *Sensors (Basel), 23*(16), 7128. Advance online publication. DOI: 10.3390/s23167128 PMID: 37631663

Peleg, Z., & Blumwald, E. (2011). Hormone balance and abiotic stress tolerance in crop plants. *Current Opinion in Plant Biology, 14*(3), 290–295. DOI: 10.1016/j.pbi.2011.02.001 PMID: 21377404

Peng, J.-S., & Gong, J.-M. (2014). Vacuolar sequestration capacity and long-distance metal transport in plants. *Frontiers in Plant Science, 5*. Advance online publication. DOI: 10.3389/fpls.2014.00019 PMID: 24550927

Peng, Y., Du, K., Yue, H., Li, H., Li, H., Liu, M., Shangguan, S., He, X., Li, X., & Chang, Y. (2025). Integrated deep eutectic system enrichment and AI-assisted high-throughput visual detection for Hg2+ in environmental samples. *Journal of Advanced Research*. Advance online publication. DOI: 10.1016/j.jare.2025.04.011 PMID: 40220898

Pereira, D., Barros, M. L., & Arruda, M. A. Z. (2021a). Evaluation of cadmium bioaccumulation-related physiological effects in *Salvinia biloba. Plants, 10*(12), 2679. DOI: 10.3390/plants10122679 PMID: 34961150

Pereira, P., Fernando, A. L., & Caeiro, S. (2021c). Assessment of trace metal contamination and bioavailability in sediments: Implications for ecosystem health. *Environmental Pollution, 277*, 116842. DOI: 10.1016/j.envpol.2021.116842

Perrin, J. E., Jernigan, S. R., Thayer, J. D., Howell, A. W., Leary, J. K., & Buckner, G. D. (2022). Sensor Fusion with Deep Learning for Autonomous Classification and Management of Aquatic Invasive Plant Species. *Robotics (Basel, Switzerland), 11*(4), 68. DOI: 10.3390/robotics11040068

Pietta, P. G. (2000). Flavonoids as antioxidants. *Journal of Natural Products, 63*(7), 1035–1042. DOI: 10.1021/np9904509 PMID: 10924197

Pilon-Smits, E. (2005). Phytoremediation. *Annual Review of Plant Biology, 56*(1), 15–39. DOI: 10.1146/annurev.arplant.56.032604.144214 PMID: 15862088

Pivetz, B. E. (2001). *Phytoremediation of contaminated soil and ground water at hazardous waste sites.* US Environmental Protection Agency, Office of Research and Development, Office of Solid Waste and Emergency Response.

Pivetz, B. E. (2001). *Phytoremediation of contaminated soil and groundwater at hazardous waste sites (EPA Ground Water Issue EPA/540/S-01/500).* U.S. Environmental Protection Agency.

Polechońska, L., Klink, A., & Sokołowska, K. (2024). Rising temperature impacts the trace metal uptake and toxicity in aquatic plants – A case study of Ni and Co in *Elodea canadensis* Michx. *The Science of the Total Environment, 955*, 177307. DOI: 10.1016/j.scitotenv.2024.177307 PMID: 39481559

Pollack Porter, K. M., Rutkow, L., & McGinty, E. E. (2018). The Importance of Policy Change for Addressing Public Health Problems. *Public Health Reports, 133*(1, suppl), 9S–14S. DOI: 10.1177/0033354918788880 PMID: 30426876

Ponnuru, A., Madhuri, J. V., Saravanan, S., Vijayakumar, T., Manimegalai, V., & Das, A. (2025). Data-Driven Approaches to Water Quality Monitoring: Leveraging AI, Machine Learning, and Management Strategies for Environmental Protection. *Journal of Neonatal Surgery, 14*(5S), 664–675. DOI: 10.52783/jns.v14.2107

Pourrut, B.. (2011). Lead and cadmium toxicity: Oxidative stress in plant cells. *Plant Physiology and Biochemistry, 49*(6), 633–638. DOI: 10.1016/j.plaphy.2011.03.004

Pourrut, B., Shahid, M., Dumat, C., Winterton, P., & Pinelli, E. (2011). Lead uptake, toxicity, and detoxification in plants. *Reviews of Environmental Contamination and Toxicology, 213*, 113–136. PMID: 21541849

Prabha, P. S., Sadhana, S., Karthik, C., & Caroline, D. G. (2020). *In vitro* evaluation of *Hydrilla verticillata* for anti-adipogenesis activity on 3T3 L1 cell lines. *Pharmacognosy Magazine, 16*(5), 498–505. DOI: 10.4103/pm.pm_234_20

Prasad, M. N. V. (2004). *Heavy metal stress in plants: From biomolecules to ecosystems* (2nd ed.). Springer., DOI: 10.1007/978-3-662-07743-6

Prasad, M. N. V., & De Oliveira Freitas, H. M. (2003). Metal hyperaccumulation in plants—Biodiversity prospecting for phytoremediation technology. *Electronic Journal of Biotechnology*, *6*(3), 110–146. DOI: 10.2225/vol6-issue3-fulltext-6

Priya, A. K., Gnanasekaran, L., Dutta, K., Rajendran, S., Balakrishnan, D., & Soto-Moscoso, M. (2022). Biosorption of heavy metals by microorganisms: Evaluation of different underlying mechanisms. *Chemosphere*, *307*, 135957. DOI: 10.1016/j.chemosphere.2022.135957 PMID: 35985378

Priya, A. K., Muruganandam, M., Ali, S. S., & Kornaros, M. (2023). Clean-Up of Heavy Metals from Contaminated Soil by Phytoremediation: A Multidisciplinary and Eco-Friendly Approach. *Toxics*, *11*(5), 422. DOI: 10.3390/toxics11050422 PMID: 37235237

Pushpamala, E., Dhanalakshmi, S., & Govindarajan, V. S. (2014). Effect of mercury and cadmium on biochemical parameters of *Hydrilla verticillata*. *Life Science–Environmental Biology*, *3*(2), L58–L66.

Qayoom, I., & Jaies, I. (2023). Review on the phytoremediation potential of aquatic macrophytes. *Environment Conservation Journal*, *24*(1), 273–280. DOI: 10.36953/ECJ.12102318

Raghavendra, R., & Kumar, N. (2023). Application of WEKA Machine Learning Tools in Phytoremediation for the Removal of Heavy Metals in Water Treatment Processes. *2023 International Conference on Communication, Security and Artificial Intelligence, ICCSAI 2023*, 219–223. DOI: 10.1109/ICCSAI59793.2023.10421571

Raghuvanshi, R., Sharma, T., Rena, R., & Rai, R. (2024). Artificial Intelligence for Smart City Development in BRICS Economies: Opportunities and Challenges. *Handbook of Artificial Intelligence for Smart City Development*, 173-208.

Rahman, M. A., Reichman, S. M., De Filippis, L., Tavakoly Sany, S. B., & Hasegawa, H. (2016). Phytoremediation of toxic metals in soils and wetlands: concepts and applications. *Environmental remediation technologies for metal-contaminated soils*, 161-195.

Rahman, M. A., & Hasegawa, H. (2011). Aquatic arsenic phytoremediation using floating macrophytes: A review. *Chemosphere*, *83*(5), 633–646. DOI: 10.1016/j.chemosphere.2011.02.045 PMID: 21435676

Rai, K. K., Pandey, N., Meena, R. P., & Rai, S. P. (2021). Biotechnological strategies for enhancing heavy metal tolerance in neglected and underutilized legume crops: A comprehensive review. *Ecotoxicology and Environmental Safety*, *208*, 111750. DOI: 10.1016/j.ecoenv.2020.111750 PMID: 33396075

Rai, P. K. (2008). Heavy metal phytoremediation by aquatic macrophytes. *International Journal of Phytoremediation*, *10*(2), 133–160. DOI: 10.1080/15226510801913918 PMID: 18709926

Rai, P. K. (2009). Heavy metal phytoremediation from aquatic ecosystems with special reference to macrophytes. *Critical Reviews in Environmental Science and Technology*, *39*(9), 697–753. DOI: 10.1080/10643380801910058

Rai, P. K., Lee, S. S., Zhang, M., Tsang, Y. F., & Kim, K. H. (2019). Heavy metals in food crops: Health risks, fate, mechanisms, and management. *Environment International*, *125*, 365–385. DOI: 10.1016/j.envint.2019.01.067 PMID: 30743144

Rai, P. K., Lee, S. S., Zhang, M., Tsang, Y. F., & Kim, K. H. (2022). Heavy metals in aquatic ecosystems and their phytoremediation using wetland plants: An eco-environmental perspective. *Environmental Geochemistry and Health*, *44*, 1–28.

Rai, P. K., & Nongtri, E. S. (2024). Heavy metals/-metalloids (As) phytoremediation with Landoltia punctata and Lemna sp.(duckweeds): Coupling with biorefinery prospects for sustainable phytotechnologies. *Environmental Science and Pollution Research International*, *31*(11), 16216–16240. DOI: 10.1007/s11356-024-32177-5 PMID: 38334920

Rajasulochana, P., & Preethy, V. (2016). Comparison on efficiency of various techniques in treatment of waste and sewage water – A comprehensive review. *Resource-Efficient Technologies*, *2*(4), 175–184. DOI: 10.1016/j.reffit.2016.09.004

Rajitha, A., Aravinda, K., Nagpal, A., Kalra, R., Maan, P., Kumar, A., & Abdul-Zahra, D. S. (2024). Machine Learning and AI-Driven Water Quality Monitoring and Treatment. *E3S Web of Conferences*, *505*. DOI: 10.1051/e3sconf/202450503012

Rajkumar, M., Ma, Y., & Freitas, H. (2012). Improvement of Ni phytostabilization by inoculation of Ni-resistant *Bacillus megaterium* SR28C. *Journal of Environmental Management*, *95*, S231–S236. DOI: 10.1016/j.jenvman.2011.06.034 PMID: 23895909

Raj, M. (2021). A Review on the Effect of Heavy Metal Contamination and its Impact on the Environment. *International Journal of Zoological Investigations.*, *7*(2), 762–771. DOI: 10.33745/ijzi.2021.v07i02.061

Rajput, J. D., & Singh, S. P. (2023). Water chestnut (*Trapa natans* L.): Functional characteristics, nutritional properties and applications in food industry—A review. *Journal of Phytopharmacology, 12*(2), 209. DOI: 10.31254/phyto.2023.12209

Rajput, V. D., Minkina, T., Sushkova, S., Behal, A., & Mandzhieva, S. (2018). Interaction of nanoparticles with plants: Impacts on growth and phytotoxicity. *Nanomaterials (Basel, Switzerland), 8*(9), 745.

Raklami, A., Oufdou, K., Tahiri, A. I., Mateos-Naranjo, E., Navarro-Torre, S., Rodríguez-Llorente, I. D., Meddich, A., Redondo-Gómez, S., & Pajuelo, E. (2019). Safe cultivation of Medicago sativa in metal-polluted soils from semi-arid regions assisted by heat-and metallo-resistant PGPR. *Microorganisms, 7*(7), 212. DOI: 10.3390/microorganisms7070212 PMID: 31336693

Ramsar Convention Secretariat. (2021). *Ramsar Wetland Conservation Awards.* Ramsar Convention.

Rane, N. L., Paramesha, M., Choudhary, S. P., & Rane, J. (2024). Machine learning and deep learning for big data analytics: A review of methods and applications. *SSRN, 2*(3), 172–197. DOI: 10.2139/ssrn.4835655

Rashid, A., Schutte, B. J., Ulery, A., Deyholos, M. K., Sanogo, S., Lehnhoff, E. A., & Beck, L. (2023). Heavy Metal Contamination in Agricultural Soil: Environmental Pollutants Affecting Crop Health. *Agronomy (Basel), 13*(6), 1521. DOI: 10.3390/agronomy13061521

Rastogi, A., Zivcak, M., Sytar, O., Kalaji, H. M., He, X., Mbarki, S., & Brestic, M. (2019). Nanoparticles and plants: Interaction at the organismal level. *Plant Physiology and Biochemistry, 134*, 1–15. DOI: 10.1016/j.plaphy.2018.10.014

Rayu, S., Karpouzas, D. G., & Singh, B. K. (2012). Emerging technologies in bioremediation: Constraints and opportunities. *Biodegradation, 23*(6), 917–926. DOI: 10.1007/s10532-012-9576-3 PMID: 22836784

Raza, A., Tabassum, J., Zahid, Z., Charagh, S., Bashir, S., Barmukh, R., Khan, R. S. A., Barbosa, F.Jr, Zhang, C., Chen, H., Zhuang, W., & Varshney, R. K. (2022). Advances in 'Omics' approaches for improving toxic metal tolerance in plants. *Frontiers in Plant Science, 13*, 8764127. DOI: 10.3389/fpls.2021.794373

Reddy, C. N.Chalamalla Nikhitha Reddy. (2025). Explainable Artificial Intelligence (XAI) for Climate Hazard Assessment: Enhancing Predictive Accuracy and Transparency in Drought, Flood, and Landslide Modeling. *International Journal on Science and Technology, 16*(1), 1309. Advance online publication. DOI: 10.71097/IJSAT.v16.i1.1309

Reddy, K. R., & D'Angelo, E. M. (1997). Biogeochemical indicators to evaluate pollutant removal efficiency in constructed wetlands. *Water Science and Technology*, *35*(5), 1–10. DOI: 10.2166/wst.1997.0152

Reddy, K. R., & DeBusk, T. A. (1984). Growth characteristics of aquatic macrophytes cultured in nutrient-enriched water: II. *Pistia stratiotes* L. *Journal of Environmental Quality*, *13*(1), 61–65.

Reddy, K. R., & Delaune, R. D. (2008). *Biogeochemistry of wetlands: Science and applications*. CRC Press. DOI: 10.1201/9780203491454

Rehman, A. U., Khan, A. U., Sohaib, M., & Rehman, H. (2024). Comparative analysis of nutritional properties, phytochemical profile, and antioxidant activities between red and green water chestnut (*Trapa natans*) fruits. *Foods*, *13*(12), 1883. DOI: 10.3390/foods13121883 PMID: 38928824

Ren, X., Zeng, G., Tang, L., Wang, J., Wan, J., Wang, J., Deng, Y., Liu, Y., & Peng, B. (2018). The potential impact on the biodegradation of organic pollutants from composting technology for soil remediation. *Waste Management (New York, N.Y.)*, *72*, 138–149. DOI: 10.1016/j.wasman.2017.11.032 PMID: 29183697

Rezania, S.. (2016). Perspectives of phytoremediation using water hyacinth for the removal of heavy metals. *Process Safety and Environmental Protection*, *102*, 128–138. DOI: 10.1016/j.psep.2016.03.017

Rezania, S., Park, J., & Rupani, P. F.. (2016a). Phytoremediation potential and cadmium uptake by water hyacinth from wastewater: A laboratory to field study. *Environmental Science and Pollution Research International*, *23*, 13401–13411.

Rezania, S., Ponraj, M., Din, M. F. M., Taib, S. M., Song, S. T., & Sairan, F. M. (2015). The diverse applications of water hyacinth with main focus on sustainable energy and production for new era: An overview. *Renewable & Sustainable Energy Reviews*, *41*, 943–954. DOI: 10.1016/j.rser.2014.09.006

Rezania, S., Taib, S. M., Din, M. F. M., Dahalan, F. A., & Kamyab, H. (2016b). Comprehensive review on phytotechnology: Heavy metals removal by diverse aquatic plant species from wastewater. *Journal of Hazardous Materials*, *318*, 587–599. DOI: 10.1016/j.jhazmat.2016.07.053 PMID: 27474848

Ribeiro, R., Pilastri, A., Moura, C., Morgado, J., & Cortez, P. (2023). A data-driven intelligent decision support system that combines predictive and prescriptive analytics for the design of new textile fabrics. *Neural Computing & Applications*, *35*(23), 17375–17395. DOI: 10.1007/s00521-023-08596-9

Ricachenevsky, F. K., Menguer, P. K., Sperotto, R. A., & Fett, J. P. (2013). Got to hide your Zn away: Molecular control of Zn accumulation and biotechnological applications. *Plant Science*, *211*, 1–12. PMID: 26025516

Rodrigo, M. A. (2021). Wetland Restoration with Hydrophytes: A Review. *Plants*, *10*(6), 1035. DOI: 10.3390/plants10061035 PMID: 34063930

Rolnick, D., Donti, P. L., Kaack, L. H., Kochanski, K., Lacoste, A., Sankaran, K., Ross, A. S., Milojevic-Dupont, N., Jaques, N., Waldman-Brown, A., Luccioni, A. S., Maharaj, T., Sherwin, E. D., Mukkavilli, S. K., Kording, K. P., Gomes, C. P., Ng, A. Y., Hassabis, D., Platt, J. C., & Bengio, Y. (2022). Tackling climate change with machine learning. *ACM Computing Surveys*, *55*(2), 1–96. DOI: 10.1145/3485128

Ronzhina, D. A., Ivanov, L. A., Lambers, G., & P'yankov, V. I. (2009). Changes in chemical composition of hydrophyte leaves during adaptation to aquatic environment. *Russian Journal of Plant Physiology: a Comprehensive Russian Journal on Modern Phytophysiology*, *56*(3), 355–362. DOI: 10.1134/S102144370903008X

Rosati, R., Romeo, L., Cecchini, G., Tonetto, F., Viti, P., Mancini, A., & Frontoni, E. (2023). From knowledge-based to big data analytic model: A novel IoT and machine learning based decision support system for predictive maintenance in Industry 4.0. *Journal of Intelligent Manufacturing*, *34*(1), 107–121. DOI: 10.1007/s10845-022-01960-x

Royal Society of Chemistry. (2023). *Contamination of heavy metals in aquatic media: Transport, toxicity and technologies for remediation*. Royal Society of Chemistry., DOI: 10.1039/9781839165111

Russ, D., Fitzpatrick, C. R., Teixeira, P. J., & Dangl, J. L. (2023). Deep discovery informs difficult deployment in plant microbiome science. *Cell*, *186*(21), 4496–4513. DOI: 10.1016/j.cell.2023.08.035 PMID: 37832524

Saba, B., Batool, S., Ahmad, M. I., & Kalsoom, U. (2021). Bioelectrochemical wetlands: An emerging hybrid technology for wastewater treatment and energy generation. *Chemosphere*, *263*, 127951. DOI: 10.1016/j.chemosphere.2020.127951

Sabreena, H. S., Hassan, S., Bhat, S. A., Kumar, V., Ganai, B. A., & Ameen, F. (2022). Phytoremediation of Heavy Metals: An Indispensable Contrivance in Green Remediation Technology. *Plants*, *11*(9), 1255. DOI: 10.3390/plants11091255 PMID: 35567256

Saeed, T., & Sun, G. (2012). A review on nitrogen and organics removal mechanisms in subsurface flow constructed wetlands: Dependency on environmental parameters, operating conditions and supporting media. *Journal of Environmental Management, 112*, 429–448. DOI: 10.1016/j.jenvman.2012.08.011 PMID: 23032989

Saha, A., Mukherjee, P., Roy, K., Sen, K., & Sanyal, T. (2022). A review on phytoremediation by aquatic macrophytes: A natural promising tool for sustainable management of ecosystem. *International Journal of Experimental Research and Review, 27*, 9–31. DOI: 10.52756/ijerr.2022.v27.002

Saha, S., Huang, L., Khoso, M. A., Wu, H., Han, D., Ma, X., Poudel, T. R., Li, B., Zhu, M., Lan, Q., Sakib, N., Wei, R., Islam, M., Zhang, P., & Shen, H. (2023). Fine root decomposition in forest ecosystems: An ecological perspective. *Frontiers in Plant Science, 14*, 1277510. Advance online publication. DOI: 10.3389/fpls.2023.1277510 PMID: 38023858

Sahoo, A., Sharma, S., Pattnaik, S., Kumar, M. R., & Shrivastava, V. K. (2024). AI-Driven Crop Health Monitoring and Efficient Irrigation Management System. *2024 IEEE 4th International Conference on Applied Electromagnetics, Signal Processing, and Communication, AESPC 2024*, 1–6. DOI: 10.1109/AESPC63931.2024.10872127

Sahoo, A., Chhotaray, S. P., Meher, I., Behera, S. P., Pal, A., Meena, M., Swapnil, P., Yadav, A., & Bhardwaj, R. (2025). Phytoremediation for a sustainable future: Integrating plant based strategies in soil and wastewater remediation. *Bioresource Technology Reports, 31*, 102266. DOI: 10.1016/j.biteb.2025.102266

Saikia, K., Dey, S., Hazarika, S. N., Handique, G. K., Thakur, D., & Handique, A. K. (2023). Chemical and biochemical characterization of *Ipomoea aquatica*: Genoprotective potential and inhibitory mechanism of its phytochemicals against α-amylase and α-glucosidase. *Frontiers in Nutrition, 10*, 1304903. DOI: 10.3389/fnut.2023.1304903 PMID: 38192648

Sakakibara, M., Watanabe, A., Inoue, M., Sano, S., & Kaise, T. (2010). Phytoextraction and phytovolatilization of arsenic from As-contaminated soils by Pteris vittata. In *Proceedings of the annual international conference on soils, sediments, water and energy*. 12(1) 26.

Salehizadeh, H., & Shojaosadati, S. A. (2003). Removal of metal ions from aqueous solution by polysaccharide produced from Bacillus firmus. *Water Research, 37*(17), 4231–4235. DOI: 10.1016/S0043-1354(03)00418-4 PMID: 12946905

Sales da Silva, I. G., Gomes de Almeida, F. C., Padilha da Rocha e Silva, N. M., Casazza, A. A., Converti, A., & Asfora Sarubbo, L. (2020). Soil bioremediation: Overview of technologies and trends. *Energies, 13*(18), 4664. DOI: 10.3390/en13184664

Salt, D. E., Baxter, I., & Lahner, B. (2018). Ionomics and the study of the plant ionome. *Annual Review of Plant Biology*, *69*, 481–504. DOI: 10.1146/annurev-arplant-042817-040212 PMID: 18251712

Salt, D. E., Blaylock, M., Kumar, N. P. B. A., Dushenkov, V., Ensley, B. D., Chet, I., & Raskin, I. (1995). Phytoremediation: A novel strategy for the removal of toxic metals from the environment using plants. *Nature Biotechnology*, *13*(5), 468–474. DOI: 10.1038/nbt0595-468 PMID: 9634787

Salt, D. E., Smith, R. D., & Raskin, I. (1998). Phytoremediation. *Annual Review of Plant Physiology and Plant Molecular Biology*, *49*(1), 643–668. DOI: 10.1146/annurev.arplant.49.1.643 PMID: 15012249

Samstag, R. W., Ducoste, J. J., Griborio, A., Nopens, I., Batstone, D. J., Wicks, J. D., & Laurent, J. (2016). CFD for wastewater treatment: An overview. *Water Science and Technology*, *74*(3), 549–563. DOI: 10.2166/wst.2016.249 PMID: 27508360

Sanders, D., Pelloux, J., Brownlee, C., & Harper, J. F. (2002). Calcium at the crossroads of signaling. *The Plant Cell*, *14*(Suppl. 1), S401–S417. DOI: 10.1105/tpc.002899 PMID: 12045291

Sand-Jensen, K. (2003). Drag forces on common plant species in temperate streams: Consequences for growth. *Hydrobiologia*, *481*(1-3), 25–30. DOI: 10.1023/A:1021299014644

Santos, D., Duarte, B., & Caçador, I. (2015). Biochemical and photochemical feedbacks of acute Cd toxicity in Juncus acutus seedlings: The role of non-functional Cd-chlorophylls. *Estuarine, Coastal and Shelf Science*, *167*, 228–239. DOI: 10.1016/j.ecss.2015.10.005

Sarah Lee. (2025). *Circular economy in environmental remediation.* https://www.numberanalytics.com/blog/circular-economy-environmental-remediation

Sari, M., Yalcin, I. E., Taner, M., Cosgun, T., & Ozyigit, I. I. (2023). An investigation on environmental pollution due to essential heavy metals: A prediction model through multilayer perceptrons. *International Journal of Phytoremediation*, *25*(1), 89–97. DOI: 10.1080/15226514.2022.2059056 PMID: 35400247

Sarma, H., Sahariah, G., Bharadwaj, A., Sharma, D., Porasar, P., & Dutta, K. N. (2024). A systematic review on botanical background, phytochemical and pharmacological properties of *Nymphaea nouchali. Journal of Applied Pharmaceutical Research*, *12*(6), 21–36. DOI: 10.69857/joapr.v12i6.714

Satrio, R. D., Fendiyanto, M. H., & Miftahudin, M. (2024). Tools and techniques used at global scale through genomics, transcriptomics, proteomics, and metabolomics to investigate plant stress responses at the molecular level. In *Molecular Dynamics of Plant Stress and its Management* (pp. 555–607). Springer Nature Singapore. DOI: 10.1007/978-981-97-1699-9_25

Schmidt, R., Hrynkiewicz, K., & Baum, C. (2008). Management of plant biomass in metal-polluted areas: Challenges and opportunities. *Environmental Pollution, 153*(3), 567–576. PMID: 18571819

Schooler, S. S., Julien, M. H., & Walsh, G. C. (2006). Predicting the response of *Pistia stratiotes* to biological control agents using a population model. *Biological Control, 38*(3), 391–400.

Sekomo, C. B., Rousseau, D. P. L., Saleh, S. A., & Lens, P. N. L. (2012). Heavy metal removal in duckweed-based stabilization ponds as a polishing step for UASB treated domestic wastewater. *Ecological Engineering, 44*, 102–110. DOI: 10.1016/j.ecoleng.2012.03.003

Seregin, I. V., & Kozhevnikova, A. D. (2023). Phytochelatins: Sulfur-Containing Metal(loid)-Chelating Ligands in Plants. *International Journal of Molecular Sciences, 24*(3), 2430. DOI: 10.3390/ijms24032430 PMID: 36768751

Shahid, M., Dumat, C., Khalid, S., Schreck, E., Xiong, T., & Niazi, N. K. (2017). Foliar heavy metal uptake, toxicity and detoxification in plants: A comparison of foliar and root metal uptake. *Journal of Hazardous Materials, 325*, 36–58. DOI: 10.1016/j.jhazmat.2016.11.063 PMID: 27915099

Shahid, M., Pourrut, B., Dumat, C., Nadeem, M., Aslam, M., & Pinelli, E. (2014). Heavy-metal-induced reactive oxygen species: Phytotoxicity and physicochemical changes in plants. *Reviews of Environmental Contamination and Toxicology, 232*, 1–44. PMID: 24984833

Shahid, S. A., Zaman, M., & Heng, L. (2018). *Soil Salinity: Assessment, Management, and Environmental Impact*. Springer.

Shanker, A. K., Cervantes, C., Loza-Tavera, H., & Avudainayagam, S. (2005). Chromium toxicity in plants. *Environment International, 31*(5), 739–753. DOI: 10.1016/j.envint.2005.02.003 PMID: 15878200

Sharma, J. K., Kumar, N., Singh, N. P., & Santal, A. R. (2023). Phytoremediation technologies and their mechanism for removal of heavy metal from contaminated soil: An approach for a sustainable environment. *Frontiers in Plant Science, 14*, 1076876. DOI: 10.3389/fpls.2023.1076876 PMID: 36778693

Sharma, M., Kant, R., Sharma, A. K., & Sharma, A. K. (2025). Exploring the impact of heavy metals toxicity in the aquatic ecosystem. *International Journal of Energy and Water Resources*, *9*(1), 267–280. DOI: 10.1007/s42108-024-00284-1

Sharma, M., Rawat, S., & Rautela, A. (2024). Phytoremediation in sustainable wastewater management: An eco-friendly review of current techniques and future prospects. *AQUA - Water Infrastructure. Aqua (London, England)*, *73*(9), 1946–1975. DOI: 10.2166/aqua.2024.427

Sharma, P., & Dubey, R. S. (2005). Lead toxicity in plants. *Brazilian Journal of Plant Physiology*, *17*(1), 35–52. DOI: 10.1590/S1677-04202005000100004

Sharma, P., Singh, S. P., Parakh, S. K., & Tong, Y. W. (2022). Health hazards of hexavalent chromium (Cr (VI)) and its microbial reduction. *Bioengineered*, *13*(3), 4923–4938. DOI: 10.1080/21655979.2022.2037273 PMID: 35164635

Sharma, R. K., Tripathi, R. D., & Srivastava, S. (2020). Synergistic effects of PGPR and nanoparticles on phytoremediation of heavy metals. *Environmental Nanotechnology, Monitoring & Management*, *13*, 100275.

Sharma, R., & Lenaghan, S. C. (2022). Duckweed: A potential phytosensor for heavy metals. *Plant Cell Reports*, *41*(12), 2231–2243. DOI: 10.1007/s00299-022-02913-7 PMID: 35980444

Sharma, S. S., & Dietz, K. J. (2009). The relationship between metal toxicity and cellular redox imbalance. *Trends in Plant Science*, *14*(1), 43–50. DOI: 10.1016/j.tplants.2008.10.007 PMID: 19070530

Sharma, S., Singh, S., & Goyal, P. (2019). Heavy metal removal from wastewater using chemical and biological methods. *Water Science and Technology*, *80*(12), 2155–2169.

Shen, T., Peng, H., Yuan, X., Liang, Y., Liu, S., Wu, Z., Leng, L., & Qin, P. (2024). Feature engineering for improved machine-learning-aided studying heavy metal adsorption on biochar. *Journal of Hazardous Materials*, *466*, 133442. DOI: 10.1016/j.jhazmat.2024.133442 PMID: 38244458

Sheoran, A. S., & Sheoran, V. (2006). Heavy metal removal mechanism of acid mine drainage in wetlands: A critical review. *Minerals Engineering*, *19*(2), 105–116. DOI: 10.1016/j.mineng.2005.08.006

Shilla, D. A., Asaeda, T., & Fujino, T. (2006). Effects of floating macrophytes on the nutrient dynamics in a shallow lake. *Limnology*, *7*(1), 27–34.

Shourie, A. (2022). Effect of cadmium and lead stress on seed germination in *Jatropha curcas*. *Biosciences Biotechnology Research Asia*, *19*(3), 671–678. Advance online publication. DOI: 10.13005/bbra/3019

Shuaib, M., Azam, N., Bahadur, S., Romman, M., Yu, Q., & Xuexiu, C. (2021). Variation and succession of microbial communities under the conditions of persistent heavy metal and their survival mechanism. *Microbial Pathogenesis*, *150*, 104713. DOI: 10.1016/j.micpath.2020.104713 PMID: 33387608

Singh R (2013). Microbial waste management. Lap lambert Academic Publishing

Singh, A., Prasad, S. M., Singh, S., & Singh, M. (2016). Phytoremediation potential of weed plants' oxidative biomarker and antioxidant responses. *Chemistry and Ecology*, *32*(7), 684–706. DOI: 10.1080/02757540.2016.1182994

Singh, A., Shahid, M., Hussain, S., & Tripathi, R. D. (2016). Signaling in plants under heavy metal stress. In *Plant Metal Interaction* (pp. 299–309). Elsevier.

Singh, A., Sharma, R. K., Agrawal, M., & Marshall, F. M. (2011). Health risk assessment of heavy metals via dietary intake of foodstuffs from the wastewater irrigated site of a dry tropical area of India. *Food and Chemical Toxicology*, *48*(2), 611–619. DOI: 10.1016/j.fct.2009.11.041 PMID: 19941927

Singh, H. P., Mahajan, P., Kaur, S., Batish, D. R., & Kohli, R. K. (2013). Chromium toxicity and tolerance in plants. *Environmental Chemistry Letters*, *11*(3), 229–254. DOI: 10.1007/s10311-013-0407-5

Singh, H., & Chandra, R. (2014). Performance of *Pistia stratiotes* and *Eichhornia crassipes* for phytoremediation of pulp and paper mill wastewater in field conditions. *International Journal of Phytoremediation*, *16*(12), 1181–1191.

Singh, H., Kumar, A., & Sharma, A. (2020). Secondary metabolites of plants in phytoremediation: Mechanisms and significance. *Plant Physiology Reports*, *25*, 710–725. DOI: 10.1007/s40502-020-00512-3

Singh, J., Kumar, V., Kumar, P., & Kumar, P. (2022). Kinetics and prediction modeling of heavy metal phytoremediation from glass industry effluent by water hyacinth (Eichhornia crassipes). *International Journal of Environmental Science and Technology*, *19*(6), 5481–5492. DOI: 10.1007/s13762-021-03433-9

Singh, J., Lee, B. K., & Kumar, V. (2022). Role of nanomaterials in phytoremediation: Recent advances and future outlook. *Chemosphere*, *287*, 131996. DOI: 10.1016/j.chemosphere.2021.131996 PMID: 34455120

Singh, M., Muller, G., & Singh, I. B. (2005). Heavy metals in freshly deposited sediments of the Gomati River (a tributary of the Ganga River): Effects of human activities. *Environmental Geology*, *48*(3), 410–420. DOI: 10.1007/s00254-005-1297-9

Singh, P., Pani, A., Mujumdar, A. S., & Shirkole, S. S. (2023). New strategies on the application of artificial intelligence in the field of phytoremediation. *International Journal of Phytoremediation*, *25*(4), 505–523. DOI: 10.1080/15226514.2022.2090500 PMID: 35802802

Singh, R. P., & Tewari, R. K. (2003). Cadmium toxicity induced changes in plant water relations and oxidative metabolism of Brassica juncea L. plants. *Journal of Environmental Biology*, *24*(1), 107–112. PMID: 12974420

Singh, R.. (2023). Heavy metal removal efficiency of duckweed species in wastewater treatment. *Environmental Nanotechnology, Monitoring & Management*, *20*, 100830. DOI: 10.1016/j.enmm.2023.100830

Singh, R., Singh, P. K., Madheshiya, P., Khare, A. K., & Tiwari, S. (2024). Heavy metal contamination in the wastewater irrigated soil and bioaccumulation in cultivated vegetables: Assessment of human health risk. *Journal of Food Composition and Analysis*, *128*, 106054. DOI: 10.1016/j.jfca.2024.106054

Singh, S., Parihar, P., Singh, R., Singh, V. P., & Prasad, S. M. (2015). Heavy metal tolerance in plants: Role of transcriptomics, proteomics, metabolomics, and ionomics. *Frontiers in Plant Science*, *6*, 1143. DOI: 10.3389/fpls.2015.01143 PMID: 26904030

Sinha, S., & Paul, D. (2014). Enhanced phytoremediation of heavy metals by *Lemna minor* and associated microbes from industrial effluent. *International Journal of Phytoremediation*, *16*(1), 1–12. DOI: 10.1080/15226514.2012.669801 PMID: 24912211

Sirianuntapiboon, S., Chularat, P., & Sansak, J. (2006). Removal of nitrogen and phosphorus from wastewater using vertical surface-flow constructed wetland planted with narrow leaf cattails. *ScienceAsia*, *32*(3), 319–326. DOI: 10.2306/scienceasia1513-1874.2006.32.319

Skuza, L., Szućko-Kociuba, I., Filip, E., & Bożek, I. (2022). Natural Molecular Mechanisms of Plant Hyperaccumulation and Hypertolerance towards Heavy Metals. *International Journal of Molecular Sciences*, *23*(16), 9335. DOI: 10.3390/ijms23169335 PMID: 36012598

Soda, S., Mishima, D., Inoue, D., & Ike, M. (2012). Removal of toxic metals using aquatic plants. *Bioresource Technology*, *121*, 432–439. DOI: 10.1016/j.biortech.2012.06.036

Soliman, M. I., Ibrahim, A. A., Rizk, R. M., & Naser, N. S. (2019). Phytoremediation, Biochemical and Molecular Studies of Some Selected Hydrophytes in Egypt. *Journal of Applied Sciences (Faisalabad), 19*(7), 708–717. DOI: 10.3923/jas.2019.708.717

Sompura, Y., Patel, K., & Joshi, N. (2024). Combined microplastic and heavy metal effects on biomass and physiology of water spinach. *Journal of Current Opinion in Crop Science, 5*(1), 48–61. DOI: 10.62773/jcocs.v5i1.233

Sondergaard, M., & Moss, B. (1998). *Impact of Submerged Macrophytes on Phytoplankton in Shallow Freshwater Lakes* (Vol. 131). Ecological Studies. DOI: 10.1007/978-1-4612-0695-8_6

Song, W. Y., Sohn, E. J., Martinoia, E., & Lee, Y. J. (2003). Engineering tolerance and accumulation of lead and cadmium in transgenic plants. *Nature Biotechnology, 21*(8), 914–919. DOI: 10.1038/nbt850 PMID: 12872132

Song, Y., Zhan, D., He, Z., Li, W., Duan, W., Yang, Z., & Lu, M. (2023). HPO-empowered machine learning with multiple environment variables enables spatial prediction of soil heavy metals in coastal delta farmland of China. *Computers and Electronics in Agriculture, 213*, 108254. DOI: 10.1016/j.compag.2023.108254

Song, Y., Zhang, L.-L., Li, J., He, X.-J., Chen, M., & Deng, Y. (2018). High-potential accumulation and tolerance in the submerged hydrophyte Hydrilla verticillata (Lf) Royle for nickel-contaminated water. *Ecotoxicology and Environmental Safety, 161*, 553–562. DOI: 10.1016/j.ecoenv.2018.06.032 PMID: 29929131

Soo, A., Kim, J., & Shon, H. K. (2024). Technologies for the wastewater circular economy: A review. *Desalination and Water Treatment, 317*, 100205. DOI: 10.1016/j.dwt.2024.100205

Sooknah, R. D., & Wilkie, A. C. (2004). Nutrient removal by floating aquatic macrophytes cultured in anaerobically digested flushed dairy manure wastewater. *Ecological Engineering, 22*(1), 27–42. DOI: 10.1016/j.ecoleng.2004.01.004

Sosa, D., Alves, F. M., Prieto, M. A., Pedrosa, M. C., Heleno, S. A., Barros, L., Feliciano, M., & Carocho, M. (2024). *Lemna minor*: Unlocking the value of this duckweed for the food and feed industry. *Foods, 13*(10), 1435. DOI: 10.3390/foods13101435 PMID: 38790736

Srivastav, A. L., Markandeya, , Patel, N., Pandey, M., Pandey, A. K., Dubey, A. K., Kumar, A., Bhardwaj, A. K., & Chaudhary, V. K. (2023). Concepts of circular economy for sustainable management of electronic wastes: Challenges and management options. *Environmental Science and Pollution Research International, 30*(17), 48654–48675. DOI: 10.1007/s11356-023-26052-y PMID: 36849690

Srivastava, S.. (2008). Phytoremediation of chromium from industrial wastewater by aquatic macrophytes. *Environmental Technology, 29*(3), 299–307. DOI: 10.1080/09593330802095083

Stottmeister, U., Wießner, A., Kuschk, P., Kappelmeyer, U., Kästner, M., Bederski, O., Müller, R. A., & Moormann, H. (2003). Effects of plants and microorganisms in constructed wetlands for wastewater treatment. *Biotechnology Advances, 22*(1–2), 93–117. DOI: 10.1016/j.biotechadv.2003.08.010 PMID: 14623046

Stumm, W., & Morgan, J. J. (1996). *Aquatic chemistry: Chemical equilibria and rates in natural waters* (3rd ed.). Wiley-Interscience.

Su, C., Jiang, L., Zhang, W., & Zhou, Q. (2012). Uptake and accumulation of heavy metals in plants: A review focusing on phytoremediation applications. *Environmental Science and Pollution Research International, 19*(7), 2251–2261. DOI: 10.1007/s11356-011-0679-0

Suman, J., Uhlik, O., Viktorova, J., & Macek, T. (2018). Phytoextraction of heavy metals: A promising tool for clean-up of polluted environment? *Frontiers in Plant Science, 9*, 1476. DOI: 10.3389/fpls.2018.01476 PMID: 30459775

Sun, Y., Chen, S., Jiang, H., Qin, B., Li, D., Jia, K., & Wang, C. (2024). Towards interpretable machine learning for observational quantification of soil heavy metal concentrations under environmental constraints. *The Science of the Total Environment, 926*, 171931. DOI: 10.1016/j.scitotenv.2024.171931 PMID: 38531447

Sutherland, D. L., & Ralph, P. J. (2019). Microalgal bioremediation of emerging contaminants– opportunities and challenges. *Water Research, 164*, 114921. DOI: 10.1016/j.watres.2019.114921 PMID: 31382151

Suyal, D. C., Soni, R., Singh, D. K., & Goel, R. (2021). Microbiome change of agricultural soil under organic farming practices. *Biologia, 76*(4), 1315–1325. DOI: 10.2478/s11756-021-00680-6

Sytar, O., Kumar, A., Latowski, D., Kuczynska, P., Strzałka, K., & Prasad, M. N. V. (2013). Heavy metal-induced oxidative damage, defense reactions, and detoxification mechanisms in plants. *Acta Physiologiae Plantarum, 35*(4), 985–999. DOI: 10.1007/s11738-012-1169-6

Szalai, G.. (2015). Phytochelatin induction in *Lemna minor* exposed to cadmium. *Ecotoxicology (London, England), 24*(2), 304–316. DOI: 10.1007/s10646-014-1375-6

Tak, H. I., Ahmad, F., & Babalola, O. O. (2013). Advances in the application of plant growth-promoting rhizobacteria in phytoremediation of heavy metals. *Reviews of Environmental Contamination and Toxicology, 223*, 33–52. PMID: 23149811

Tandon, P. K., & Singh, S. B. (2016). Redox processes in water remediation. *Environmental Chemistry Letters, 14*(1), 15–25. DOI: 10.1007/s10311-015-0540-4

Tangahu, B. V., Abdullah, S. R. S., Basri, H., Idris, M., Anuar, N., & Mukhlisin, M. (2011). A review on heavy metals (As, Pb, and Hg) uptake by plants through phytoremediation. *International Journal of Chemical Engineering, 2011*(1), 1–31. DOI: 10.1155/2011/939161

Tang, K. H. D. (2024). Phytoremediation of petroleum hydrocarbons: An update of its recent progress. *Tropical Environment, Biology, and Technology, 2*(2), 106–123. DOI: 10.53623/tebt.v2i2.532

Tan, H. W.. (2023). Phytoremediation of heavy-metal-contaminated waters: Challenges including secondary pollution and biomass handling. *The Science of the Total Environment, 872*, 162411. DOI: 10.1016/j.scitotenv.2023.162411

Tanko, W. S., Albalwe, F. M., Bakr, A. A., Ali, A. M., Ibrahim, R. I. H., Abdel-Mageed, A. M., Shahhat, I. M. A. M., Elsheikh, S. Y. S., Hussein, H.-A. A., Qaoud, E.-S. M., Ezzat, S. M., Elshamy, M. M., Abdein, M. A., & Nafea, E. M. (2025). Submerged hydrophytes as a tool for the removal of heavy metals. *Asian Journal of Water, Environment and Pollution, 22*(4), 64–74. DOI: 10.36922/AJWEP025150105

Tanner, C. C. (2001). Plants as ecosystem engineers in subsurface-flow treatment wetlands. *Water Science and Technology, 44*(11–12), 9–17. DOI: 10.2166/wst.2001.0804 PMID: 11804163

Tanner, C. C., & Headley, T. R. (2011). Components of floating emergent macrophyte treatment wetlands influencing removal of stormwater pollutants. *Ecological Engineering, 37*(3), 474–486. DOI: 10.1016/j.ecoleng.2010.12.012

Tao, M., Zhang, W., Fan, H., Huang, L., & Xu, Q. (2020). *E3S Web of Conferences, 143,* 02020.

Tchounwou, P. B., Yedjou, C. G., Patlolla, A. K., & Sutton, D. J. (2012). Heavy metal toxicity and the environment. In *Molecular, Clinical and Environmental Toxicology* (Vol. 101, pp. 133–164). Springer. https://doi.org/DOI: 10.1007/978-3-7643-8340-4_6

Tebo, B. M., Bargar, J. R., Clement, B. G., Dick, G. J., Murray, K. J., Parker, D., Verity, R., & Webb, S. M. (2004). Biogenic manganese oxides: Properties and mechanisms of formation. *Annual Review of Earth and Planetary Sciences, 32*(1), 287–328. DOI: 10.1146/annurev.earth.32.101802.120213

Thakur, S., Dangi, P., Pandey, A., & Shukla, R. (2016). Engineered nanomaterials for water treatment and remediation: Challenges and opportunities. *Environmental Technology & Innovation*, *5*, 161–173.

Thakur, S., Mathur, S., Patel, S., & Paital, B. (2022). Microplastic accumulation and degradation in environment via biotechnological approaches. *Water (Basel)*, *14*(24), 4053. DOI: 10.3390/w14244053

Thakur, T. K., Barya, M. P., Dutta, J., Mukherjee, P., Thakur, A., Swamy, S. L., & Anderson, J. T. (2023). Integrated Phytobial Remediation of Dissolved Pollutants from Domestic Wastewater through Constructed Wetlands: An Interactive Macrophyte-Microbe-Based Green and Low-Cost Decontamination Technology with Prospective Resource Recovery. *Water (Basel)*, *15*(22), 3877. Advance online publication. DOI: 10.3390/w15223877

Thapa, G., Sadhukhan, A., Panda, S. K., & Sahoo, L. (2021). Molecular mechanistic model of heavy metal tolerance in plants: Interactions between signaling pathways. *Plant Signaling & Behavior*, *16*(4), 1870063.

Thayyib, P. V., Mamilla, R., Khan, M., Fatima, H., Asim, M., Anwar, I., Shamsudheen, M. K., & Khan, M. A. (2023). State-of-the-art of artificial intelligence and big data analytics reviews in five different domains: A bibliometric summary. *Sustainability (Basel)*, *15*(5), 4026. DOI: 10.3390/su15054026

Thomaz, S. M., & Bini, L. M. (2003). Ecological responses of aquatic vegetation to nutrient enrichment. *Acta Limnologica Brasiliensia*, *15*(1), 23–29.

Thomaz, S. M., & Cunha, E. R. (2010). The role of macrophytes in habitat structuring in aquatic ecosystems: Methods of measurement, causes and consequences on animal assemblages' composition and biodiversity. *Acta Limnologica Brasiliensia*, *22*(2), 218–236. DOI: 10.4322/actalb.02202011

Tian, S., Xie, R., Wang, H., Hu, Y., Hou, D., Liao, X., Brown, P. H., Yang, H., Lin, X., Labavitch, J. M., & Lu, L. (2017). Uptake, sequestration and tolerance of cadmium at cellular levels in the hyperaccumulator plant species Sedum alfredii. *Journal of Experimental Botany*, *68*(9), 2387–2398. DOI: 10.1093/jxb/erx112 PMID: 28407073

Tripathi, D. K., Singh, S., Singh, S., Mishra, S., Chauhan, D. K., & Dubey, N. K. (2021). Nanomaterials in the environment: Fate, transport, toxicity and interactions with plants. *Environmental Technology & Innovation*, *21*, 101200.

Tront, J. M., Reinhold, D. M., Bragg, A. W., & Saunders, F. M. (2007). Uptake of Halogenated Phenols by Aquatic Plants. *Journal of Environmental Engineering*, *133*(10), 955–961. DOI: 10.1061/(ASCE)0733-9372(2007)133:10(955)

Tsai, C.-H., Shen, Y.-H., & Tsai, W.-T. (2021). Sustainable material management of industrial hazardous waste in Taiwan: Case studies in circular economy. *Sustainability (Basel), 13*(16), 9410. DOI: 10.3390/su13169410

Turetsky, M. R., Benscoter, B., Page, S., Rein, G., Van Der Werf, G. R., & Watts, A. (2020). Global vulnerability of peatlands to fire and carbon loss. *Nature Geoscience, 13*(1), 1–6. DOI: 10.1038/s41561-019-0468-3

U.S. Environmental Protection Agency. (2018). *Green remediation: Best management practices for phytoremediation.* US EPA.

Uddin, M. M., Zakeel, M. C. M., Zavahir, J. S., Marikar, F. M. M. T., & Jahan, I. (2021). Heavy Metal Accumulation in Rice and Aquatic Plants Used as Human Food: A General Review. *Toxics, 9*(12), 360. DOI: 10.3390/toxics9120360 PMID: 34941794

Ullrich, S. M., Tanton, T. W., & Abdrashitova, S. A. (2001). Mercury in the aquatic environment: A review of factors affecting methylation. *Critical Reviews in Environmental Science and Technology, 31*(3), 241–293. DOI: 10.1080/20016491089226

United Nations (UN). (2015). *Transforming our world: The 2030 Agenda for Sustainable Development.* UN.

United Nations Development Programme (UNDP). (2023). *Sustainable Development Goals.*

United Nations Environment Programme (UNEP). (2018). *Nature-based solutions for water.* UNEP.

United Nations. (2015). Transforming our world: The 2030 agenda for sustainable development. United Nations General Assembly. https://sdgs.un.org/2030agenda

Upadhyay, S. K., Srivastava, A. K., Rajput, V. D., Chauhan, P. K., Bhojiya, A. A., Jain, D., Chaubey, G., Dwivedi, P., Sharma, B., & Minkina, T. (2022). Root Exudates: Mechanistic Insight of Plant Growth Promoting Rhizobacteria for Sustainable Crop Production. *Frontiers in Microbiology, 13*, 916488. Advance online publication. DOI: 10.3389/fmicb.2022.916488 PMID: 35910633

van Oosterhout, C. (2024). AI-informed conservation genomics. *Heredity, 132*(1), 1–4. DOI: 10.1038/s41437-023-00666-x PMID: 38151537

Vangronsveld, J.. (2009). Phytoremediation of contaminated soils and sediments: A review. *Environmental Science and Pollution Research International, 16*(6), 715–731.

Varol, M. (2011). Assessment of heavy metal contamination in sediments of the Tigris River (Turkey) using pollution indices and multivariate statistical techniques. *Journal of Hazardous Materials, 195*, 355–364. DOI: 10.1016/j.jhazmat.2011.08.051 PMID: 21890271

Verdouw, C., Tekinerdogan, B., Beulens, A., & Wolfert, S. (2021). Digital twins in smart farming. *Agricultural Systems, 189*, 103046. Advance online publication. DOI: 10.1016/j.agsy.2020.103046

Verhoeven, J. T. A., & Meuleman, A. F. M. (1999). Wetlands for wastewater treatment: Opportunities and limitations. *Ecological Engineering, 12*(1–2), 5–12. DOI: 10.1016/S0925-8574(98)00050-0

Verma, P. K., Sinha, A., & Singh, V. K. (2020a). Role of phytochemicals in phytoremediation of heavy metals: Mechanistic insights and future perspectives. *Environmental Technology & Innovation, 18*, 100684. DOI: 10.1016/j.eti.2020.100684

Verma, R., Sinha, S., & Chauhan, P. (2020b). Recent trends in phytoremediation: A critical review. *Environmental Science and Pollution Research International, 27*, 13855–13868. DOI: 10.1007/s11356-020-07904-3

Verma, S., & Dubey, R. S. (2003). Lead toxicity induces lipid peroxidation and alters the activities of antioxidant enzymes in growing rice plants. *Plant Science, 164*(4), 645–655. DOI: 10.1016/S0168-9452(03)00022-0

Verma, S., Gupta, R., & Singh, N. (2020c). Mechanisms of heavy metal removal by aquatic plants and their future prospects. *Environmental Science and Pollution Research International, 27*(25), 31725–31737. DOI: 10.1007/s11356-020-09512-9

Verstraete, W., Yanuka-Golub, K., Driesen, N., & De Vrieze, J. (2022). Engineering microbial technologies for environmental sustainability: Choices to make. *Microbial Biotechnology, 15*(1), 215–227. DOI: 10.1111/1751-7915.13986 PMID: 34875143

Vestena, S. A., Ribeiro, C., & Cambraia, J. (2017). Cadmium-induced oxidative stress and antioxidative enzyme response in water hyacinth and *Salvinia. Brazilian Journal of Plant Physiology, 23*(2), 131–139. DOI: 10.1590/S1677-04202011000200005

Vestena, S., Cambraia, J., & Ribeiro, C. (2016). Lead-induced heavy metal toxicity alters growth and antioxidative enzyme levels in *Eichhornia crassipes. Botanical Studies, 55*(1), 54. DOI: 10.1186/s40529-014-0054-6 PMID: 28597420

Vikraman Meera, A., Thampatti, K. C. M., John, J., Sudha, B., & Sajeena, A. (2022). Aquatic plants as bioremediators in pollution abatement of heavy metals. In Sharma, P., Pandey, A., Tong, Y. W., & Ngo, H. H. (Eds.), *Current developments in biotechnology and bioengineering* (pp. 223–239). IntechOpen., DOI: 10.5772/intechopen.99627

Villamagna, A. M., & Murphy, B. R. (2010). Ecological and socio-economic impacts of invasive water hyacinth (*Eichhornia crassipes*): A review. *Freshwater Biology*, *55*(2), 282–298. DOI: 10.1111/j.1365-2427.2009.02294.x

Visioli, F., Bellomo, G., & Galli, C. (1998). Free radical-scavenging properties of olive oil polyphenols. *Biochemical and Biophysical Research Communications*, *247*(1), 60–64. Advance online publication. DOI: 10.1006/bbrc.1998.8735 PMID: 9636654

Vlot, A. C., Dempsey, D. A., & Klessig, D. F. (2009). Salicylic acid, a multifaceted hormone to combat disease. *Annual Review of Phytopathology*, *47*(1), 177–206. DOI: 10.1146/annurev.phyto.050908.135202 PMID: 19400653

Voesenek, L. A. C. J.. (2016). Plant responses to flooding stress. *Current Opinion in Plant Biology*, *33*, 64–71. DOI: 10.1016/j.pbi.2016.06.005 PMID: 27322538

Vögeli-Lange, R., & Wagner, G. J. (1996). Relationship between cadmium, glutathione and cadmium-binding peptides (phytochelatins) in leaves of intact tobacco seedlings. *Plant Science*, *114*(1), 11–18. DOI: 10.1016/0168-9452(95)04299-7

Vögeli-Lange, R., & Wagner, G. J. (1996). Relationship between cadmium-induced phytochelatin accumulation and the response of protein synthesis to cadmium in suspension-cultured plant cells. *Plant Science*, *114*(2), 147–155.

Vulpe, C.-B., Toplicean, I.-M., Agachi, B.-V., & Datcu, A.-D. (2025). A Review on Uses of Lemna minor, a Beneficial Plant for Sustainable Water Treatments, in Relation to Bioeconomy Aspects. *Water (Basel)*, *17*(9), 1400. DOI: 10.3390/w17091400

Vymazal, J. (2007). Removal of nutrients in various types of constructed wetlands. *The Science of the Total Environment*, *380*(1–3), 48–65. DOI: 10.1016/j.scitotenv.2006.09.014 PMID: 17078997

Vymazal, J. (2011). Constructed wetlands for wastewater treatment. *Environmental Science & Technology*, *45*(1), 61–69. DOI: 10.1021/es101403q PMID: 20795704

Vymazal, J. (2011). Plants used in constructed wetlands with horizontal subsurface flow: A review. *Hydrobiologia*, *674*(1), 133–156. DOI: 10.1007/s10750-011-0738-9

Vymazal, J. (2014). Constructed wetlands for treatment of industrial wastewaters: A review. *Ecological Engineering*, *73*, 724–751. DOI: 10.1016/j.ecoleng.2014.09.034

Vymazal, J., & Březinová, T. (2015). The use of constructed wetlands for removal of pesticides from agricultural runoff and drainage: A review. *Environment International*, *75*, 11–20. DOI: 10.1016/j.envint.2014.10.026 PMID: 25461411

Vymazal, J., & Kröpfelová, L. (2011). *A three-stage constructed wetland for wastewater treatment: Performance studies*. Springer.

Wagh, M. S., Sivarajan, S., & Osborne, J. W. (2024). Deciphering the enhanced translocation of Pb, Ni and Cd from artificially polluted soil to Chrysopogon zizanioides augmented with Bacillus xiamenensis VITMSJ3. *3 Biotech, 14*(7), 180. DOI: 10.1007/s13205-024-04001-x

Wang, C., & Keller, A. A. (2008). Natural and engineered nano-scale iron for environmental remediation. *Trends in Analytical Chemistry*, *27*(2), 152–158. DOI: 10.1016/j.trac.2007.10.010

Wang, H., Cui, S., Ma, L., Wang, Z., & Wang, H. (2021). Variations of arsenic forms and the role of arsenate reductase in three hydrophytes exposed to different arsenic species. *Ecotoxicology and Environmental Safety*, *221*, 112415. DOI: 10.1016/j.ecoenv.2021.112415 PMID: 34171691

Wang, J., & Aghajani Delavar, M. (2023). Techno-economic analysis of phytoremediation: A strategic rethinking. *The Science of the Total Environment*, *902*, 165949. DOI: 10.1016/j.scitotenv.2023.165949 PMID: 37536595

Wang, J., Zhang, C., Dai, H., & Zhang, Z. (2017). Advances in the genetic engineering of aquatic plants for phytoremediation. *Environmental Science and Pollution Research International*, *24*, 17118–17131. DOI: 10.1007/s11356-017-9351-6

Wang, W., Wang, Y., Zhang, Y., & Gao, F. (2012). Bioaccumulation and antioxidative responses of *Eichhornia crassipes* to heavy metals. *Ecotoxicology (London, England)*, *21*(3), 681–689. DOI: 10.1007/s10646-012-0852-4 PMID: 22113456

Wang, W.-X., & Chen, X. (2019). Bioavailability and bioaccumulation of heavy metals in aquatic organisms from China. *Environmental Science and Pollution Research International*, *26*(19), 19023–19041. DOI: 10.1007/s11356-019-04988-7 PMID: 31090002

Wang, X., Liu, C., & Huang, Y. (2008). Cadmium responses in *Vallisneria spiralis*: A freshwater submerged macrophyte. *Chemosphere*, *71*(2), 297–303. DOI: 10.1016/j.chemosphere.2007.10.008

Wani, K. A., Manzoor, J., Dar, A. A., Shuab, R., & Lone, R. (2020). Heavy Metal Intrusion and Accumulation in Aquatic Ecosystems. *Fresh Water Pollution Dynamics and Remediation*, 83–104. DOI: 10.1007/978-981-13-8277-2_6

Wan, Y., Liu, J., Zhuang, Z., Wang, Q., & Li, H. (2024). Heavy metals in agricultural soils: Sources, influencing factors, and remediation strategies. *Toxics*, *12*(1), 63. DOI: 10.3390/toxics12010063 PMID: 38251018

Weis, J. S., & Weis, P. (2004). Metal uptake, transport and release by wetland plants: Implications for phytoremediation and restoration. *Environment International*, *30*(5), 685–700. DOI: 10.1016/j.envint.2003.11.002 PMID: 15051245

Wei, X., Liu, Y., Shen, L., Lu, Z., Ai, Y., & Wang, X. (2024). Machine learning insights in predicting heavy metals interaction with biochar. *Biochar*, *6*(1), 1–11. DOI: 10.1007/s42773-024-00304-7

Wentzell, B. M. (2025). Phytoremediation for water quality improvement: Current advances and future prospects. *Biotechnology for the Environment*, *2*(1), 7. DOI: 10.1186/s44314-025-00022-9

Wetzel, R. G. (2001). *Limnology: Lake and river ecosystems* (3rd ed.). Academic Press.

Wetzel, R. G. (2001). *Limnology: Lake and River Ecosystems* (3rd ed.). Academic Press.

Whang, S. E., Roh, Y., Song, H., & Lee, J. G. (2023). Data collection and quality challenges in deep learning: A data-centric ai perspective. *The VLDB Journal*, *32*(4), 791–813. DOI: 10.1007/s00778-022-00775-9

WHO. (2018). Climate change and health. World Health Organization. https://www.who.int/news-room/fact-sheets/detail/climate-change-and-health

Williams, A., Aguilar, M. R., Pattiya Arachchillage, K. G. G., Chandra, S., Rangan, S., Ghosal Gupta, S., & Artes Vivancos, J. M. (2024). Biosensors for public health and environmental monitoring: The case for sustainable biosensing. *ACS Sustainable Chemistry & Engineering*, *12*(28), 10296–10312. DOI: 10.1021/acssuschemeng.3c06112 PMID: 39027730

Williams, L. E., & Mills, R. F. (2005). P1B-ATPases—An ancient family of transition metal pumps with diverse functions in plants. *Trends in Plant Science*, *10*(10), 491–502. DOI: 10.1016/j.tplants.2005.08.008 PMID: 16154798

Wolverton, B. C., & McDonald, R. C. (1981). The water hyacinth: From prolific pest to potential provider. *Ambio*, *10*(2–3), 82–87.

World Health Organization (WHO). (2021). *Green and blue spaces and mental health: new evidence and perspectives*. WHO Regional Office for Europe.

World Health Organization (WHO). (2021). *Water, Sanitation and Hygiene: Guidelines on Heavy Metal Limits*. WHO Press.

Wróbel, M., Śliwakowski, W., Kowalczyk, P., Kramkowski, K., & Dobrzyński, J. (2023). Bioremediation of heavy metals by the genus Bacillus. *International Journal of Environmental Research and Public Health, 20*(6), 4964. DOI: 10.3390/ijerph20064964 PMID: 36981874

Wu, B., Zhang, M., Zeng, H., Tian, F., Potgieter, A. B., Qin, X., Yan, N., Chang, S., Zhao, Y., Dong, Q., Boken, V., Plotnikov, D., Guo, H., Wu, F., Zhao, H., Deronde, B., Tits, L., & Loupian, E. (2023). Challenges and opportunities in remote sensing-based crop monitoring: A review. *National Science Review, 10*(4), nwac290. DOI: 10.1093/nsr/nwac290 PMID: 36960224

Wu, S., Kuschk, P., Brix, H., Vymazal, J., & Dong, R. (2014). Development of constructed wetlands in performance intensifications for wastewater treatment: A nitrogen and organic matter targeted review. *Water Research, 57*, 40–55. DOI: 10.1016/j.watres.2014.03.020 PMID: 24704903

Wu, S., Wallace, S., Brix, H., Kuschk, P., Kirui, W. K., Masi, F., & Dong, R. (2014). Treatment of industrial effluents in constructed wetlands: Challenges, operational strategies and overall performance. *Environmental Pollution, 192*, 245–261. DOI: 10.1016/j.envpol.2014.05.007 PMID: 25792030

Wu, Y., Sun, J., Hu, B., Zhang, G., & Rousseau, A. N. (2023). Wetland-based solutions against extreme flood and severe drought: Efficiency evaluation of risk mitigation. *Climate Risk Management, 40*, 100505. DOI: 10.1016/j.crm.2023.100505

WWAP (United Nations World Water Assessment Programme). (2020). The United Nations World Water Development Report 2020: Water and Climate Change. UNESCO. https://unesdoc.unesco.org/ark:/48223/pf0000372985

Xie, P., Zahoor, F., Iqbal, S. S., Ullah, S., Noman, M., Din, Z. U., & Yang, W. (2022). Elimination of toxic heavy metals from industrial polluted water by using hydrophytes. *Journal of Cleaner Production, 352*, 131358. DOI: 10.1016/j.jclepro.2022.131358

Xu, C., Zhou, T., Wang, D., & Sun, X. (2014). Proteomic insights into Cd tolerance in *Eichhornia crassipes*. *Environmental Science & Technology, 48*(12), 7012–7020. DOI: 10.1021/es500914n PMID: 24819143

Xue, Y., Wang, J., Huang, J., Li, F., & Wang, M. (2018). The Response of Duckweed (*Lemna minor* L.) Roots to Cd and Its Chemical Forms. *Journal of Chemistry, 2018*, 1–9. DOI: 10.1155/2018/7274020

Xu, Q. S., Hu, J. Z., Xie, K. B., Yang, H. Y., Du, K. H., & Shi, G. X. (2010). Accumulation and acute toxicity of silver in Potamogeton crispus L. *Journal of Hazardous Materials*, *173*(1–3), 186–193. DOI: 10.1016/j.jhazmat.2009.08.067 PMID: 19740606

Xu, S., & Sun, M. (2025). The interpretable machine learning model for depression associated with heavy metals via EMR mining method. *Scientific Reports*, *15*(1), 10811. Advance online publication. DOI: 10.1038/s41598-025-95938-3 PMID: 40155474

Xu, Y., Sun, X., Zhang, Q., Li, X., & Yan, Z. (2018). Iron plaque formation and heavy metal uptake in Spartina alterniflora at different tidal levels and waterlogging conditions. *Ecotoxicology and Environmental Safety*, *153*, 91–100. DOI: 10.1016/j.ecoenv.2018.02.008 PMID: 29407744

Yadav, A. K., Dash, P., Mohanty, A., Abbassi, R., & Mishra, B. K. (2012). Performance assessment of innovative constructed wetland-microbial fuel cell for electricity production and dye removal. *Ecological Engineering*, *47*, 126–131. DOI: 10.1016/j.ecoleng.2012.06.029

Yadav, A., Hasan, M., Kour, S., Chandra, R., Singh, V., Noor, I. M., Yahya, M. Z. A., Singh, H. M., & Tripathi, A. (2025). Heavy Metal Contamination in Plants: Sources, Monitoring, and Data-Driven Insights. *Macromolecular Symposia*, *414*(1), 2400239. Advance online publication. DOI: 10.1002/masy.202400239

Yadav, B. K., Siebel, M. A., & van Bruggen, J. J. (2011). Rhizofiltration of a heavy metal (lead) containing wastewater using the wetland plant Carex pendula. *Clean (Weinheim)*, *39*(5), 467–474. DOI: 10.1002/clen.201000385

Yadav, R. K., Saini, R., & Sharma, A. (2022). Decision support tools for sustainable phytoremediation: A review of frameworks and practical applications. *Ecotoxicology and Environmental Safety*, *238*, 113597. DOI: 10.1016/j.ecoenv.2022.113597

Yadav, S. K. (2010). Heavy metals toxicity in plants: An overview on the role of glutathione and phytochelatins in heavy metal stress tolerance of plants. *South African Journal of Botany*, *76*(2), 167–179. DOI: 10.1016/j.sajb.2009.10.007

Yadav, V., Arif, N., Kováč, J., Singh, V. P., Tripathi, D. K., Chauhan, D. K., & Vaculík, M. (2021). Structural modifications of plant organs and tissues by metals and metalloids in the environment: A review. *Plant Physiology and Biochemistry*, *159*, 100–112. DOI: 10.1016/j.plaphy.2020.11.047 PMID: 33359959

Yamashita, T., Ishikawa, M., & Fujiwara, T. (2021). The effects of temperature and seasonal changes on pollutant removal in constructed wetlands: A review. *Journal of Environmental Chemical Engineering*, *9*(1), 104680. DOI: 10.1016/j.jece.2020.104680

Yan, A., Wang, Y., Tan, S. N., Mohd Yusof, M. L., Ghosh, S., & Chen, Z. (2020). Phytoremediation: A promising approach for revegetation of heavy metal-polluted land. *Frontiers in Plant Science*, *11*, 513099. DOI: 10.3389/fpls.2020.00359 PMID: 32425957

Yang, K. M. (2025). Recent trend in phytoremediation of petroleum hydrocarbon contaminated soil: A bibliometric review. *International Journal of Phytoremediation*, •••, 1–9. DOI: 10.1080/15226514.2025.2539429 PMID: 40735952

Yang, R., Roshani, D., Gao, B., Li, P., & Shang, N. (2024). Metallothionein: A Comprehensive Review of Its Classification, Structure, Biological Functions, and Applications. *Antioxidants*, *13*(7), 825. DOI: 10.3390/antiox13070825 PMID: 39061894

Yang, Z., Yang, F., Liu, J. L., Wu, H. T., Yang, H., Shi, Y., Liu, J., Zhang, Y. F., Luo, Y. R., & Chen, K. M. (2022). Heavy metal transporters: Functional mechanisms, regulation, and application in phytoremediation. *The Science of the Total Environment*, *809*, 151099. DOI: 10.1016/j.scitotenv.2021.151099 PMID: 34688763

Yaseen, Z. M., & Alhalimi, F. L. (2025). Heavy metal adsorption efficiency prediction using biochar properties: A comparative analysis for ensemble machine learning models. *Scientific Reports*, *15*(1), 13434. Advance online publication. DOI: 10.1038/s41598-025-96271-5 PMID: 40251173

Yeh, C. M., Chang, I. F., Lu, C. A., Wu, Y. M., Lu, M. Y., & Shih, M. C. (2007). A novel family of transcription factors involved in plant defense response. *Plant Molecular Biology*, *64*(6), 771–787. DOI: 10.1007/s11103-007-9227-4

Yuan, X., Li, J., Lim, J. Y., Zolfaghari, A., Alessi, D. S., Wang, Y., Wang, X., & Ok, Y. S. (2024). Machine Learning for Heavy Metal Removal from Water: Recent Advances and Challenges. *ACS ES&T Water*, *4*(3), 820–836. DOI: 10.1021/acsestwater.3c00215

Yuan, X., Xue, N., & Han, Z. (2021). A meta-analysis of heavy metals pollution in farmland and urban soils in China over the past 20 years. *Journal of Environmental Sciences (China)*, *101*, 217–226. DOI: 10.1016/j.jes.2020.08.013 PMID: 33334517

Yu, K., Fang, S., & Zhao, Y. (2021). Heavy metal Hg stress detection in tobacco plant using hyperspectral sensing and data-driven machine learning methods. *Spectrochimica Acta. Part A: Molecular and Biomolecular Spectroscopy*, *245*, 118917. DOI: 10.1016/j.saa.2020.118917 PMID: 32949945

Zahari, N. Z., Tuah, P. M., & Rahim, S. A. (2021, September). Inoculation of Bacillus cereus enhance phytoremediation efficiency of Pistia stratiotes and Eichhornia crassipes in removing heavy metal Pb. [). IOP Publishing.]. *IOP Conference Series. Earth and Environmental Science*, *847*(1), 012012. DOI: 10.1088/1755-1315/847/1/012012

Zahoor, I., & Mushtaq, A. (2023). Water pollution from agricultural activities: A critical global review. *International Journal of Chemical and Biochemical Science*, *23*(1), 164–176.

Zayed, A., Gowthaman, S., & Terry, N. (1998). Phytoaccumulation of trace elements by wetland plants: I. Duckweed. *Journal of Environmental Quality*, *27*(3), 715–721. DOI: 10.2134/jeq1998.00472425002700030032x

Zedler, J. B., & Kercher, S. (2004). Causes and consequences of invasive plants in wetlands: Opportunities, opportunists, and outcomes. *Critical Reviews in Plant Sciences*, *23*(5), 431–452. DOI: 10.1080/07352680490514673

Zedler, J. B., & Kercher, S. (2005). Wetland resources: Status, trends, ecosystem services, and restorability. *Annual Review of Environment and Resources*, *30*(1), 39–74. DOI: 10.1146/annurev.energy.30.050504.144248

Zeid, I., Ghaly, E. K., & Shedeed, Z. A. (2024). Azolla pinnata as a phytoremediator: Improves germination, growth and yield of maize irrigated with Ni-polluted water. *Scientific Reports*, *14*(1), 22284. DOI: 10.1038/s41598-024-72651-1 PMID: 39333677

Zhai, Y., Zhou, L., Qi, H., Gao, P., & Zhang, C. (2023). Application of Visible/Near-Infrared Spectroscopy and Hyperspectral Imaging with Machine Learning for High-Throughput Plant Heavy Metal Stress Phenotyping: A Review. *Plant Phenomics (Washington, D.C.)*, *5*, 0124. Advance online publication. DOI: 10.34133/plantphenomics.0124 PMID: 38239738

Zhakypbek, Y., Kossalbayev, B. D., Belkozhayev, A. M., Murat, T., Tursbekov, S., Abdalimov, E., Pashkovskiy, P., Kreslavski, V., Kuznetsov, V., & Allakhverdiev, S. I. (2024). Reducing Heavy Metal Contamination in Soil and Water Using Phytoremediation. *Plants*, 13.

Zhang, D. Q., Jinadasa, K. B. S. N., Gersberg, R. M., Liu, Y., Ng, W. J., & Tan, S. K. (2020). Application of constructed wetlands for wastewater treatment in developing countries – A review of recent developments (2010–2020). *Journal of Environmental Management, 293*, 112862. DOI: 10.1016/j.jenvman.2021.112862

Zhang, D. Q., Jinadasa, K. B. S. N., Gersberg, R. M., Liu, Y., Tan, S. K., & Ng, W. J. (2014). Application of constructed wetlands for wastewater treatment in developing countries – A review of recent developments (2000–2013). *Journal of Environmental Management, 141*, 116–131. DOI: 10.1016/j.jenvman.2014.03.015 PMID: 24784754

Zhang, D.. (2024). Cadmium tolerance mechanisms in water hyacinth involving MT1 and ROS balance. *Environmental Pollution, 315*, 120316. DOI: 10.1016/j.envpol.2022.120316

Zhang, D., Gersberg, R. M., Ng, W. J., & Tan, S. K. (2014). Removal of pharmaceuticals and personal care products in aquatic plant-based systems: A review. *Environmental Pollution, 184*, 620–639. DOI: 10.1016/j.envpol.2013.09.009 PMID: 24080393

Zhang, H., Zhang, L.-L., Li, J., Chen, M., & An, R.-D. (2020). Comparative study on the bioaccumulation of lead, cadmium and nickel and their toxic effects on the growth and enzyme defence strategies of a heavy metal accumulator, Hydrilla verticillata (Lf) Royle. *Environmental Science and Pollution Research International, 27*(9), 9853–9865. DOI: 10.1007/s11356-019-06968-0 PMID: 31927739

Zhang, X., Xia, H., Li, Z., & Zhuang, P. (2014). Potential of four ornamental plants in remediation of contaminated soils with heavy metals. *Ecological Engineering, 70*, 244–252. DOI: 10.1016/j.ecoleng.2014.05.003

Zhang, X., Xia, H., Li, Z., Zhuang, P., Gao, B., & Gao, Y. (2014). Removal efficiency of heavy metals in a pilot-scale constructed wetland system and accumulation in plant tissues. *Ecological Engineering, 64*, 356–362. DOI: 10.1016/j.ecoleng.2013.12.044

Zhang, Y., Li, Z., & Wang, L. (2024). Environmental problems of emerging toxic metals and remediation techniques: Role of phytoremediation and microbes. *Environmental Pollution, 334*, 122479. DOI: 10.1016/j.envpol.2024.122479

Zhang, Y., Shao, Y., Tian, S., & He, Z. (2020). Influence of co-contaminants on heavy metal uptake and accumulation in plants: A review. *Environmental Pollution, 257*, 113613. DOI: 10.1016/j.envpol.2019.113613 PMID: 31838392

Zhang, Y., Zhang, M., Yao, Y., & Hu, R. (2020). Biochar for environmental management: A meta-analysis of its effects on soil nitrogen transformation. *The Science of the Total Environment, 731*, 139105. DOI: 10.1016/j.scitotenv.2020.139105

Zhao, F. J.. (2018). Phytoremediation of heavy metals: A review. *Environmental Pollution, 234*, 785–794.

Zhao, F. J., Ma, J. F., & Meharg, A. A. (2015a). Arsenic uptake, metabolism, and toxicity in plants. *The New Phytologist, 206*(4), 1063–1078. DOI: 10.1111/nph.13281

Zhao, F. J., Ma, Y., Zhu, Y. G., Tang, Z., & McGrath, S. P. (2015b). Soil contamination in China: Current status and mitigation strategies. *Environmental Science & Technology, 49*(2), 750–759. DOI: 10.1021/es5047099 PMID: 25514502

Zhao, H.. (2021). Floating-leaved aquatic macrophytes: Functional traits and ecosystem services. *Aquatic Botany, 172*, 103394. DOI: 10.1016/j.aquabot.2021.103394

Zhao, Y., Liu, J., & Zhang, J. (2020). Remote sensing technologies for real-time monitoring and optimization of constructed wetlands: A review. *The Science of the Total Environment, 749*, 141438. DOI: 10.1016/j.scitotenv.2020.141438

Zhao, Y., Wang, Z., Zhang, Q., & Wang, C. (2018). Enhancing Cd phytoremediation efficiency in *Pistia stratiotes* through endophytic fungi colonization. *Environmental Science and Pollution Research International, 25*(16), 15408–15416. DOI: 10.1007/s11356-018-1747-9

Zheng, J., Tan, M., Shibata, Y., Tanaka, A., Li, Y., Zhang, G., & Zhang, Y. (2007). Characteristics of lead isotope ratios and elemental concentrations in PM2.5 in Beijing, China. *Atmospheric Environment, 41*(3), 541–551. DOI: 10.1016/j.atmosenv.2006.08.040

Zhou, C., Xu, Y., Song, X., & Yang, Y. (2022). Phytoremediation of heavy metal-contaminated water and sediments using aquatic plants: A review. *Ecological Engineering, 179*, 106608. DOI: 10.1016/j.ecoleng.2022.106608

Zhou, H., & Lai, C. (2018). Chelation strategies in phytoremediation of heavy metals. *Frontiers in Plant Science, 9*, 1290.

Zhou, Q.. (2007a). Role of aquatic macrophytes in phytoremediation of heavy metals: A review. *Environmental Reviews, 15*(1), 1–12. DOI: 10.1139/a06-003

Zhou, Q.. (2022). Phytoremediation of heavy metals using aquatic macrophytes: Mechanisms and applications. *Environmental Pollution, 298*, 118800. DOI: 10.1016/j.envpol.2021.118800

Zhou, Q., Chen, Y., Gao, Y., & Xu, J. (2022). Arsenic-induced oxidative stress and the regulatory role of WRKY transcription factors in *Nelumbo nucifera. Environmental Pollution, 294*, 118646. DOI: 10.1016/j.envpol.2021.118646

Zhou, Q., Li, X., Lin, Y., Yang, C. P., Tang, W. C., Wu, S. H., Li, D. H., & Lou, W. (2019). Effects of copper ions on removal of nutrients from swine wastewater and on release of dissolved organic matter in duckweed systems. *Water Research, 158*, 171–181. DOI: 10.1016/j.watres.2019.04.036 PMID: 31035194

Zhou, Q., Sun, T., Wang, X., & Shan, Y. (2010). Mechanisms of heavy metal uptake and accumulation by aquatic macrophytes. *Journal of Environmental Sciences (China), 22*(5), 716–723. DOI: 10.1016/S1001-0742(09)60185-5 PMID: 20608508

Zhou, Q., Zhang, J., Fu, J., Shi, J., & Jiang, G. (2007b). Biomonitoring: An appealing tool for assessment of metal pollution in the aquatic ecosystem. *Analytica Chimica Acta, 606*(2), 135–150. DOI: 10.1016/j.aca.2007.11.018 PMID: 18082645

Zhu, Y. L., Zayed, A. M., Qian, J. H., De Souza, M., & Terry, N. (2014). Phytoaccumulation of trace elements by wetland plants: III. Uptake and accumulation of ten trace elements by twelve plant species. *Journal of Environmental Quality, 28*, 1448–1455. DOI: 10.2134/jeq1999.00472425002800050029x

Ziegler, P., Adelmann, K., Zimmer, S., Schmidt, C., & Appenroth, K. J. (2016). Relative in vitro growth rates of duckweeds (Lemnaceae) – The most rapidly growing higher plants. *Plant Biology, 17*(s1, S1), 33–41. DOI: 10.1111/plb.12184 PMID: 24803032

Ziegler, P., Appenroth, K. J., & Sree, K. S. (2023). Survival Strategies of Duckweeds, the World's Smallest Angiosperms. *Plants, 12*(11), 2215. DOI: 10.3390/plants12112215 PMID: 37299193

Zimmels, Y., Kirzhner, F., & Malkovskaja, A. (2006). Application of *Eichhornia crassipes* and *Pistia stratiotes* for treatment of urban sewage in Israel. *Journal of Environmental Management, 81*(4), 420–428. DOI: 10.1016/j.jenvman.2005.11.014 PMID: 16597484

Zorza, L., Bertins, M., Saleniece, K., Kizane, G., Grinbergs, A., Eismonts, U., Reinholds, I., Viksna, A., & Muter, O. (2022). Caesium-133 Accumulation by Freshwater Macrophytes: Partitioning of Translocated Ions and Enzyme Activity in Plants and Microorganisms. *Sustainability (Basel), 14*(3), 1132. DOI: 10.3390/su14031132

Zulkernain, N. H., Basant, N., Ng, C. C., Kriti, K., Salari, M., & Mallick, S. (2023). Recovery of precious metals from e-wastes through conventional and phytoremediation treatment methods: A review and prediction. *Journal of Material Cycles and Waste Management, 25*(6), 2726–2752. DOI: 10.1007/s10163-023-01717-5

About the Contributors

Mansi Sharma is an Assistant Professor in the Department of Physics and Environmental Sciences at Sharda School of Engineering and Science, Sharda University, Uttar Pradesh, India. She completed her Ph.D. in 2023 in the field of Environment Science from Panjab University, Chandigarh, India. She completed her B.Sc. in 2013 in Biomedical Sciences from Delhi University, India and M.Sc. in 2015 in Biodiversity and Conservation, University School of Environment Management from Guru Gobind Singh Indraprastha University, Delhi, India. She was a rank 1 holder and gold medalist in M.Sc. and qualified UGC-NET (JRF) in 2018. Her areas of research interest include Ecophysiology, Stress physiology, Sustainable agriculture, Biodiversity & Conservation, and Environmental remediation. She has published 12 research/review articles in high impact factor journals of National/International repute (All Scopus/SCI-indexed) and 2 book chapters. Her most cited articles in the last few years include: 1. Essential oils as anticancer agents: Potential role in malignancies, drug delivery mechanisms, and immune system enhancement (published in the Journal of Biomedicine and Pharmacotherapy; IF = 6.9) and 2. Plant essential oils as biopesticides: Applications, mechanisms, innovations, and constraints (Published in the MDPI Journal Plants; IF = 4.4). Currently, she has an h-index (Scopus/Google scholar) of 6 and i-10 index (Google scholar) of 5. She is already one of the Editors of a running book proposal accepted by Springer Nature (80% work has been accomplished). She has participated and presented her research work in a number of National and International Conferences. She is an active reviewer for the following journals: Journal of Energy Storage (by Wiley), Journal of Porous Materials (Springer Nature), Journal of Polymers and the Environment (Springer Nature), Journal of Materials Protection, and Exploration of Foods and Foodomics.

Priyanka Mahajan is currently working as an Assistant Professor in the University Center for Research and Development (UCRD), Chandigarh University,

Mohali, Punjab. She has 3 years of teaching experience and 5 years of post-doctoral research experience in the field of plant sciences. She has completed her graduation and post-graduation degree from Panjab University, Chandigarh in Botany (Hons.). She has acquired her Ph.D. degree from Department of Botany, Panjab University, Chandigarh in the field of Botany (Eco-physiology) in 2015. The potential of monoterpene (Beta-pinene)- a natural plant product was explored to ameliorate the phytotoxic effects of hexavalent chromium in the cereal crops during her doctorate and post- doctorate research. Her academic and research achievements include Gold-Medal in M.Sc., 13 research and review articles in the international journal of high impact factors, 2 edited books, and 4 book chapters. She has participated and presented her research work in number of national and international conferences.

Har Mohan Singh is working as an Assistant Professor in the Department of Physics and Environmental Sciences, Sharda University, Greater Noida, India. He was Guest Faculty at Department of Environmental Sciences, BBA University (A Central University), Lucknow, U.P., India and Guest Lecturer at Institute of Newand Renewable Energy, University of Lucknow. He completed his Ph.D. from Shri Mata Vaishno Devi University, Katra, Jammu, India. He did M. Tech. (Energy and Environment) and M.Sc. (Environmental Science) from Department of Environmental Sciences, BBA University (A Central University), Lucknow, U.P., India. His main area of research is to develop algal cultivation systems (photobioreactors) and value-added products from algal biomass. He is actively involved in waste to energy and treatment of wastewater, phyco-remediation, and wastewater treatment technologies. He has published 30 articles (Research/Review) in reputed International Journals with good impact factor and also 15 book chapters in the edited book published by reputed publishing houses. He was awarded Senior Research Fellowship (SRF) under the National Renewable Energy-Fellowship Program by the Ministry of New and Renewable Energy, Govt. of India in 2017. Dr. Singh is a Lifetime member of Biotech Research Society of India [BRSI]. He is serving as an active reviewer for various national and international journals. He is awarded Best Oral Presentation for Photobioreactors as Green Factories for Food, Fuel, Feed and Fertilizers at National Conference on Sustainable Environment: Water-Health Agriculture-Climate Nexus-2022 (SEWHACN-2022), Department of Environmental Sciences, Central University of Jammu, India. He is one of the Editors of a running book proposal accepted by Springer Nature (80% work has been accomplished).

Ajay Kumar holds a B.Tech in Biotechnology, an M.Tech in Environmental Science and Engineering, and a Ph.D in Life Sciences (Environmental Sciences) from Guru Nanak Dev University, Amritsar. He was a UGC-BSR Fellow and qualified the National Eligibility Test (NET-UGC) in 2015. Previously, he served as a Senior Project Associate at Punjab Biotechnology Incubator, Mohali, under Mission Tandrust Punjab. He also worked as a Postdoctoral fellow at the University of Wisconsin, Madison, USA. Currently, he is an Assistant Professor at the University Centre for Research and Development (UCRD), Biotechnology Engineering and Food Technology, Chandigarh University, Mohali. His research focuses on Cancer Chemoprevention using Natural Plant Products. He has authored more than 45 publications in high-impact journals such as Antioxidants, Journal of Functional Foods, Environmental Science and Pollution Research, Frontiers in Pharmacology, Molecules (MDPI), Biomedicine and Pharmacotherapy, Food and Chemical Toxicology, among others. Additionally, he has published two books, 9 book chapters, and has more than 750 citations, an h-index of 15, and an i10-index of 20 (Google Scholar). He has presented his research at various international conferences and is a life member of the Environmental Mutagen Society of India.

Richa Arora is faculty in the Department of Microbiology, Scottish Church College, Kolkata. Presently she is pursuing her research in JIS University, Agarpara, Kolkata.She has over 15 years of undergraduate teaching experience in microbiology, specializing in environmental microbiology, virology, and immunological techniques. Her research includes studies on microbial diversity, heavy metal resistance, and water quality assessment, with several publications in national and international seminars. She has actively participated in research projects funded by the United Board of Christian Higher Education in Asia. As Head of the Department from 2021- 2024, she has also contributed to various academic committees and mentors student research.

Bakare Mariam Oyinade is a passionate aquaculture advocate and a graduate of Fisheries Production from Olabisi Onabanjo University, Ogun State, Nigeria. She currently serves as the World Aquaculture Society (WAS) Global Student Director (2024/2025), where she contributes to student engagement and leadership in the sector. Mariam has a strong record of leadership, having served as both Vice President and President of the Fisheries Student Association at her university, and as the Student Secretary for the WAS West African Chapter (WAS-WAR). She is deeply committed to personal growth, knowledge sharing, and the advancement of aquaculture across Africa. Mariam has volunteered with several aquaculture-related initiatives and organizations, where she has supported awareness, training, and community outreach programs. With plans to pursue a master(tm)s degree, she remains focused on building a career centered on sustainable aquaculture development, youth mentorship,and capacity building in the fisheries and aquaculture sector.

Priyanka Devi is an Assistant Professor in the Department of Agronomy at Lovely Professional University, Punjab, India. Her research focuses on crop physiology, stress tolerance, and precision agriculture techniques.

Titiksha Kamal currently serves as Assistant Professor in the Department of Botany at St. Bede(tm)s College, Shimla. She holds M.Sc. in Botany from St. Bede(tm)s College, Shimla. A qualified SET and GATE scholar, she brings academic excellence to her role. With over two years of teaching experience, Ms. Kamal is known for her student-centric approach and dedication to academic mentoring. She has guided several M.Sc. students in their dissertation and thesis work, encouraging original research.

Preeti Kaundal is currently serving as the Head of the Department and Assistant Professor in the Department of Botany at St. Bede ?(tm)s College, Shimla, Himachal Pradesh. She holds a Ph.D., M.Phil., and M.Sc. Botany Degree from Himachal Pradesh University, where she was awarded Dr. K. K. Jain Gold Medal for academic excellence in her M.Sc. program. She also holds a Bachelor of Education (B.Ed.) degree from Himachal Pradesh University. She has also completed a certificate course in the German language with a Gold Medal. In addition to this, she has qualified GATE and SLET examinations. With over eight years of teaching experience, Dr. Kaundal is a passionate educator and dedicated researcher. She has guided many M.Sc. students for their projects/dissertation/thesis work, mentoring them in various areas of botanical research. Her academic specialization lies in medicinal plant research and ethnobotany, with additional interests in taxonomy, environmental botany and phytochemistry. She is committed to integrating traditional knowledge with modern scientific methods, contributing to the conservation and sustainable use of plant biodiversity.

Sandeep Kaur is currently serving as an assistant professor in the Department of UCRD (University Center for Research and Development) at Chandigarh University, Mohali, (India). Dr. Kaur holds Bachelor of Science in Medical Sciences, Bachelor of Education (Life sciences), M. Sc. (Hons.) and Ph. D degrees in Botany from Guru Nanak Dev University, Amritsar, Punjab (India). She has over 12 years of research experience and 6 years of teaching experience. Her specialization includes natural plant products, cancer cell biology, isolation and characterization of compounds, chemoprevention, pharmaceutical sciences and genetic toxicology. She has also authored 56 publications, including articles in SCI-indexed journals, achieving an h-index of 18 and over 772 citations. She has served as a reviewer board member for esteemed journals, thereby contributing to the rigorous evaluation of scientific literature. Additionally, she has co-edited books on functional foods and cancer prevention and one text book in Kalyani publishers, which reflect her dedication to translating scientific insights into impactful knowledge. She has presented her research work in various national and international conferences and published numerous abstracts. She was honored with the Young Scientist Award in 2023 and had received accolades for her presentations at international conferences. She has been actively participating in workshops/Hand-On-Cum Trainings and had attended different seminars, symposia and conferences held at the National/international level. She has published research articles and book chapters in journals and books of international repute. She is an active member of professional societies, including the Environmental Mutagen Society of India, which reflects her commitment to staying at the forefront of scientific advancements. At present, she guides graduate and post-graduate students.

Vaseem Raja (Ph.D.) is working as an Assistant Professor at University Centre for Research and Development, Chandigarh University, Mohali Punjab-140413. He obtained his master?(tm)s degree (M.Sc.) from Jamia Hamdard New Delhi, and doctorate degree (Ph.D.) in Botany (plant molecular biology) from University of Kashmir, Hazratbal Srinagar (India). During his PhD he has received Senior Research fellowship from Department of Biotechnology Government of India. He has also worked as a post-doctorate research fellow in Department of Botany University of Kashmir. His main research area is stress physiology and molecular biology. He has published more than 60 research papers in peer reviewed journals. Besides these research publications he has also published eight book chapters and two review articles. Dr. Vaseem has also presented his work in various national and international conferences. He is actively involved in studying molecular and physio-biochemical responses of different plant species to environmental stresses ??. Dr. Vaseem is actively engaged in the teaching of plant physiology, biodiversity, plant ecology and taxonomy, cell and molecular biology, crop genetics and plant breeding from last 7 years.

Pushkar Singh Rawat is a dedicated biochemist with extensive expertise in clinical biochemistry and neurobiology. He is currently associated with the Department of Biochemistry, Lucknow University, Lucknow, India. He has built a strong academic and research profile based on rigorous training and impressive scientific contributions. He received his PhD in Biochemistry in 2022, focusing on lead (Pb) toxicity in children, with a focus on environmental toxicology and public health. He completed his B.Sc. in Biology (2012) and M.Sc. in Biochemistry (2014) from Lucknow University, Lucknow, India. He also holds a Diploma in Pharmacy (2008) from Government Polytechnic Dwarahat, Almora, Uttarakhand, India. Dr. Rawat has received several prestigious honors, including the ICMR-SRF (2019;

2022), and has qualified national/state exams in life sciences, such as ASRB-NET, CG-SET, HP-SET, and U-SET. Dr. Rawat is an active researcher with numerous peer-reviewed publications in renowned journals. His research primarily focuses on heavy metal toxicity, particularly the neurobiochemical effects of lead exposure in children. Additionally, Dr. Rawat has contributed several international book chapters (~20) on topics such as toxicology, nutraceuticals, medicinal plants, and plastic biodegradation, Artificial Intelligence demonstrating his interdisciplinary and applied scientific approach.

Sangeeta Sharma currently serves as Assistant Professor in the Department of Botany at St. Bede(tm)s College, Shimla. She holds a Ph.D. and M.Sc. in Botany from Panjab University, Chandigarh, and a B.Ed. from Himachal Pradesh University. A qualified UGC-NET-JRF and GATE scholar, she brings both academic excellence and research insight to her role. With over two years of teaching experience, Dr. Sharma is known for her student-centric approach and dedication to academic mentoring. She has guided several M.Sc. students in their dissertation and thesis work, encouraging original research in various areas of botany. Her core academic interests lie in plant physiology and biochemistry, complemented by a strong passion for field botany. Her research emphasizes the integration of traditional botanical knowledge with contemporary scientific methods, contributing to biodiversity conservation and sustainable resource use. Dr. Sharma is committed to fostering scientific curiosity and environmental stewardship among her students.

Brahmjot Singh has Bachelor degree in pharmacy, master degree in Pharmacology and Ph.D. in Life Sciences (Pharmaceutical Sciences) under the supervision of Prof.(Dr.) Balbir Singh from the Department of Pharmaceutical Sciences, Guru Nanak Dev University, Amritsar, Punjab, India. He qualified national competitive examination GPAT (Graduate Pharmacy Aptitude Test) in 2015. Previously, he worked as Assistant Professor in the PCTE Group of College, Baddowal, Ludhiana, Punjab, India. Currently, he is working as Assistant Professor in the Department of Pharmaceutical Sciences, Guru Nanak Dev University, Amritsar, Punjab, India. His areas of interest are role of natural compounds in the prevention and mitigation of diabetes mellitus, diabetic nephropathy and other diabetes-associated complication (tm)s impacts on the body and anti-microbial studies. He has more than 15 publications in the reputed and high-impacted journals (Phytotherapy Research, The Journal of Functional Foods, Cornea, Naunyn-Schmiedeberg's Archives of Pharmacology, Current Drug Delivery, International Immunopharmacology etc). He has presented his research at international conferences. Scopus ID: 57218321987

Shalini Singh is a distinguished academician and researcher in the field of Biochemistry and Biotechnology, currently serving as an Assistant Professor at the Institute of Biosciences and Technology, Shri Ramswaroop Memorial University, Lucknow, India. With a strong educational foundation and extensive research background, Dr. Singh has established herself as a dynamic contributor to the scientific community. She earned her Ph.D. in Biochemistry (2022) from the University of Lucknow under the supervision of Prof. Sudhir Kumar Agarwal, with her doctoral research focusing on the purification and characterization of cathepsin H from buffalo lung. Prior to that, she secured a Master (tm)s degree in Microbiology (2014) from Veer Bahadur Singh Purvanchal University, where she was a University Gold Medalist, and a Bachelor (tm)s degree in Biotechnology (2012) from Chaudhary Charan Singh University. Dr. Singh (tm)s primary areas of interest include Recombinant DNA Technology, Microbiology, Enzymology, and Bioinformatics. Her technical expertise spans various molecular biology techniques such as DNA/RNA isolation, PCR, gene cloning, enzymatic assays, protein purification, and bioinformatics tools including molecular docking and simulation. She has held notable research positions, including Junior Research Fellowships at the University of Lucknow and the Central University of Rajasthan. Her project work at NRCPB, IARI Pusa, involved cloning ectoine biosynthetic genes from Halomonas elongata in E. coli. Dr. Singh has contributed extensively to scientific literature with several peer-reviewed publications, book chapters, and editorial projects. Her work is featured in journals like Biochemical and Biophysical Reports, International Journal of Biochemistry and Physiology, Journal of Trace Elements in Medicine and Biology, and Indian Journal of Clinical Biochemistry. She is also a co-editor of upcoming academic books focusing on AI applications in bioremediation and genomics. A recipient of several prestigious awards, she qualified the GATE (2012), ASRB-NET (2016), secured the top rank in the Ph.D. entrance exam, and was honoured with the DST-INSPIRE Fellowship (2015). She also earned accolades for her presentations at national and international conferences. Beyond teaching and research, Dr. Singh actively participates in academic governance. She is a Board of Studies (BOS) member, a steering committee member of the Food Safety and Drug Administration (FSD), and contributes to several institutional and departmental committees. Dr. Singh has conducted and participated in numerous workshops and conferences and delivered both oral and poster presentations on diverse biochemical and environmental health topics. Her commitment to scientific inquiry is evident in her robust portfolio of ongoing research and collaborations. In her personal life, Dr. Shalini Singh was born on July 24, 1992, and resides in Lucknow, Uttar Pradesh. She is married and maintains strong academic ties through professional networks like ORCID, Google Scholar, and ResearchGate.

Index

A

B

C

E

H

M

N

O

P

R

S

T

W

www.ingramcontent.com/pod-product-compliance
Lightning Source LLC
LaVergne TN
LVHW062323300926
849161LV00007B/1550